POLLUTION CONTROL IN THE PLASTICS AND RUBBER INDUSTRY

Pollution Control in the Plastics and Rubber Industry

Marshall Sittig

NOYES DATA CORPORATION
Park Ridge, New Jersey London, England
1975

Library of Congress Catalog Card Number: 75-2940
ISBN: 0-8155-0572-8
Printed in the United States

Published in the United States of America by
Noyes Data Corporation
Noyes Building, Park Ridge, New Jersey 07656

FOREWORD

This book is No. 18 in our Pollution Technology Review Series. It has become increasingly evident that much can and must be done by industry during the early stages of manufacture of raw materials, intermediates, and even of end products to control environmental pollution and improve waste removal techniques.

Therefore, the present volume is not just a book on pollution control in the plastics and rubber industries, but also a comprehensive treatise on plastics and rubber manufacturing processes and product finishing. The point of view is an integrated one: New and timely reviews of industrial practice are given together with constant attention to air pollution, water pollution, and solid waste problems.

Advanced composition and production methods developed by Noyes Data are employed to bring new durably bound books to the reader in a minimum of time. Special techniques are used to close the gap between "manuscript" and "completed book." Industrial technology is progressing so rapidly that time-honored, conventional typesetting, binding and shipping methods are no longer suitable. Delays in the conventional book publishing cycle have been bypassed to provide the user with an effective and convenient means of reviewing up-to-date information in depth.

The Table of Contents is organized in such a way as to serve as a subject index and provides easy access to the information contained in this book. There is also an index of patentees and a valuable list of references to government reports which are hard to find via ordinary bibliographic search tools.

CONTENTS AND SUBJECT INDEX

INTRODUCTION

The plastics and rubber industries—the polymer industries—have been among the fastest growing sectors of our economy. With growth, however, has come the responsibility for disposal of increasing quantities of air and water pollutants and especially solid wastes.

One aspect of this which can be mentioned at the outset is that plastics (particularly thermoplastics rather than thermosets) are recyclable whereas rubber is essentially not recyclable (vulcanized rubber perhaps being considered as more analogous to thermosetting resinous materials).

Today, some 10 million tons of plastics are used annually in products ranging from automobile components to artificial heart valves. Some of these plastics products—in fact the great majority—are used in applications that require long service. Some plastics, however, are used in forms that will be discarded after short-term use. The growing use of these short-term applications has led to speculation about contributions plastics may make to pollution.

In packaging, plastics have replaced other materials because of a number of advantages. Plastics in packaging aid in display of products, eye appeal and are lightweight. Plastics are used to package many household products including cosmetics, shampoos, bleaches, liquid detergents and a wide variety of food. Packaging plastics are those which primarily are discarded after serving their useful purpose within a year of manufacture. Ecologically, it is interesting to note that the same quantity of products packaged in most materials other than plastics would mean a larger contribution from packaging to the solid waste load—larger on a weight basis and, in most cases, on a volume basis according to Ethyl Corporation (1).

However in 1970, plastic packaging waste was estimated at 1.8 million tons compared to all other plastic solid waste of 0.75 million tons according to National Industrial Pollution Control Council (2), so the problem is there—and growing.

The plastics and rubber industry is unique in that is presents fairly massive problems of air pollution, water pollution and solid waste disposal. The air pollution problems result from manufacturing processes to some extent but more from possible products of incineration of plastics and rubber wastes. The water pollution problems are primarily a product of manufacturing operations. The solid waste disposal problems are perhaps the greatest of the three.

As pointed out by Ethyl Corporation (1), the volume of plastics in all solid waste is not large, but it is growing since the uses of plastics are growing. One recent study showed that the solid waste generated in the U.S. in a year was in the range of 300 million tons. Realistic estimates indicate that all plastics account for about three million tons of solid waste a year. Based on weights of refuse collected, the percentage of plastics, on average, is between 1.5 and 1.7% while paper products, for example, account for about 33% of the total. Food wastes comprise about 16% of the solid waste load with textiles and garden trash accounting for about 18% of the total. About 85% of the solid waste is disposed of via landfills and 10% via incineration.

According to H.H. Connolly, then Acting Deputy Commissioner, Solid Waste Management Office, U.S. Environmental Protection Agency in an address delivered to Palisade Section, Society of Plastics Engineers, Cherry Hill, New Jersey, October 1970, however, there is good reason to believe that the figures for plastic wastes could be as much as 12% by 1980.

In 1969, according to the EPA (3), the rubber industry consumed 3.2 million tons of rubber. The rubber industry consumption includes new rubber as well as reclaimed rubber and old tires consumed by retreaders and tire splitters. Tire splitting is a clear case of reuse, analogous to wiping cloth production from old textiles. Retreading is a more ambiguous case and could be viewed either as recycling or reuse. The table below shows recent patterns of rubber consumption in the United States.

Rubber Usage in the United States
'000 Long Tons

	1971	1972	1973	1974
Total new rubber	2,999	3,161	3,453	3,311
Reclaim	199	191	201	151
Retread	276	259	245	na*

na = not available

Source: Rubber Manufacturers Association

In reclaiming, obsolete and new rubber wastes are reduced to a uniform size by shredding and cracking; metals and fibrous materials are removed; the rubber is mixed with reclaiming oils, softened, and mixed with compounding agents; and after further refining and milling operations, the reclaimed rubber is sold in slabs or bales.

The tire splitting industry engages in the production of various industrial products die-cut from obsolete tires. Typical products are gaskets, shims, insulators, doormats, conveyor rollers, belt pieces, and the like.

Rubber wastes, usually in the form of obsolete tires and inner tubes, are handled by secondary materials dealers; some are rubber specialists, but this is a disappearing breed because rubber waste handling is not a very profitable business. Rubber dealers, who once paid waste generators for tires, are converting their operation to a new basis—whereby the generators pay them to remove the tires. Since tires occur in relatively small concentrations at garages and filling stations, the sales price is usually insufficient to cover a dealer's costs of picking up tires, removal and disposal of unacceptable tires (excessively worn tires, studded snow tires, steel-wire reinforced tires), and delivery or freight to the reclaimer. He must be paid for the removal service in order to stay in business.

Reclaimers, who can best use rubber wastes that are already free of metals and fibers, pay considerably more for such wastes than for tires.

Conventional recycling and reuse of rubber products is thus a limited activity. Rubber waste recycling and reuse depends in large measure on retread tire sales. The bulk of rubber production reaches the discard state in the form of tires which, though largely rubber, are composites of several materials, including textiles and metals. The reprocessing of this rubber by removal of impurities is usually less costly than production of virgin rubber, but the cost advantages have not been for the most part sufficiently great in light of the technical limitations of reclaim, to cause greatly increased use of reprocessed rubber.

Rubber in mixed municipal wastes usually also occurs as a component of products—soles and heels on footwear, rubber backings on textiles, wire coatings, gaskets and insulation in small appliances, etc. Some rubber is pure (water bottles, rubber bands, toys) but this is an insignificant portion of the total. Removal of these materials from mixed wastes for recovery appears impractical. Furthermore, a critical cost and logistics problem seems to be the collection of tires to central processing locations.

The most practical use for rubber, particularly old tires, appears to be raw materials recovery by distillation or energy recovery.

PLASTICS MANUFACTURE

Some idea of the environmental impacts of the manufacture of specific plastics in terms of air pollution, water pollution and solid waste disposal (as well as indications of environmental imports due to energy consumption and transportation), are given in Tables 1 through 5. These were taken (with permission) from *Plastics: Resource & Environmental Profile Analysis,* Washington, D.C., Manufacturing Chemists Association (June 28, 1974)(4).

TABLE 1: DATA FOR MANUFACTURE OF 1,000 POUNDS OF POLYVINYL CHLORIDE

Raw Materials	
(Vinyl chloride monomer - 1,030 lb)	
Peroxides, stabilizers, antioxidants and hydrocarbons	3.0 lb
Energy	
Electric	84.0 kwhr
Natural gas	160.0 scf
Distillate oil	52.0 gal.
Water Volume	1,000.0 gal.
Process Solid Wastes	3.0 lb
Process Atmospheric Emission	
Hydrocarbons (as vcm)	30.0 lb
Particulates	0.34 lb
Waterborne Wastes	
BOD	0.45 lb
COD	2.25 lb
Suspended solids	0.45 lb
Transportation	
Rail	400 ton-miles
Truck	100 ton-miles

Source: Manufacturing Chemists Association (4)

TABLE 2: DATA FOR MANUFACTURE OF 1,000 POUNDS OF HIGH DENSITY POLYETHYLENE

Raw Materials--lb	
(Ethylene - 1,040)	
Solvent	45.7 lb
Additives	3.0 lb
Energy	
Electric	390.0 kwhr
Natural gas	2.5 scf
Water Volume	1,780.0 gal.
Process Solid Wastes	2.4 lb
Process Atmospheric Emissions	
Hydrocarbons	5.0 lb
Waterborne Wastes	
BOD	0.37 lb
COD	2.04 lb
Suspended solids	0.62 lb
Transportation	250 ton-miles

TABLE 3: DATA FOR MANUFACTURE OF 1,000 POUNDS OF LOW DENSITY POLYETHYLENE

Raw Materials--lb	
(Ethylene--1,047)	
Additives	35.0
Energy	
Electric	580.0 kwhr
Natural gas	5,000.0 scf
Diesel	0.02 gal.
Gasoline	0.05 gal.
Water Volume	262.0 gal.
Process Solid Waste	9.0 lb
Process Atmospheric Emissions	
Particulates	1.3 lb
Hydrocarbons	3.0 lb
Waterborne Wastes	
BOD	0.23 lb
COD	0.61 lb
Suspended solids	0.28 lb
Transportation	
Rail	250 ton-miles

Source: Manufacturing Chemists Association (4)

TABLE 4: DATA FOR MANUFACTURE OF 1,000 POUNDS OF ABS RESIN

Raw Materials	
(styrene - 680, acrylonitrile - 237, polybutadiene - 71)	
Material process addition	12.0 lb
Energy	
Electric	206 kwhr
Natural gas industrial heat	2,386 scf
Water Volume	2,841 gal.
Process Solid Wastes	1.0 lb
Process Atmospheric Wastes	
Hydrocarbons	2.70 lb
Waterborne Wastes	
BOD	0.46 lb
COD	2.36 lb
Oil	0.02 lb
Suspended solids	0.49 lb
Chromium	0.0016 lb
Iron	0.016 lb
Aluminum	0.016 lb
Nickel	0.008 lb
Cyanide	0.0008 lb
Transportation	
Rail	125.0 ton-miles
Truck	125.0 ton-miles

TABLE 5: DATA FOR MANUFACTURE OF 1,000 POUNDS OF HIGH IMPACT POLYSTYRENE RESIN

Raw Materials	
(Styrene - 989, Polybutadiene - 52)	
Material process addition	30.0 lb
Energy	
Electric	170.0 kwhr
Natural gas	1,471 scf
Water Volume	1,160 gal.
Solid Waste Process	5.0 lb
Atmospheric Hydrocarbons	2.9 lb
Waterborne Wastes	
BOD	0.19 lb
COD	0.87 lb
Suspended solids	0.24 lb
Oil	0.01 lb
Transportation	
Rail	250 ton-miles
Truck	250 ton-miles

Source: Manufacturing Chemists Association (4)

AIR POLLUTION

The major sources of possible air contamination in resin manufacturing are the emissions of raw materials or monomer to the atmosphere, emissions of solvent or other volatile liquids during the reaction, emissions of sublimed solids such as phthalic anhydride in alkyd production, emissions of solvents during thinning of some resins, and emissions of solvents during storage and handling of thinned resins. Table 6 lists the most probable types and sources of air contaminants from various resin-manufacturing operations.

In the formulation of polyurethane foam, a slight excess of tolylene diisocyanate (TDI) is usually added. Some of this is vaporized and emitted along with carbon dioxide during the reaction. The TDI fumes are extremely irritating to the eyes and respiratory system and are a source of local air pollution. Since the vapor pressure of TDI is small, the fumes are minute in quantity and, if exhausted from the immediate work area and discharged to the outside atmosphere, are soon diluted to a nondetectible concentration. No specific controls have been needed to prevent emission of TDI fumes to the atmosphere.

The finished solid resin represents a very small problem—chiefly some dust from crushing and grinding operations for molding powders. Generally the material is pneumatically conveyed from the grinder or pulverizer through a cyclone separator to a storage hopper. The fines escaping the cyclone outlet are collected by a baghouse-type dust collector. The collector should be designed for a filter velocity of about 4 fpm or less.

Most of the contaminants are readily condensable. In addition to these, however, small quantities of noncondensable, odorous gases similar to those from varnish cooking may be emitted. These are more prevalent in the manufacture of oil-modified alkyds where a drying oil such as tung, linseed, or soya is reacted with glycerin and phthalic anhydride. When a drying oil is heated, acrolein and other odorous materials are emitted at temperatures exceeding about 350°F (see further discussion under Varnish Cookers). The intensity of these emissions is directly proportional to maximum reaction temperatures.

TABLE 6: PRINCIPAL AIR CONTAMINANTS AND SOURCES OF EMISSION FROM RESIN-MANUFACTURING OPERATIONS

Resin	Air contaminant	Possible sources of emission
Phenolic	Aldehyde odor	Storage, leaks, condenser outlet, vacuum pump discharge
Amino	Aldehyde odor	Storage, leaks
Polyester and alkyds	Oil-cooking odors Phthalic anhydride fumes Solvent	Uncontrolled resin kettle discharge Kettle or condenser discharge
Polyvinyl acetate	Vinyl acetate odor Solvent	Storage, condenser outlet during reaction, condenser outlet during steam distillation to recover solvent and unreacted monomer
Polyvinyl chloride	Vinyl chloride odor	Leaks in pressurized system
Polystyrene	Styrene odor	Leaks in storage and reaction equipment
Petroleum and coal tar resins	Monomer odors	Leaks in storage and reaction equipment
Polyurethane resins	Tolylene diisocyanate	Emission from finished foam resulting from excess TDI in formulation

Source: EPA (5)

Thus, the intensity of noncondensable gases from resin formulation should be considerably less than that of gases from varnish cooking since the reaction temperature is approximately 100°F lower.

Control of monomer and volatile solvent emissions during storage before the reaction and of solvent emissions during thinning and storage after the polymerization of the resin is relatively simple. It involves care in maintaining gastight containers for gases or liquefied gases stored under pressure, and condensers or cooling coils on other vessels handling liquids that might vaporize. Since most resins are thinned at elevated temperatures near the boiling point of the thinner, resin-thinning tanks, especially, require adequate condensers. Aside from the necessity for control of air pollution, these steps are needed to prevent the loss of valuable products.

Heated tanks used for storage of liquid phthalic and maleic anhydrides should be equipped with condensation devices to prevent losses of sublimed material. An excellent device is a water-jacketed, vertical condenser with provisions for admitting steam to the jacket and provisions for a pressure relief valve at the condenser outlet set at perhaps 4 ounces pressure.

During storage the tank is kept under a slight pressure of about 2 ounces, an inert gas making the tank completely closed. During filling, the displaced gas, with any sublimed phthalic anhydride, is forced through the cooled condenser where the phthalic anhydride is deposited on the condenser walls. After filling is completed, the condensed phthalic anhydride is remelted by passing steam through the condenser jacket.

Addition of solids such as phthalic anhydride to other ingredients that are above the sublimation temperature of the phthalic anhydride causes temporary emissions that violate most air pollution standards regarding opacity of smoke or fumes. These emissions subside somewhat as soon as the solid is completely dissolved but remain in evidence at a reduced opacity until the reaction has been completed. The emissions can be controlled fairly easily with simple scrubbing devices. Various types of scrubbers can be used.

A common system that has been proved effective consists of a settling chamber, commonly called a resin slop tank, followed by an exhaust stack equipped with water sprays. The spray system should provide for at least 2 gallons per 1,000 scf at a velocity of 5 fps. The settling chamber can consist of an enclosed vessel partially filled with water capable of being circulated with gas connections from the reaction vessel and to the exhaust stack. Some solids and water of reaction are collected in the settling tank, the remainder being knocked down by the water sprays in the stack. Here the vapors from a polyester resin process kettle are first passed through a spray chamber type precleaner followed by a venturi scrubber.

This system effectively reduces visible emissions. Scrubber water may be recirculated or used on a once-through basis, depending primarily upon the available wastewater disposal system. The scrubber water can be odorous and should be discharged to a sanitary sewer.

Many resin polymerization reactions, for example, polyvinyl acetate by the solution method, require refluxing of ingredients during the reaction. Thus, all reactors for this or other reactions involving the vaporization of portions of the reactor contents must be equipped with suitable reflux- or horizontal-type condensers or a combination of both. The only problems involved here are proper sizing of the condensers and maintaining the cooling medium at the temperature necessary to effect complete condensation.

When noncondensable, odor-bearing gases are emitted during the reaction, especially with alkyd resin production as already mentioned, and these gases are in sufficient concentration to create a public nuisance, more extensive air pollution control equipment is necessary. It includes equipment for absorption and chemical oxidation, adsorption, and combustion, both catalytic and direct-flame type.

As regards the intermedia aspects of the plastics manufacturing industry, water pollution control can give rise to some air pollution problems. If incineration of waste sludges is employed, for example, there is potential for air pollution, principally particulates and possibly nitrogen oxides, although the latter should be minimal because incineration of sludges does not normally take place at temperature levels where the greatest amounts of nitrogen oxide are generated. Odors from the wastewater treatment plants may cause occasional problems since wastewaters are sometimes such that heavy, stable, foams occur on aerated basins and septicity is present. But, in general, odors are not expected to be a significant problem when compared with odor emissions possible from other plant sources.

Polyvinyl Chloride Manufacture

The general types of emissions, emission control techniques and control costs have been described by B.H. Carpenter of Research Triangle Institute (6).

The problems of vinyl chloride monomer (VCM) emission have of course reached almost panic proportions in the public press due to disclosures of the involvement of vinyl chloride in deaths from liver cancer (angiosarcoma). As a consequence, the U.S. Occupational Safety and Health Administration (OSHA) has proposed rules for vinyl chloride exposure. However, as pointed out in *Chemical Week* for January 15, 1975, the proposed OSHA rules are already in litigation in the courts.

The Society of the Plastics Industry (SPI) and several of its member companies are opposing OSHA's recently issued regulation stating that employees must not be exposed to air in which the vinyl chloride content is more than one part per million. The new standard was stated to go into effect in January 1975 but the Second Circuit Court of Appeals in New York granted a stay pending a decision on the merits of the SPI objection. Meanwhile an earlier standard of 50 ppm remains in effect.

In PVC production, as reported by Carpenter (6), the pounds of VCM losses per pound of PVC are at least an order of magnitude higher than those from VCM production. Most producers report PVC production as 3 to 4% lower than VCM intake, and some producers report 1 to 1.5% of PVC production lost as fines to the atmosphere. Thus, the actual VCM losses (2 to 3%) result from the batch operation of the polymerization and in the filtration and the drying of the polymer. Direct reduction to 50% of this level seems possible, but a 90% reduction without process changes in some existing PVC plants is likely to be beyond control techniques and acceptable costs. However, intensive stripping at the end of the suspension reaction may achieve the 90% reduction at acceptable costs.

Controls appraised for VCM production sources include recycling of vent streams, condensation with refrigeration, compression, adsorption with carbon, incineration, oxidation with ozone, absorption (scrubbing), and venting to flares. Monomer loading (and unloading) involves special controls: vapor collection adapters with recycling, thermal level detectors with recycling, and magnetic gauges. The PVC production sources can benefit from the same controls plus vacuum stripping of crude product, steam stripping, and carrier airstream recycling. Fugitive sources require use of better valves, packings, etc. A typical PVC plant includes the following operations:

[1] Receiving and storage of VCM and catalysts.
[2] Polymerization of VCM: measuring and charging, and reaction.
[3] Stripping and recovery: reactor blowdown and recovery, and slurry handling and storage.
[4] Centrifugation or filtration.
[5] Drying.
[6] Pneumatic conveying and storage.
[7] Packaging and shipping.
[8] Blending.
[9] Waste treatment.

Compounding, which may be done in a PVC resin plant or in separate facilities, is not included in this study.

Total PVC output for 1972-73 is attributed to four process types: suspension polymerization, 78%; emulsion polymerization, 12%; bulk polymerization, 6%; and solution polymerization, 4%. All polymerizations in the United States are batch operations, which probably contribute strongly to the severity of the VCM emissions problem.

Suspension polymerization (Figure 1) uses a water media, a suspension agent such as polyvinyl alcohol, and an oil-soluble catalyst. A typical recipe is 100 parts VCM, 200 of water, 0.1 to 0.2 of catalyst, and 0.005 to 0.0 of suspension agent. The VCM-water ratio can be varied over a wide range; the limiting factor is sufficient fluidity of the final PVC-water slurry, for adequate dissipation of the heat evolved during polymerization. Oil-soluble catalysts such as dibenzoyl peroxide, acetyl benzoyl peroxide, or dibutyl peroxide are preferred to the water soluble persulfate catalyst. Persulfate is slower, and it produces particles too fine to be handled in conventional filtering and drying equipment.

Commercial suspension resins show little or no retention on a 100-mesh (U.S. standard) screen and from 30 to 80% retention on a 200-mesh screen. Such a particle size is achieved by rapid agitation in glass-lined autoclaves at 35° to 45°C. Vigorous agitation suspends VCM as fine droplets in water, controls size, and gives a granular polymer for better filtering and drying.

During polymerization, alkaline buffers such as sodium carbonate, bicarbonate, and phosphate have been used to maintain the pH at 5 to 8 to prevent formation of hydrochloric acid. Use of an inert substance such as nitrogen increases reaction rate, reduces the amount of hydrochloric acid evolved, and thereby increases the stability of the polymer.

There is little published information on molecular weight distribution and chain branching of vinyl chloride polymers. However, chloride polymers in general have narrow molecular distributions, especially compared to those of polystyrene.

The major advantage of suspension polymerization is the excellent heat transfer rate attainable in conventional equipment. The major disadvantage is that water must be separated from the polymer, usually by centrifugation and drying.

In Table 7 the VCM emissions rates are given for the sources shown in Figure 1; the table is restricted to those plants for which data have been supplied.

Fugitive emissions account for 12 to 46% of total emissions. Vacuum stripping of the crude product would reduce these emissions 50 to 80%; carbon adsorption, 50 to 99%; condensation with refrigeration, 40 to 60%; incineration, 50 to 99%; and adsorption, 50 to 90%.

Collectively, the dryer vents, the air conveyor vent, the storage bin vent, and the centrifuge vent provide 35% of the total emissions. Vacuum stripping could achieve 50 to 80% reduction, and either carbon adsorption, ozone oxidation, incineration, steam stripping, or recycling to compressors would reduce emissions 40 to 99%. Air recycling should reduce these emissions 40 to 80%.

Blend surge tanks account for 11% of total emissions. Carbon adsorption, ozone oxidation, and incineration should give 50 to 99% reductions. Vacuum stripping and absorption are expected to give 50 to 80%; recycles to compressors, 40 to 60%.

The emulsion process (Figure 2) is similar to the suspension process (Figure 1). In emulsion, the particle size is smaller. Emulsifiers and additives maintain a nonsettling emulsion during polymerization. At least four components are needed–water, VCM, water-soluble initiators, and an emulsifying agent. Water is the continuous phase; VCM is the discontinuous phase. Low concentrations of emulsifiers (either anionic or cationic) behave essentially as electrolytes because they are uniformly distributed between the water-VCM phases. When the concentration is increased, the surface tension between the phases decreases and the conductivity of the mixture increases.

FIGURE 1: SIMPLIFIED FLOW DIAGRAM OF PVC SUSPENSION PROCESS

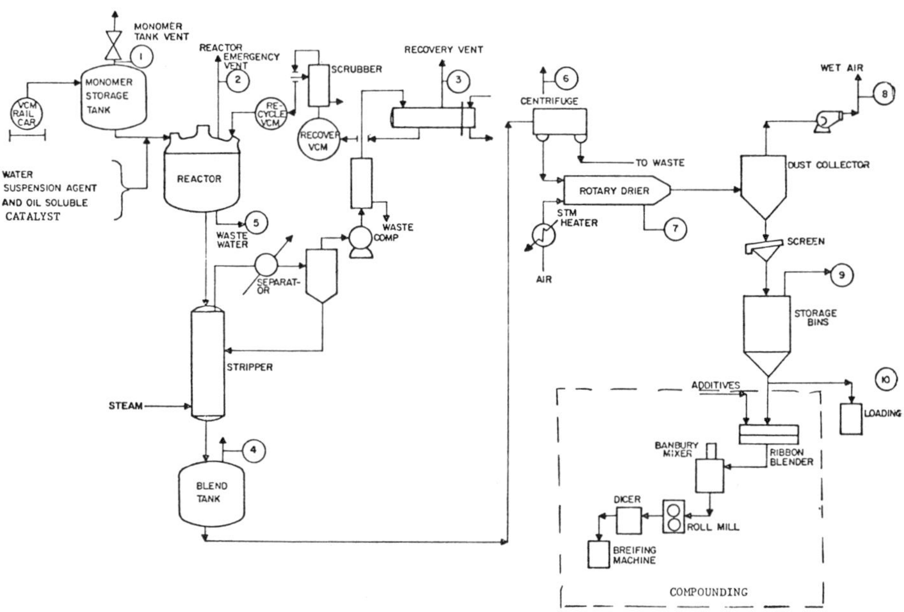

FIGURE 2: SIMPLIFIED FLOW DIAGRAM FOR BATCH EMULSION POLYMERIZATION OF VINYL CHLORIDE

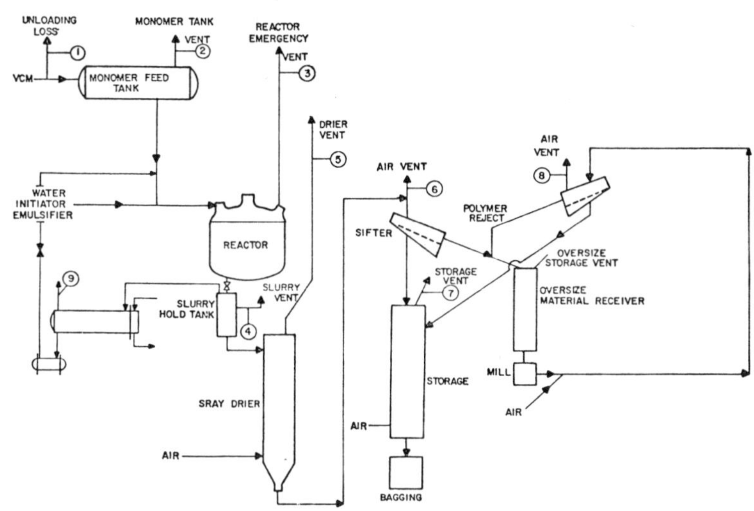

Source: B.H. Carpenter (6)

TABLE 7: EMISSION RATES FOR PVC SUSPENSION PROCESS

PROCESS EMISSION SOURCE[a]	PLANT A lb VCM/MM lb produced[b]	PLANT A % VCM emissions	PLANT B lb VCM/MM lb produced[b]	PLANT B % VCM emissions
1. VCM tank vent, C	Part of vent 15		Part of vent 15	
2. Reactor emergency vent, I	Part of vent 15		Part of vent 15	
3. VCM recovery still vent, C	Part of vent 15		Part of vent 15	
4. Blend or surge tank vent, C	4280(48.2)	11.0	3447(34.5)	11.5
5. Wastewater, VCM recovery, C				
6. Centrifuge vent, C	Part of vent 9		Part of vent 9	
7. Dryer vent, C	Part of vent 9		Part of vent 9	
8. Air conveyor vent, C	Part of vent 9		Part of vent 9	
9. Storage bin vent, C	13,498(152.0)	35.0	7872(78.8)	23
10. VCM loading, storage vents, I	Part of vent 15		Part of vent 15	
11. VCM charging, I	Part of vent 15		Part of vent 15	
12. Polymerizer cleaning, I	3019(34.0)	7.8	3447(34.5)	10-12
13. Packing/hopper car, I	1243(14.0)	3.2	300(3.0)	1
14. System cleaning, I	Part of vent 15		Part of vent 15	
15. Fugitive losses, C&I	17,467(196.7)	40-46	18,691(187.1)	50-58
16. Abnormal losses, I				
Total	39,507(444.9) or 3.5% VCM in PVC		33,757(337.9) or 3.0% VCM in PVC	

[a]C is continuous, I is intermittent. [b]Value in parentheses is lb VCM/hr.

Note: Plant C with a capacity of 175 MM lb/yr emitted 7079 lb VCM/MM lb(155 lb VCM/hr) from vent 1 and 2283 lb VCM/MM lb(50 lb VCM/hr) from vent 4; emissions totaled 24.2% for vent 1 and 7.8% for vent 4.

Source: B.H. Carpenter (6)

Above a critical concentration, the surface tension and conductivity change less rapidly and the emulsifier begins to agglomerate into groups of 20 to 30 molecules called micelles. In the water phase, the initiator forms a free radical which migrates to the micelles, combines with the VCM molecules, and begins the polymer chain. That is, a polymer particle starts to form, the emulsifier collects at the surface, and the VCM molecules diffuse (from the dispersed VCM droplets) through the water phase and through the emulsifier to the growing polymer chain.

As polymerization of VCM progresses and as the other polymer particles grow, more emulsifier is needed at the particle surfaces. In some cases when 1 to 2% conversion is reached, the emulsifier micelles disappear and all of the emulsifier is concentrated at the particle surfaces; at higher conversions, perhaps 60%, all of the VCM is in the PVC phase. The number of PVC particles formed (i.e., their size) seems to be controlled in the early stages of polymerization. Termination steps involving coupling or disproportionation are rare because the number of PVC chains for each particle is small. Hence, PVC produced by the emulsion process tends to have high molecular weights. In principle, the process can be continuous; however, only the batch operation is practiced in the United States.

In Table 8, the VCM emissions are given as pounds of VCM per million pounds of PVC produced and as total pounds of VCM released per hour. The VCM emissions sources and rates are restricted to those plants for which data have been received. The data show that VCM emissions are about 4% of the VCM feed; this emissions level is similar to that observed for suspension polymerization. However, data from additional plants are needed to draw a firm conclusion regarding the range and reliability of the emission levels.

The dryer vent, the air conveyor vent, the storage vent, and the VCM recovery vent appear to account collectively for up to 85% of total emissions. Carbon adsorption, ozone oxidation, and steam stripping could reduce these emissions 50 to 99%. Absorption might achieve 90%. The recycle of airstreams could be 40 to 80% efficient.

Fugitive losses contribute 3% of the emissions and the resin receiver vents contribute another 7%. For either source, vacuum stripping, would effect 50 to 80% reduction; carbon adsorption, 50 to 99%. Condensation with refrigeration would reduce either source 40 to 60%; absorption, 50 to 80%. Preventative maintenance, would reduce fugitive losses 25 to 50%. The surge tank vents could be recycled (40 to 60% reduction) or oxidized with ozone (50 to 99% reduction).

Emissions may be reduced by process modifications. Of all the process modifications, vacuum stripping after polymerization is one of the most important. At least two companies are practicing this method in emulsion PVC polymerization, just as in suspension polymerization. This should be considered a state of the art control. The cost for retrofitting vacuum stripping is guesstimated at $250,000 to $500,000 per unit.

Emissions may be combined; the combined airstreams may be treated at less cost than individual ones. Control methods may be combined and/or used in sequence; for example, ozonation is expensive, but may be efficient as a final step to destroy VCM. All of these should be considered in the developmental stage and in need of evaluation.

In bulk polymerization (Figure 3), only VCM and small amounts of additives are charged to the reactors; inert diluents such as water or solvents are not used. The Pechiney-St. Gobain bulk polymerization process consists of two stages. In the prepolymerizer, the additives are slurried in monomer, which is then about 10% converted to insoluble polymer. In the autoclave, VCM is polymerized in a "paste" phase to obtain 90% conversion.

The advantages claimed are [1] no impurities in the final PVC and [2] low capital and operating costs because separation of an inert diluent is not required. Data from only one company is available. In Table 9, the VCM emissions are given in pounds of VCM per million pounds of PVC produced and as total pounds of VCM released per hour.

TABLE 8: EMISSION RATES FOR PVC EMULSION PROCESS

PROCESS EMISSION SOURCE[a]	PLANT A		PLANT B	
	lb VCM/MM lb produced[b]	% VCM emissions	lb VCM/MM lb produced[b]	% VCM emissions
1. Unloading losses, I	Part of vent 9		Part of vent 9	
2. VCM feed tank vent, I	Part of vent 9		Part of vent 9	
3. Reactor emergency vent, I				
4. Slurry hold tank vent, C	219(1.0)	0.6	3026(11.4)	7.2
5. Dryer vent, C	10,293(490)	25.6	35,570(134.0)	84.0
6. Air conveyor vent, C				
7. Storage vent, C				
8. Resin-screening vent, C				
9. VCM recovery vent, C	26,958(1231)	76.0	796(3.0)	1.9
10. Polymerizer cleaning, I	2738(12.5)	6.8		
11. Packing/hopper car, I				
12. System cleaning, I				
13. Fugitive losses, C&I			2920(11.0)	6.9
14. Abnormal losses, I				
Total	40,208(183.6)		42,312(159.4)	

[a]C is continuous, I is intermittent.

[b]Value in parentheses is lb VCM/hr.

Source: B.H. Carpenter (6)

FIGURE 3: SIMPLIFIED FLOW DIAGRAM FOR BULK POLYMERIZATION OF VINYL CHLORIDE

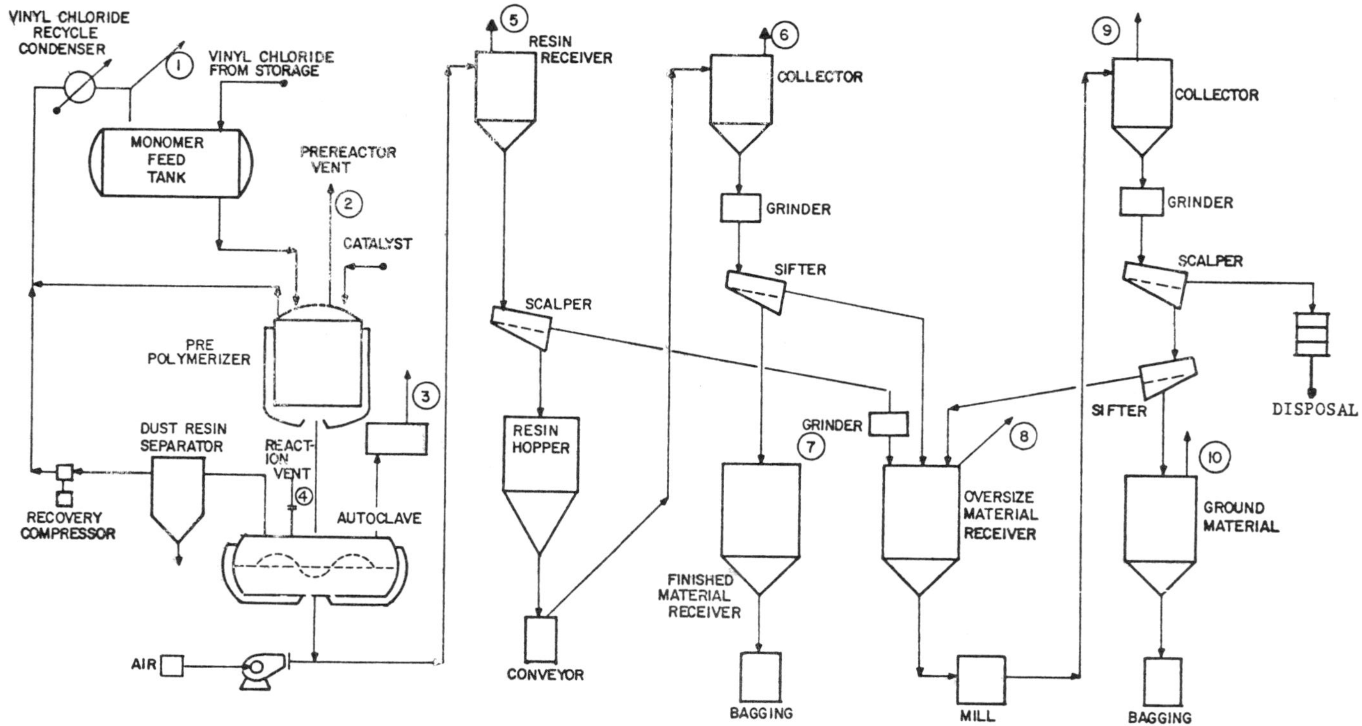

Source: B.H. Carpenter (6)

TABLE 9: EMISSION RATES FOR PVC BULK PROCESS

PROCESS EMISSION SOURCE[a]	PLANT A		
	lb VCM/MM lb produced	lb VCM/hr emitted	% VCM emissions
1. VCM feed tank vent, I	7900	29.1	33
2. Prereactor vent, I	Part of vent 3	Part of vent 3	
3. Reactor, I	2800	10.3	12
4. Reactor emergency vent, I	0	0	0
5. Resin receiver vent, C	2000	7.4	
6. Resin collector vent, C	Part of vent 5	Part of vent 5	
7. Finished product storage vent, C	Part of vent 5	Part of vent 5	20
8. Oversized PVC storage vent, C	Part of vent 5	Part of vent 5	
9. Second collector vent, C	Part of vent 5	Part of vent 5	
10. Ground material vent, C	Part of vent 5	Part of vent 5	
11. Fugitive losses, C&I	800	29.4	35
Total	13,500	76.2[b,c]	

[a]C is continuous, I is intermittent.

[b]PVC emitted at 2800 lb VCM/MM lb produced (10.3 lb VCM/hr emitted).

[c]Or 3.1% VCM in PVC.

Source: B.H. Carpenter (6)

The economically attractive bulk polymerization may offer more resistance to a reduction of emissions than suspension polymerization. Dry steam may enhance the vacuum stripping action and reduce strongly the VCM emission.

The VCM reactor vent (12%), fugitive emissions (35%), and the combined resin receiver, collector, and storage (20%) are the major emission sources. For the reactor vent, adsorption could achieve 50 to 90% reduction. Intensive maintenance is believed to be capable of giving 50 to 75% reduction of the diverse, ill-defined fugitive emissions (which need further study). The product-collection systems vents could be water washed and then adsorbed (50 to 90% reduction), oxidized with ozone (90% reduction), or incinerated (50 to 90% reduction).

Solution polymerization (Figure 4) is often conducted in systems where PVC is slightly soluble in the solvent and VCM is soluble. The products tend to have low molecular weights because the solvents act to some extent as chain transfer agents; therefore, straight comparisons with the other polymerization methods are not possible.

Absence of water simplifies the recovery of the final product and is the major advantage for this process. If a volatile solvent like n-butane is used, unreacted VCM and solvent are easy to remove.

No data are available on solution polymerization. Total losses are expected to be approximately 2 to 3% VCM and 1% PVC fines on PVC produced. Reduction should be by control steps similar to those indicated for the suspension process.

The solution process probably leads to a PVC recovery of 96 to 97% of VCM intake. Most likely, the actual VCM losses total 2 to 3%. These losses are not specified as to different vent streams.

While no data are available at present, solution polymerization is expected to have the emission characteristics of the suspension polymerization process and to respond similarly to the same controls.

FIGURE 4: SIMPLIFIED FLOW DIAGRAM OF SOLUTION POLYMERIZATION PROCESS FOR PVC

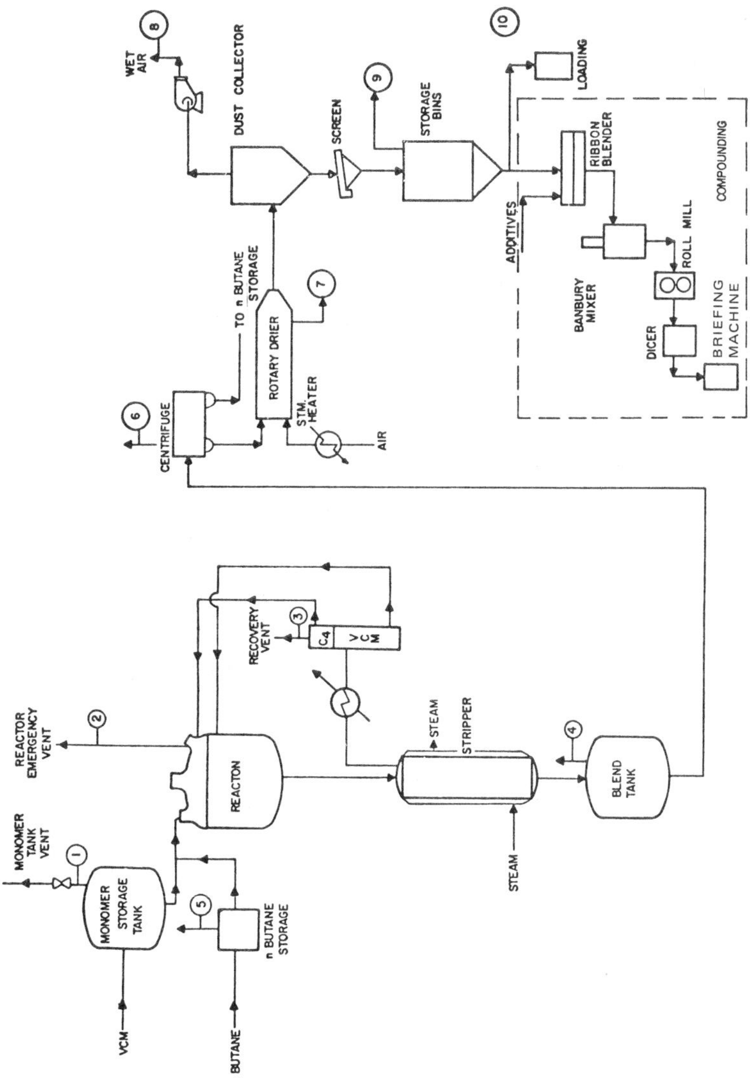

Source: B.H. Carpenter (6)

A process developed by *A.A. Bellisio; U.S. Patent 3,807,138; April 30, 1974; assigned to GAF Corporation* relates to the recovery of the polyvinyl chloride vapors vented from polyvinyl chloride reactors. More particularly, it relates to the recovery and purification of the vinyl chloride vapors vented from polyvinyl chloride reactors and accumulated with associated purge gases.

Figure 5 shows the essential features of this process. Referring to the drawing, vinyl chloride-containing gaseous mixtures from polyvinyl chloride reactors, not shown, are vented to an atmospheric gas-holder accumulator **1** in the vinyl chloride recovery section of the overall operation.

FIGURE 5: LACTAM SCRUBBING PROCESS FOR RECOVERY AND PURIFICATION OF VINYL CHLORIDE VENTING FROM PVC REACTORS

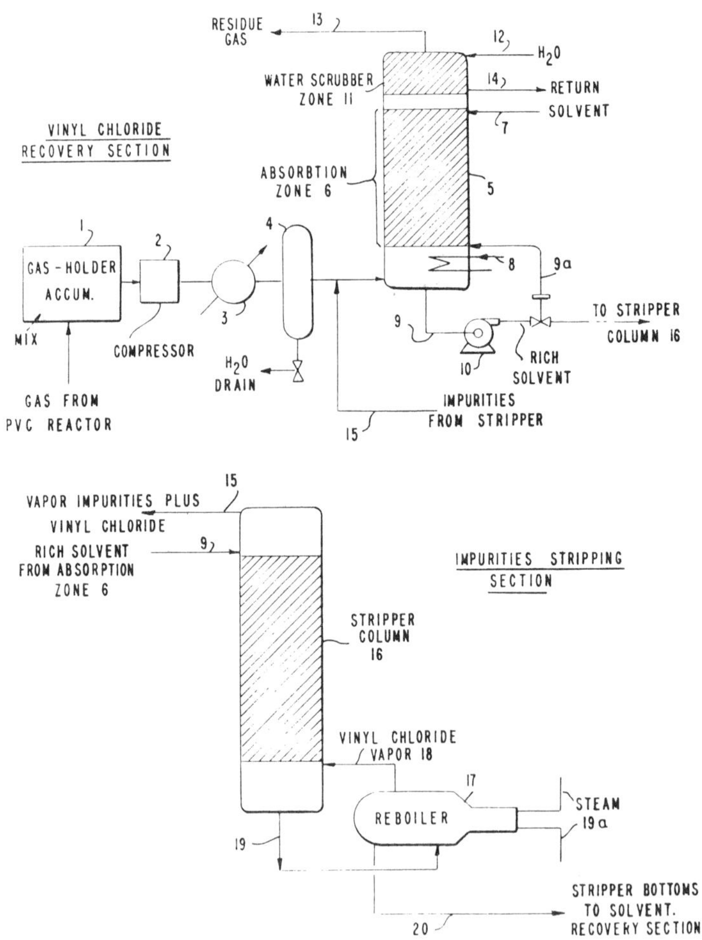

Source: A.A. Bellisio; U.S. Patent 3,807,138; April 30, 1974

Inert purge gas employed to assure that all of the vinyl chloride vapors are removed from the reactors is likewise passed to the gas-holder **1** in which a propeller mixer may be employed to produce a gaseous mixture of uniform composition. A vinyl chloride-containing gas stream to be treated is passed from gas-holder **1** to compressor unit **2** in order to bring the gas stream up to a suitable pressure for the efficient absorption of vinyl chloride vapors therefrom and to reduce the water vapor content thereof.

The gas stream will generally be compressed to a pressure of 25 to 100 psig. The gas stream from compressor **2** is passed through a suitable heat exchange unit **3** in order to reduce the temperature of the gas stream from the order of 90° to 110°C to from 10° to 40°C. The thus pressurized and cooled gas stream passes into water knock-out tank **4** from which liquid water is drained prior to the passage of the gas stream into absorption zone **6** of column **5** for ascending passage therethrough in countercurrent contact with a descending stream of liquid lactam solvent composition introduced into the upper portion of the absorption zone **6** through line **7**.

Cooling water may be circulated through the lower portion of column **5**, as illustrated by line **8**, for indirect heat exchange to assure that absorption zone **6** is maintained at a temperature within the desired range. A rich lactam solvent composition stream having vinyl chloride absorbed therein is withdrawn from the bottom of column **5** through line **9** for passage by means of pump **10** to impurity stripper section shown in the lower part of Figure 5. A side stream **9a** of the rich solvent stream in line **9** may be recycled to the lower portion of absorption zone **6** for maintaining the desired temperature control in the absorption zone.

The treated gas stream emerging from absorption zone **6** of column **5** passes upwardly directly into water scrubber zone **11** of the column **5**. In zone **11**, the ascending stream of gas passes in countercurrent contact with a descending stream of water introduced into the upper portion of scrubber zone **11** through line **12**. Upon contact with the ascending stream of gas, the water in scrubber zone **11**, serves to effectively remove from the gas stream essentially all of the lactam that is either vaporized or entrained in the gas stream leaving absorption zone **6** of column **5**. The treated residue gas exhausted from column **5** through line **13**, therefore, not only has a substantial portion of its vinyl chloride content removed therefrom, but is also substantially free of vaporized or entrained lactam solvent.

This latter feature is highly significant from a practical operating viewpoint, in that any appreciable loss of the lactam solvent in the gas stream exhausted from column **5** would constitute a major operating disadvantage adversely effecting the overall economic evaluation of the vinyl chloride recovery method of this process.

Water having recovered lactam therein is shown being withdrawn from scrubber zone **11** through line **14**. It will be understood that this stream will thereafter be treated to recover the lactam content thereof for subsequent use in the absorption of vinyl chloride from additional quantities of the gaseous mixture vented from polyvinyl chloride reactors. This subsequent lactam recovery operation can be accomplished in a number of ways as will be readily appreciated by those skilled in the art.

The rich lactam solvent liquid stream in line **9** is passed to stripping column **16** having a reboiler **17**. In the stripping column **16**, the rich lactam solvent stream passes downwardly in countercurrent contact with an ascending stream of reboiling vapors entering column **16** from reboiler **17** through line **18**. Steam for the operation of reboiler **17** is furnished to line **19a**. The stripper reboiler vapors are essentially vinyl chloride, with the vapors being essentially free of acetylene, butadiene and methyl chloride.

Most of the lighter impurities present in the rich lactam solvent stream in column **16** are stripped therefrom, with the vapor impurities stripped from the rich solvent together with recycled vinyl chloride vapors passing from the column **16** through line **15** for recycle to absorption zone **6** as indicated above. Very little of the vinyl chloride is stripped from the rich solvent in column **16** since the stripping agent is itself vinyl chloride. Stripper bottoms from column **16** are passed through line **19** to reboiler **17**, from which a liquid stream is withdrawn through line **20**. This stripper bottoms liquid stream from reboiler **17** comprises the lactam solvent having the recovered vinyl chloride absorbed therein, but with essentially all of the vaporous impurities in the gaseous mixture accumulated in gasholder **1** removed therefrom. Also present in the liquid stream in line **20** are any acetaldehyde, sulfur and nonvolatile materials accompanying the gaseous mixture vented and purged from the polyvinyl chloride reactors.

The residue gas is thus discharged to the stack with the following composition.

Component	Pound Mols Per Day
Nitrogen	97.654
Acetylene	0.00004
Butadiene	0.00063
Methyl chloride	0.0063
Vinyl chloride	4.5
Water	1.03
Other impurities	Trace

Polyurethane Resin Manufacture

A report from Midwest Research Institute (25) characterizes the air pollutant emissions from a polyurethane resin manufacturing plant. TDI emissions were found to be at a maximum in the first part of the resin formation reaction. The report concluded that there was no emissions hazard. A scrubber used as an air pollution control device had negligible effect on the TDI emissions.

WATER POLLUTION FROM SPECIFIC PROCESSES

Acrylic Resin Manufacture

Polymers have been produced commercially from a wide variety of different esters of both acrylic acid and methacrylic acid. Specialty plastics have been produced from 2-haloacrylic esters and 2-cyanoacrylic esters. In recent years the production of methacrylate polymers has exceeded production of acrylic polymers. Since production methods are similar, methyl methacrylate polymers will be discussed here as representative of the acrylate resins.

Methacrylic acid, $CH_2{=}C(CH_3)COOH$, can be considered as the parent substance from which methyl methacrylate monomer and poly(methyl methacrylate) and all other methacrylate compounds are derived. In one process, methyl methacrylate monomer is manufactured starting with acetone cyanohydrin and 98% sulfuric acid. The methacrylamide sulfate formed as an intermediate is not isolated but reacts with methanol to produce methyl methacrylate. Both steps are carried out continuously.

Polymerization of methyl methacrylate produces poly(methyl methacrylate). The clarity, outstanding weather resistance, light weight, formability, and strength of poly(methyl methacrylate) have led to extensive use of acrylic plastics in aircraft glazing, signs, lighting, construction, transportation, appliances, and merchandizing. Because of excellent suspending, rheological, and durability characteristics, acrylic emulsions have wide use in paints for exterior and interior applications on wood, masonry, metals, etc.

Monomer is delivered to the polymerization manufacturer usually in tank car quantities. Low concentration (5 to 15 ppm) of inhibitor is often adequate for safe handling and storage of methyl methacrylate. Low inhibitor content is desirable, since subsequent polymerization without first removing the inhibitor is possible. The polymerization of methyl methacrylate may be conducted in a variety of ways. In commercial casting of sheets of poly(methyl methacrylate) each sheet is cast in a mold assembled from two sheets of plate glass spaced apart at the edges by a gasket (see Figure 6). The mold is filled by pouring in a charge of monomer with exact amounts of catalyst, colorant if desired, or other additives for special effects. The closed mold then goes through a controlled temperature cycle, generally between 40° and 75°C (113° and 158°F). Annealing often follows the casting process.

Homo- and copolymers can also be conveniently prepared by an emulsion technique. In a representative procedure shown in Figure 7, methyl methacrylate is emulsified with an anionic emulsifier and deionized water containing a little ferrous sulfate and ammonium peroxysulfate. The emulsion is flushed with nitrogen and then treated at 20°C (68°F) with small quantities of sodium metabisulfite and tert-butyl hydroperoxide. The temperature rises spontaneously to 70°C (158°F) and the polymerization is completed at this temperature in about 15 minutes. After the batch is cooled to room temperature, the product typically contains 35% total solids with a viscosity of about 6 cp and a pH of 2.5.

FIGURE 6: ACRYLIC RESIN PRODUCTION—BULK PROCESS

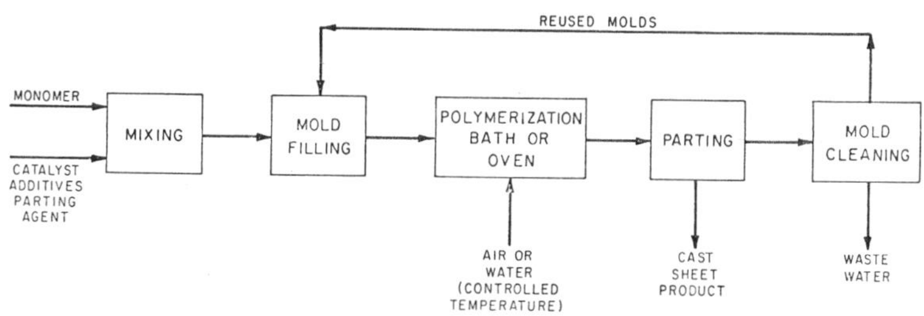

FIGURE 7: ACRYLIC RESIN PRODUCTION—EMULSION PROCESS

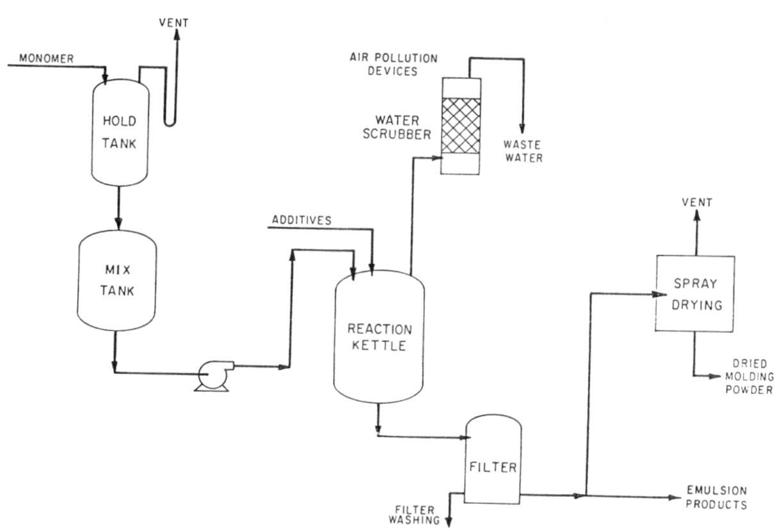

Source: EPA (7)

In suspension as well as in emulsion polymerization, water is used as a heat transfer medium in the reaction zone, but the methods differ in the state in which the polymer is obtained. Suspension polymerization (see Figure 8) yields discrete beads, granules, or particles ranging in size from a few microns to a fraction of an inch in diameter.

With water as the reaction medium, the following factors have to be controlled for successful suspension polymerization methods:

[1] The initiator should be soluble in the monomer and insoluble in water. This prevents the polymerization from occurring in the aqueous phase and is especially necessary when the monomer is appreciably soluble in water.

[2] Suspending agents may be used to prevent droplet contact and merging and to aid in the suspension of polymerizing droplets. Such agents may be soluble products such as cellulose derivatives, starches, gums, and salts of acrylic polymers; polyvinyl alcohol, or they may be insoluble materials such as clay or talc. Thickeners such as polyoxyalkylene derivatives may also be present to prevent droplet contact.

[3] Inorganic salts are often added to increase the density of the aqueous medium, to reduce the water solubility of the monomers, and to increase the interfacial tension of the system.

FIGURE 8: ACRYLIC RESIN PRODUCTION–SUSPENSION PROCESS

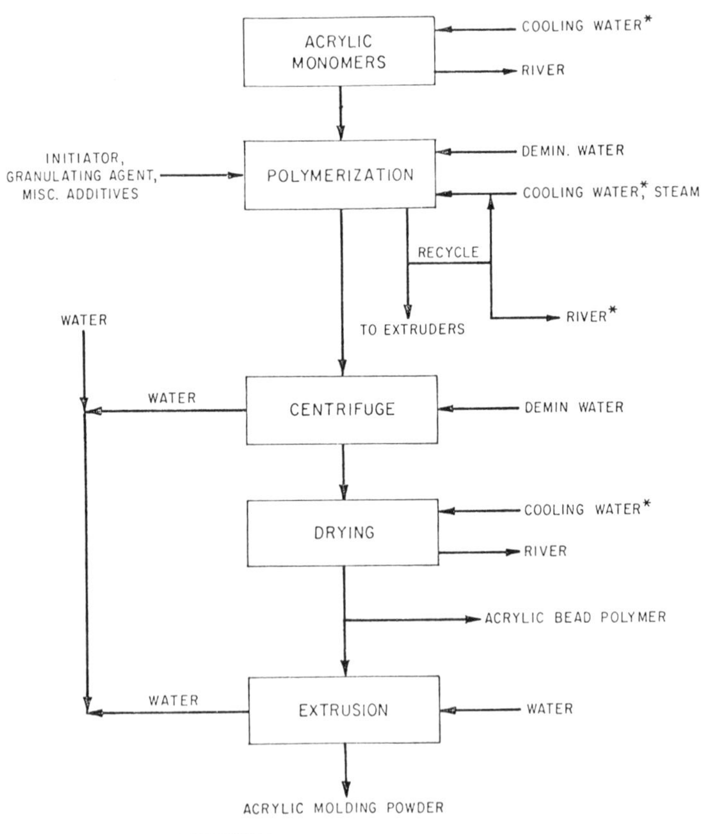

Source: EPA (8)

The primary wastewater streams are obvious from inspection of the process schematic diagram. Upon cooling, the polymer product is washed, usually within the cooling vessel, and the water or brine leaving this vessel will be contaminated with some monomer, some polymer, and the various stabilizing, emulsifying, and chain-regulating agents as well as the catalyst. Final dewatering occurs in the centrifuge, and the wastewater stream from this equipment will contain the same type of contaminants listed above.

If the monomer is assumed to be present at its saturation concentration in the wash water, it will comprise approximately 1.5% in the waste stream. If two volumes of water per volume polymer are used to wash the product beads, then 1,335 mass units of water will be released per 1,000 mass units of polymer product (SG = 1.5).

There is no water of reaction for the polymerization of methyl methacrylate. Oil and grease are due to the presence of lubricants used for the extrusion process. Depending on the specific wastewater chemical conditions and the analytical methods used, cyanides may be detected due to the presence of cyanoacrylic esters and acetone cyanohydrin.

According to a process developed by *J. Greenberg; U.S. Patent 3,766,087; October 16, 1973; assigned to Anti-Pollution Systems, Inc.*, wastewater comprising an emulsion of acrylic residues may be treated by a process wherein the polluted water is fed onto the surface of a brine solution maintained at the boiling point, the emulsion being broken instantly at the surface of the brine and accumulated as an organic residue, while the water is boiled off. The organic residue is skimmed off for separate handling.

Figure 9 shows a suitable form of apparatus for the conduct of the process. Referring to the drawing, there is shown a container **10** in which is maintained a brine solution **11** having a level **12**. The solution is preferably a concentrated brine solution of water and salt. The process is nonspecific as to the salt, and may be worked with good results with ordinary NaCl, chlorates, nitrates and carbonates. In a preferred embodiment, the brine is a saturated solution of NaCl (39.8% by weight), with some small amount of salt settling at the bottom of container **10.**

FIGURE 9: PROCESS FOR TREATING WATER CONTAINING ACRYLIC RESIDUES

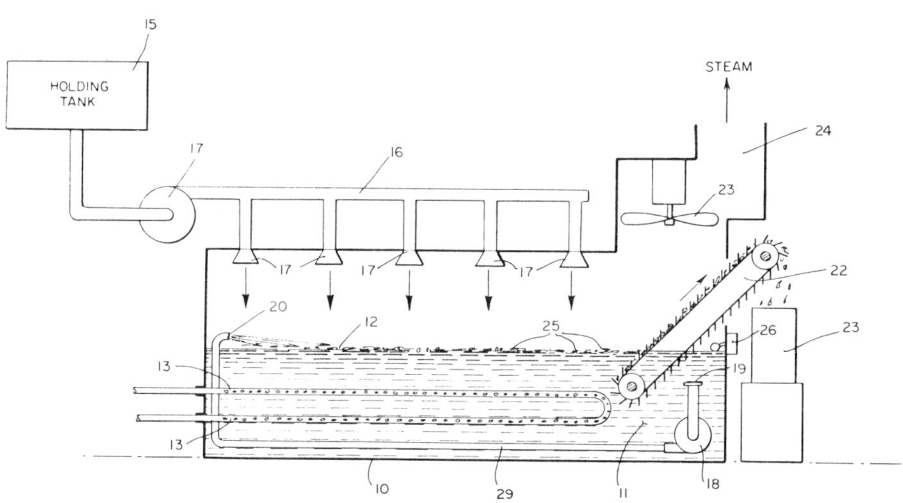

Source: J. Greenberg; U.S. Patent 3,766,087; October 16, 1973

A gas fired immersion tube **13** is positioned within container **10** below the brine surface **12**, and provides a source of heating energy to maintain the brine at or just above the boiling level. In the preferred embodiment, the brine solution is maintained at the boiling point, i.e., above 100°C. It is, of course, understood that any suitable means of heating may be employed.

The polluted water solution is fed from a holding tank **15** through a spigot feed **16**, having a plurality of spigots **17** from which the polluted solution is entrained or dropped onto the surface **12** of the brine. A pump **17** is used for transporting the solution into the spigot feed.

A second pump **18** takes brine collected from an inlet **19** near the surface **12**, and pumps it through pipe **29** to an ejecting orifice **20** positioned so that the emitted brine pushes the accumulated organic residue (noted at **25**) on the surface **12** from left to right as seen in the drawing. There is thus, in the illustration, a clockwise circulation of the brine, transporting the surface accumulation of organic residue, or scum, toward a conventional skimmer **22** which removes the scum and deposits it into a collection drum **23**. This skimming system is exemplary only, and any type of conventional skimming apparatus and/or system may be utilized which has the function of transporting the scum on the surface to a point where it may be skimmed or otherwise removed into a suitable collection instrument.

In operation, the polluted solution is spread from spigot **17** and distributed substantially uniformly onto the surface **12** of the brine. The brine, which is maintained at the boiling point, causes the emulsion to be broken instantly at the surface, such that it accumulates as organic residue. This effect in breaking the emulsion is also known as "salting-out." At the same time that the emulsion is being broken, the water is being boiled off, and is drawn by exhaust fan **23** through port **24** where it is passed into the atmosphere in a safe and substantially pure form. The remaining organic residue, or scum, is pushed and/or drawn toward the skimmer and removed into the collection drum.

Since there is about a 5% carryout of the salt with the curdled scum, or organic matter, the salt must be periodically replenished. In practice, the pump **18** and skimmer **22** operate continuously, drawing off the unburned residue. The system has been operated successfully at a rate of pressing 400 gal/hour of polluted water having a 2% emulsion content. It is, of course, understood that the system capacity may be enlarged by appropriate engineering design, and the method is not limited by the system rate.

A water balance is maintained in the system, by adapting the rate of feed from the holding tank to the vapor exhaust rate, such that the amount of water passing through port **24** substantially equals that being fed in through spigot feed **16**. The surface **12** may be monitored, by conventional level detecting device **26**, with a signal derived to control the operation of pump **17** so as to provide a closed loop servo-type system for maintaining surface **12** within any prescribed range.

As noted before, the salt that is used in the process is nonspecific. The process disclosed herein may be performed with excellent results using NaCl, chlorides, nitrates and carbonates. It is to be understood that the process may be practiced at higher temperatures, by maintaining a salt bath of higher concentration. However, maximum efficiency is obtained by running the system relatively cool, i.e., no higher than 230°F, within which temperature efficient breaking of the emulsion is obtained, as well as efficient boiling off of the water. Indeed, it is a primary feature of this process that it can be carried out at a low temperature, contrasted with prior art schemes requiring high temperatures and high levels of energy consumption.

Amino Resin Manufacture (Urea and Melamine)

The term "amino resins" is used to describe a broad group of polymers formed from formaldehyde and various nitrogen containing organic chemicals. The nitrogen groups is in the form of the NH_4 radical. Although called amino resins most of the compounds used are more of the nature of amides than true amines. The resins are characterized as being thermosetting, amorphous materials which are insoluble in most solvents. Although many amino compounds are used in the formation of amino resins, the two of primary commercial significance are urea and melamine. Specialty materials are formed from other amino compounds such as thiourea, acrylamide or aniline. These, however, are produced only in small volumes and have little significance in the total amino resin market.

Formaldehyde, the common raw material in all types of amino resins, is normally a gas but is handled industrially as an aqueous solution. It is infinitely miscible with water. Urea, a solid under normal conditions, is highly soluble in water. Melamine could be described as sparingly soluble and is also a solid under the usual conditions, melting at the high temperature of 355°C (671°F).

Another characteristic of the group of amino resins is that the polymerization reaction proceeds in two distinct stages. In the first of these, as indicated in Figure 10, Equations (1) and (2) urea (depending upon the mol ratio of the reactants) forms materials such as monomethylol urea and dimethylol urea which are the reactive monomers involved in the final polymer. As indicated in Equation (3), these materials may react among themselves to form dimers. Although the structure of just one dimer is shown, a consideration of the active hydrogen groups involved shows that many other dimers containing both methylene and ether linkages are possible. The initial reaction is an addition reaction with no water formed as a result of the combination. The condensation reaction, as indicated by Equation (3), involves the formation of one mol of water for each linkage formed.

FIGURE 10: TYPICAL POLYMERIZATION FOR UREA AND FORMALDEHYDE

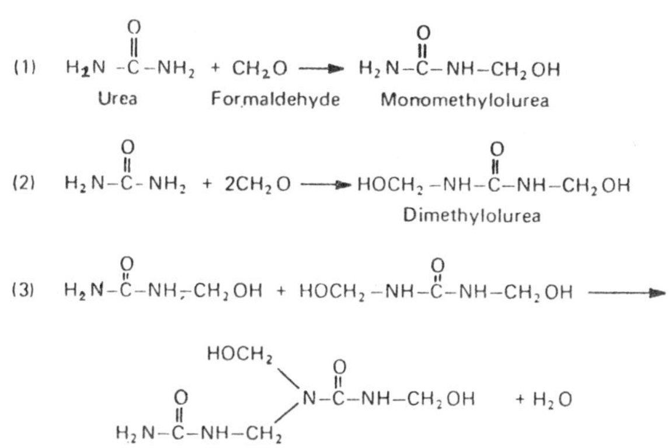

Source: EPA (8)

As shown in Figure 11, the reactions in the case of melamine and formaldehyde are entirely analogous to those shown for urea-formaldehyde. It should be noted, however, that since melamine contains three NH_2 groups, permutations are much greater than is the case for urea. Again, the first two reactions indicate the initial step of the polymerization. This consists of the formation of reactive monomers between melamine and formaldehyde. The further reactions, as indicated schematically by Equation (3), can involve the reaction of an additional mol of melamine with one of the monomers, shown in this case as

FIGURE 11: TYPICAL POLYMERIZATION REACTIONS FOR MELAMINE AND FORMALDEHYDE

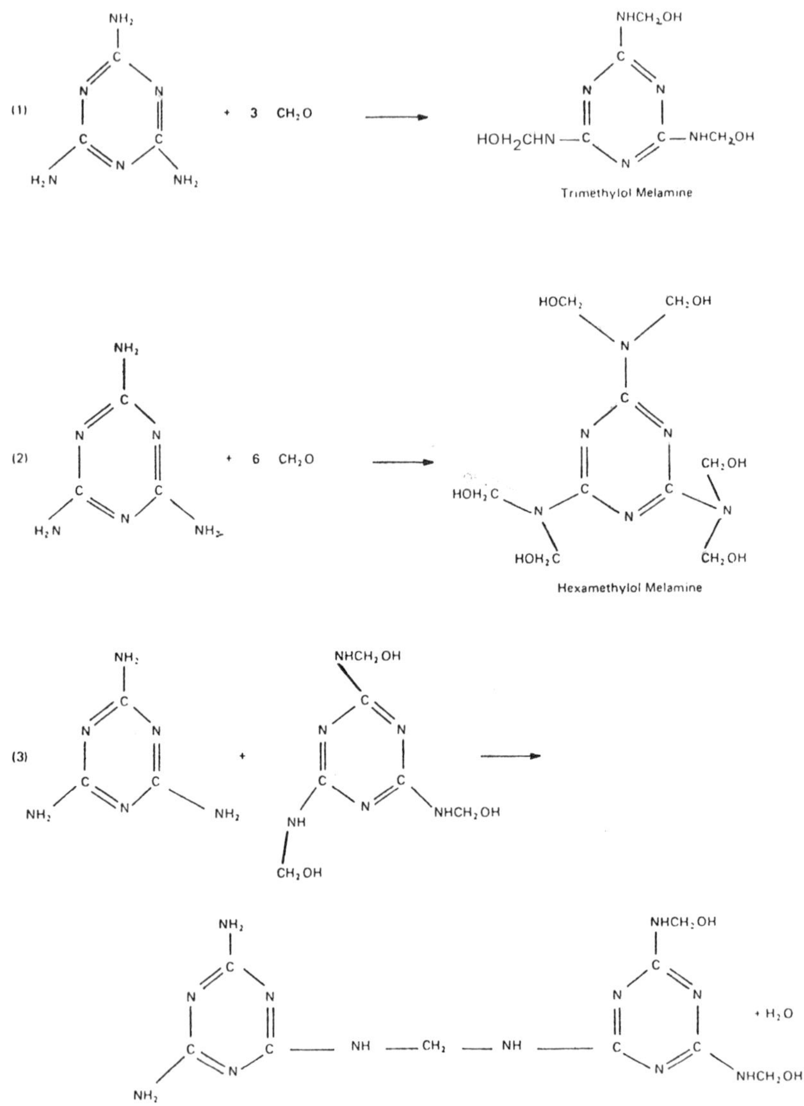

Source: EPA (8)

trimethylol melamine, to form condensation compounds which involve the elimination of water of reaction. Although not shown, it can be readily visualized that a mol of trimethylolamine could react with an additional mol of trimethylolamine to eliminate water and form an ether linkage as contrasted to the methylene linkage formed between the trimethylolamine and another molecule of melamine.

These reactions are catalyzed by hydrogen ions and, in general, are moderated or slowed down by hydroxyl ions. Thus, the proper pH selection is an important consideration in determining the structure of the ultimate polymer formed.

The basic amino resin manufacturing process is generally stopped with the formation of a predetermined amount of monomers, dimers and trimers depending upon the specifications desired for the ultimate resin. This mixture of materials is then utilized by the customer to form the final thermoset resin which is an insoluble, heat resistant material. This is contrasted with the mixture of very low molecular weight materials produced by the basic manufacturer which are usually water soluble, very heat sensitive materials.

Consideration of the equations presented above will show there are numerous possibilities for cross-linking the various monomers, dimers and trimers which would be involved in the initial stages of the reaction. The ultimate customer forms these cross-links between the molecules by the application of heat and pressure, sometimes with the aid of a catalyst depending upon the nature of the application.

For most of these applications the resin is used in the form of either an aqueous solution or a mixture of an aqueous and alcohol solution, ethanol being the usual alcohol. For molding compounds and some of the others, a solid material is utilized. In nearly all of these applications, the melamine part of the amino resin family has superior properties. Because of its higher cost, however, it is utilized principally where these superior properties are necessary. The urea-formaldehyde resins, which are lower cost, are equally applicable in other instances.

Since, as mentioned above, the reactive monomers, polymers, trimers and low molecular weight material formed by the basic resin manufacturer contain all of the reactive groups necessary to further crosslink, the solution materials have a limited shelf life, in the order of 60 days. Thus the users who have a large volume requirement for solution forms, such as paper mills, textile mills, and the like, may purchase material made in solution form by the manufacturer since they will utilize it quickly and not have a residual inventory. Other users, where the shelf life of the product is of considerable importance, will purchase the material in an anhydrous solid form which has a relatively indefinite shelf life. Often, before the final use, the solid may be redissolved in either water or alcohol or mixtures thereof if a solution form is utilized in the application.

Since amino resins are produced in many specialty grades with each run being a relatively modest volume, continuous processes are not in general use in the industry. The typical process is a standard batch polymer kettle arrangement. As shown in Figure 12, the normal arrangement consists of a jacketed polymer kettle ranged in size from 7.6 to 38 m^3 (2,000 to 10,000 gal). The larger sizes contain internal coils for additional heating and cooling surface in order to provide a reasonable surface-to-volume ratio. The kettles are agitated and can operate under either pressure or vacuum conditions.

The kettle is equipped with a water-cooled condenser and tied into a vacuum system so that the operating temperature can be controlled through the use of both reflux and cooling or heating in the jacket and coils of the kettle. The feed system consists generally of weigh tanks for the batch operation of the kettle.

The techniques used are very similar for both melamine or urea types of formaldehyde amino resins. As a typical example, the production of a plywood adhesive grade urea-formaldehyde resin is as follows. Formaldehyde as a 30% solution is added to the kettle and the pH adjusted to about 7 to 7.8. Boric acid, the catalyst, is then added, and then

FIGURE 12: AMINO-FORMALDEHYDE RESIN PRODUCTION

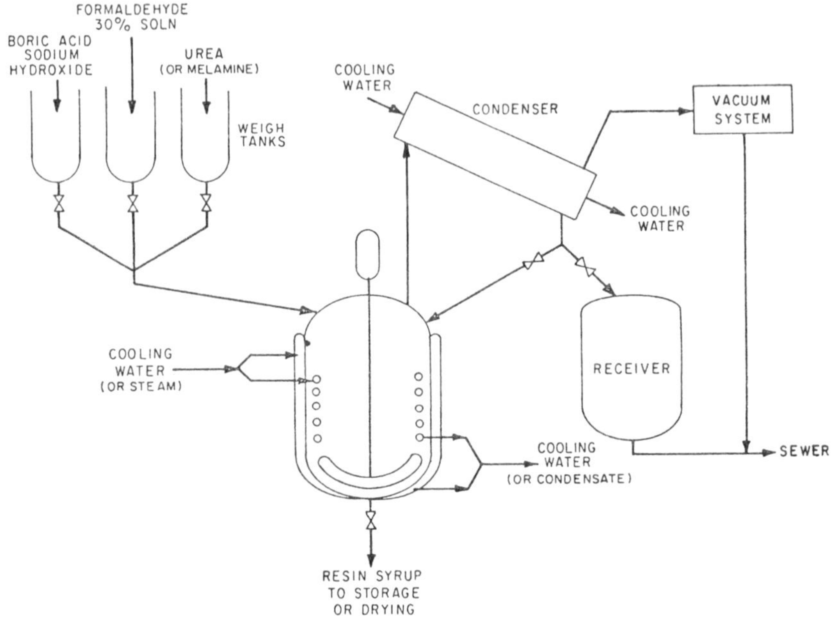

Source: EPA (8)

urea in the form of a solid is fed into the reaction vessel. The pH of the mixture is again brought back to approximately neutral and the mixture heated to 100°C (212°F) under atmospheric reflux conditions.

During this initial heating period the pH drops to about 4 as the reaction between urea and formaldehyde takes place to form di- and trimethylol urea. Atmospheric reflux is maintained for a period of about 2 hours. Then the vacuum is applied, and the system temperature drops to approximately 40°C (104°F). It is maintained at this level for approximately five hours. During this period of time there is a small amount of condensation reactions taking place between the various monomers formed earlier. Simultaneously with this further reaction, water is removed from the system so that the final water content, in the case of this particular adhesive formulation, is about 50%.

The water in the system comes from two sources, that introduced with the 30% formaldehyde solution used as a raw material, and that produced by the reaction between the monomers, which eliminates a mol of water for each pair of monomers or trimers reacting.

At the end of the vacuum reflux period, the system is put on total reflux and the pH adjusted to slightly alkaline conditions. The reactor is then returned to atmospheric pressure, and the product is ready to be removed. The total cycle time is about 10 hours.

The mixture, at this point in the form of a thick syrup, is drained to storage where quality checks are made to determine the exact condition of the polymers. The material may be shipped in this form for further polymerization by the customer or it may be dried to be shipped as a solid which, as mentioned earlier, has a much longer shelf life. If the material is to be dried, it is fed to either a belt drier or a spray drier where the remaining water is removed at low temperature in order to prevent further polymerization. As mentioned earlier, the final adjustment of the pH also helps prevent further condensation reaction and polymerization of the monomers. The water removed during these final drying operations is vented to the atmosphere.

Depending upon the end-use requirements, the final solid product may be milled with pigments, dyes and fillers to provide a molding compound suitable for the particular end-use desired.

The equipment used for the production of the first-step amino resins is often used for other materials, such as phenolics. Between these different uses, and indeed between production batches of melamine and urea resins or between batches of significantly different resins, it is customary to clean the equipment by utilizing a hot dilute caustic solution. This material is drained as process waste.

Cellulose Acetate Resin Manufacture

Cellulose acetate resin (flake) is produced by a batch type operation shown schematically in Figure 13. Purchased, dissolving grade wood pulp is defiberized in attrition mills, pretreated with acetic acid to activate the cellulose, and charged to acetylation reactors where it is reacted with acetic anhydride in the presence of glacial acetic acid. Sulfuric acid is added as a catalyst. The acetic acid/anhydride mixture is prechilled by refrigeration to aid in removing heat of reaction. The reaction is carried to nearly complete acetylation of the cellulose.

The clear, viscous solution is then transferred to a hydrolysis reactor where dilute aqueous acetic acid is added, and the acetate hydrolyzed back to the specified acetyl content. Some magnesium acetate may be added to adjust the concentration of sulfuric acid which also serves as a catalyst for the hydrolysis. The hydrolysis step is necessary to remove sulfate ester groups and to provide close control of the final acetyl content. At the desired point, the reaction is stopped by adding magnesium acetate to neutralize remaining sulfuric acid.

The overall reactions involved in the production of cellulose acetate and triacetate may be represented as follows:

Acetylation

$$\underset{}{R(cell)(OH)_3} + 3(CH_3CO)_2O \longrightarrow \underset{\text{cellulose triacetate}}{R(cell)(OCOCH_3)_3} + 3CH_3COOH$$

Hydrolysis

$$R(cell)(OCOCH_3)_3 + XH_2O \longrightarrow \underset{\text{cellulose acetate}}{R(cell)(OH)_x(OCOCH_3)_{3-x}} + XCH_3COOH$$

Cellulose acetate flake is recovered from the reaction solution on a continuous precipitator by precipitation with weak acetic acid solution from the countercurrent wash step that follows. In the flake washing process, fresh water enters the second stage washer, and flows countercurrent to the flake through the second and first-stage washers. As noted above, the water from the first-stage washer is used for the flake precipitation step. This water, which is separated from the flake on the vibrating screens, is sent to acid recovery. Entrained fines are collected in filters and recycled to the process.

Process wastewaters from acetate resin production are treated in an acid recovery plant to recover acetic acid. A recovery plant flowsheet is shown in Figure 13. The process waste streams are filtered and held in a filtered acid tank. Miscellaneous streams with sufficient acid value may also be collected in this tank.

Acetic acid is separated from the process water by solvent extraction and distillation. Glacial acetic acid is recovered from the bottom of the still. Reportedly about 99.8% of

FIGURE 13: CELLULOSE ACETATE RESIN PRODUCTION

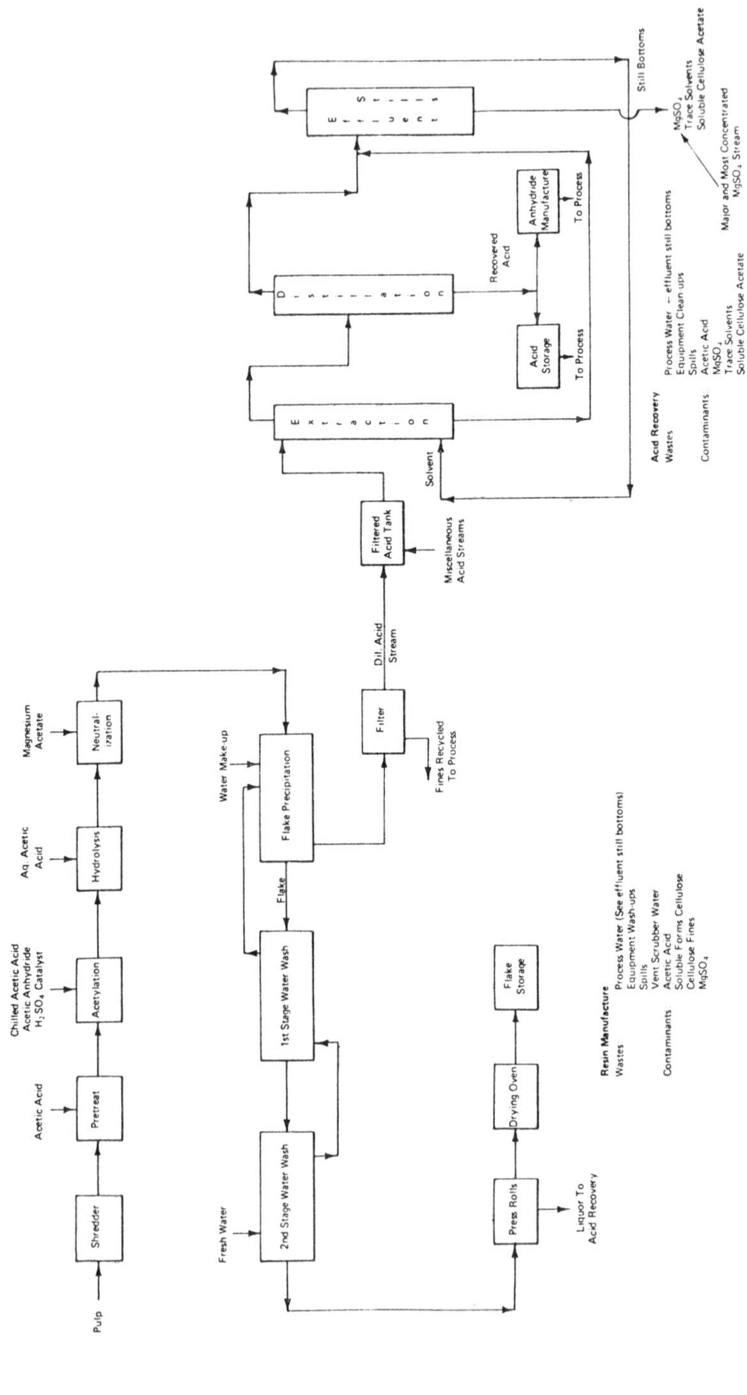

Source: EPA (9)

the acetic acid in the collected process water is recovered in this operation for reuse. A portion of the acid is converted by a catalytic pyrolysis process to anhydride for the acetylation reaction.

The solvent and water mixture from the top of the acid recovery stills is sent to effluent stills where the solvent is recovered and recycled to the extraction column. The water removed from the bottom of the effluent stills flows to waste treatment. This stream represents the major source of dissolved solids (magnesium sulfate) in the plant discharge.

Cellulose Ether Manufacture

This group of materials includes ethylcellulose, methylcellulose, carboxymethylcellulose, and and hydroxyethylcellulose. All are ethers of cellulose.

The first three of these derivatives are made by reaction of an alkyl chloride – ethyl chloride, methyl chloride, and chloracetic acid – with cellulose. Hydroxyethylcellulose is made by reaction of ethylene oxide with cellulose. Several of the commercial grades of methylcellulose are mixed ethers, made by reaction of propylene oxide as well as methyl chloride with cellulose.

All of the reactions are run using alkali cellulose, a mixture of cellulose with sodium hydroxide. In the course of the reactions involving alkyl halides, the alkali is neutralized by formation of sodium chloride. This salt, and excess alkali, must be removed from some of the products to provide materials that are usable.

Figure 14 shows a block flow diagram for production of cellulose ethers. The use of a solvent in the process maintains the cellulose as a relatively easy-to-handle slurry. Depending on the cellulose derivative involved, the solvent may be either an alcohol or a hydrocarbon.

FIGURE 14: CELLULOSE ETHERS PRODUCTION

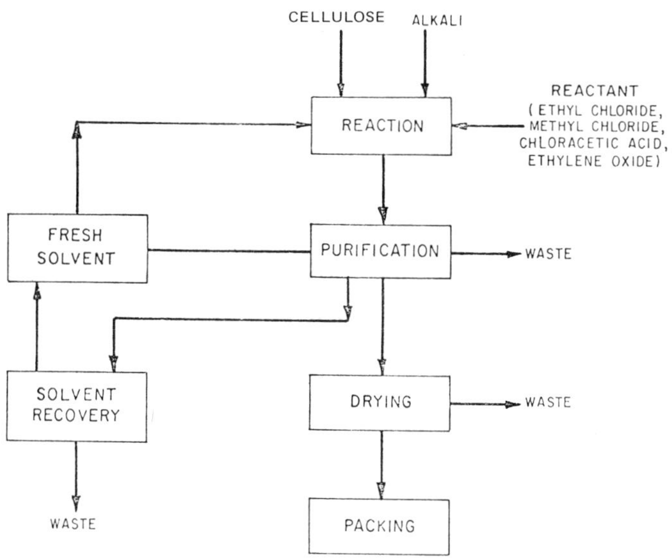

Source: EPA (7)

Proprietary processes appear to be widely used in production of cellulose ethers. Manufacturers refuse to discuss the processes in any detail. Uses of the cellulose ethers are varied. Ethylcellulose is a plastic. Methylcellulose, carboxymethylcellulose, and hydroxyethylcellulose are water soluble and are generally used in applications involving water solubility. Carboxymethylcellulose may be used as a suspending agent. Methylcellulose is used as a film former. All three ethers are used in certain foodstuffs.

Wastes generated in production of cellulose derivatives constitute alkali, salt, solvent residues, pulp, and treatment chemicals. These wastes indicate relatively high BOD, COD, and dissolved solids levels. Organic nitrogen may be present in the wastewaters of facilities where nitrogen containing cellulose derivatives (other than cellulose nitrate) are produced.

Cellulose Nitrate Manufacture

Cellulose nitrate is produced by reaction of fibrous cellulose with a mixture of nitric and sulfuric acids. One may calculate that cellulose nitrate contains 14.1% nitrogen. The commercial product contains about 12% nitrogen; this level is attained by using mixtures containing carefully controlled amounts of nitric acid, sulfuric acid and water. These liquids are present in the mix in the approximate proportion 1:3:0.75. The fibrous nature of the original cellulose is essentially unaltered by the reaction. In order to provide a commercially useful product, the initially obtained nitrate is taken through the following processes:

[1] Washing with water to remove all acid.
[2] Stabilization by boiling with water to remove small amounts of combined sulfuric acid.
[3] Digestion (heating in the presence of water) to reduce viscosity of the product to a useful level.
[4] Dehydration or exchange of the water, by alcohol.

The steps listed above are shown in Figure 15 which also indicates wastewater streams.

FIGURE 15: CELLULOSE NITRATE PRODUCTION

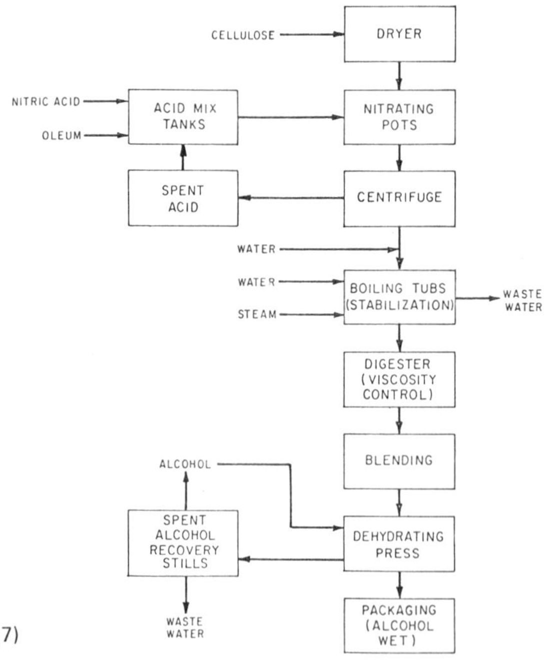

Source: EPA (7)

Aqueous wastes generated in the manufacture of nitrocellulose constitute primarily acids (both nitric and sulfuric) and alcohol lost in the dehydration process. Spent acids are recovered as far as possible, but some are inevitably lost. Alcohol is recovered and recycled. Suspended solids in the wastes include a small amount of cellulosic material. The strongly acidic wastes are handled by neutralization with lime. The calcium sulfate which is formed may be removed by settling.

Epoxy Resin Manufacture

Epoxy resins are characterized by the presence of the epoxy group within their structure. Rather than an end resin in itself, the epoxy family should be regarded as intermediates. They all require further reaction with a second component, or curing agent as the second material is often termed, in order to yield the final thermoset material.

Almost all of the commercially produced epoxy resins are made by the reaction between epichlorohydrin and bisphenol A. Small volumes, however, are produced from polyols other than bisphenol A, such as aliphatic glycols and novolak resins formed from phenol and formaldehyde. It is also possible to produce epoxy resins by introducing the epoxy group after the polymer has been formed. As example of this is the epoxidation of a polybutadiene material. The double bond present in these materials forms the site for the epoxy linkage. The following discussion, however, is limited to the materials produced from epichlorohydrin and bisphenol A.

Epichlorohydrin is a liquid with a boiling point of 117°C (242°F). Bisphenol A is a solid which melts at 152°C (305°F). Bisphenol A is insoluble in water, dissolving to the extent of 0.3% at 85°C (185°F) whereas epichlorohydrin is somewhat more soluble (approximately 5%). The reaction between the two raw materials takes place under alkaline conditions as shown by the equations in Figure 16. The first step, shown by Equation (1), is the condensation of the epichlorohydrin with the bisphenol A to form the chlorohydrin compound.

This compound is dehydrohalogenated with caustic soda to form epoxy linkages yielding diglycidyl ether or bisphenol A, as shown by Equation (2). Sodium chloride and water of reaction are also formed as by-products with the ether. Further reaction between the ether and additional bisphenol A results in growth in the chain length, as shown by Equation (3).

Operating conditions and type of catalyst are selected to minimize the formation of side chains and to prevent phenolic termination of the principal chain. The final resin properties are enhanced when the chain is terminated with epoxy groups, as shown in Equation (3), and when the chain is linear with minimum branching. The possibility of branching exists since epichlorohydrin could react with the hydroxyl group to start a side chain.

The product epoxy resins fall into two broad categories, the low molecular weight liquids and the high molecular weight solids.

In the liquids, n, the number of repeating units in the final chain as designated in Equation (3), is low, ranging in commercial materials from 0.1 to 0.6 as the average value. For solid materials, n ranges from 1.8 to 16. Control over chain length is exercised primarily by the ratio of the two reactants charged to the system. To produce the low molecular weight liquids, a large excess of epichlorohydrin is used so that n is close to 0 in the final product. In order to produce the high molecular weight solid resins, the ratio of epichlorohydrin per mol of bisphenol A is usually less than 2.

There are two general approaches to carrying out the synthesis of epoxy resins. In the one-step process all of the reactions described above proceed at the same time. These are usually carried out in the presence of sodium or potassium hydroxide. In the two-step process, reaction (1) is carried out by itself in the presence of a catalyst. Sodium or potassium hydroxide is then added to carry out the dehydrohalogenation and further condensation or polymerization as a second stage.

FIGURE 16: REACTIONS BETWEEN EPICHLOROHYDRIN AND BISPHENOL A

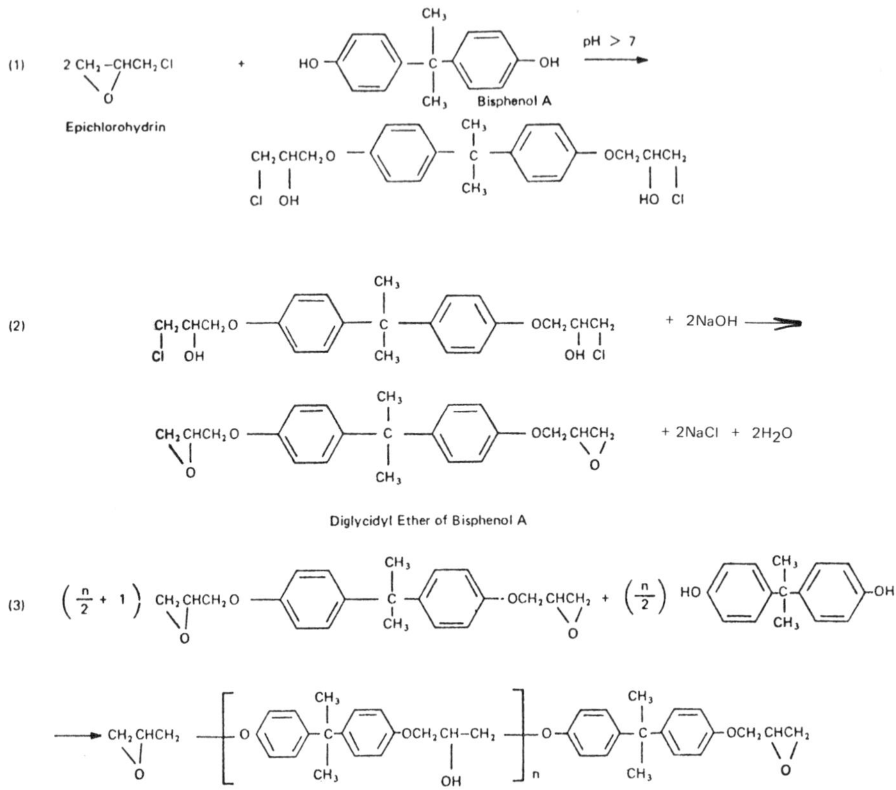

Source: EPA (8)

Regardless of which of these two approaches is used, the overall chemistry remains the same.

The product resins are utilized by the customer in conjunction with a curing agent to provide the cross-linking necessary to form a thermoset material. The curing agents used cover a broad variety of materials such as amines, polyamides, acids, acid anhydrides, resins such as phenolic, urea or melamine formaldehyde combinations, any of which is capable of reacting with either the epoxy groups or the hydroxyl groups present in the resin. The specific material picked depends upon the properties desired in the end resin.

There is substantial production of the so-called modified epoxies. Most of these are manufactured by reacting some material such as a fatty acid, tall oil or the like to form an ester with some of the epoxy groups present in the resin. The degree of esterification carried out depends upon the properties desired in the final material. Most of these modified epoxies find their way into coatings markets.

Continuous Process, Liquid Resins and Liquid Resin Solutions: The low molecular weight liquid resins can be manufactured by either batch or continuous processes. Most of the larger producers utilize a continuous process for this material as well as batch processes for the lower volume products. Figure 17 shows a schematic flowsheet of a typical continuous process. Bisphenol A, with a large mol excess of epichlorohydrin, is introduced into the polymerizer where, under the influence of the catalyst and caustic conditions, the reaction takes place. The excess epichlorohydrin is vaporized from the material and recycled.

FIGURE 17: LIQUID EPOXY RESIN PRODUCTION

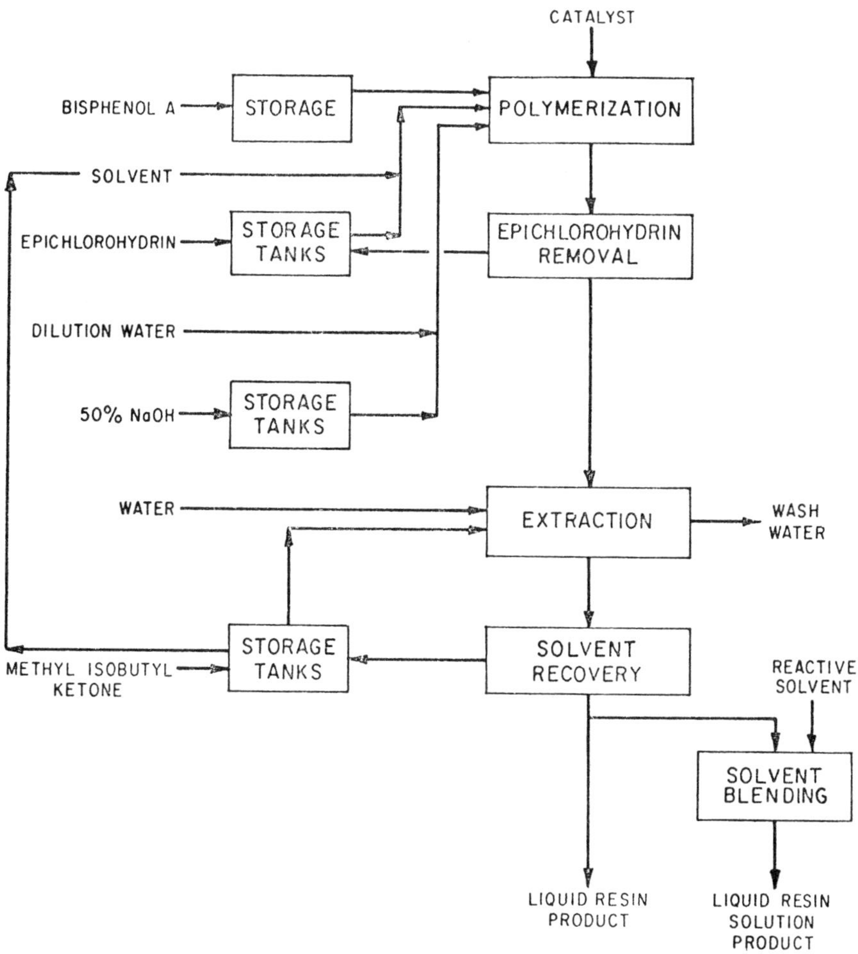

Source: EPA (8)

A solvent, usually a ketone such as methyl isobutyl ketone, is then added together with additional caustic and water. The epoxidation of the resin takes place with the formation of salt.

A solution of resin in the ketone solvent is water-washed to remove the final traces of salt, the decanted water is sent to waste, and the solvent is removed by vaporization. The liquid epoxy resin product is then sent to storage. Some resins are redissolved in solvent to produce a liquid resin solution product. Liquid resins are also produced in batch reactors, with reactant ratios similar to the above description of the continuous process, but in the equipment and processing steps as described below.

Batch Process, Liquid and Solid Resins: The solid resins, which have a high molecular weight, are produced by batch techniques in resin kettles. In producing these materials where the repeating part of the epoxy chain is a high number ranging from 1.8 to 16, the mol ratio of epichlorohydrin to bisphenol A charged to the kettle is less than 2. No excess epichlorohydrin is used in this case. The process is shown schematically in Figure 18. Aqueous sodium or potassium hydroxide is added to serve both as a catalytic agent and as one of the reactants to form the epoxy links during the polymerization reaction. Upon completion of the polymerization reaction, the water-containing salt and a very small amount of excess caustic are decanted to the process wastewater system.

FIGURE 18: BATCH EPOXY RESIN PRODUCTION

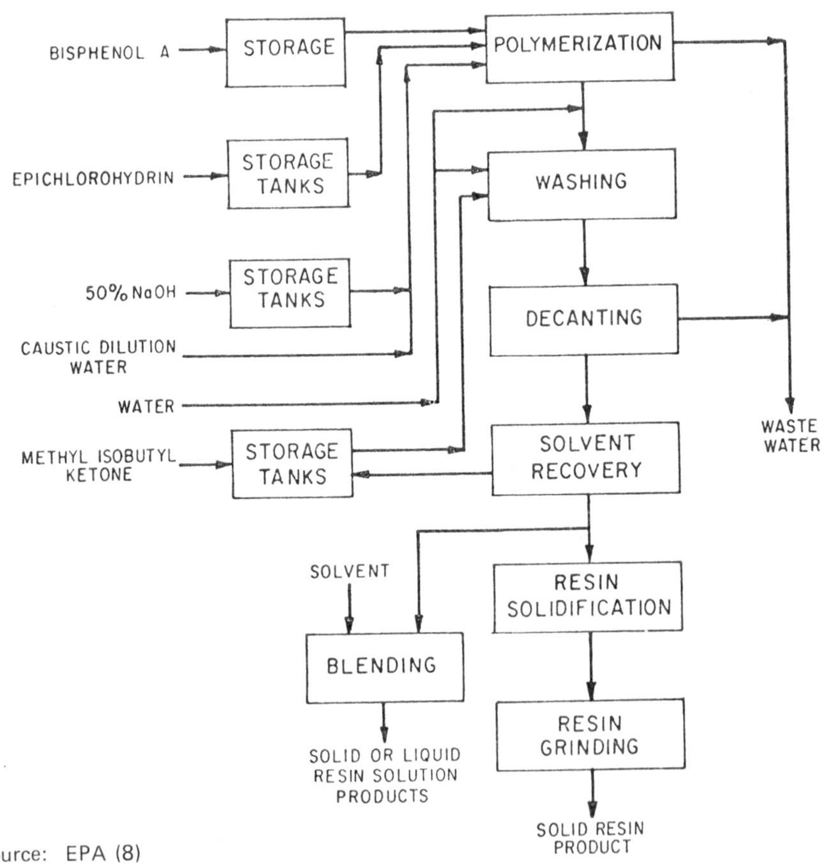

Source: EPA (8)

A solvent such as methyl isobutyl ketone is then added to dissolve the resin, and the solution is washed with water to remove the remaining amounts of sodium chloride and other salts which may be present. This water is decanted to the process waste system, and then the methyl isobutyl ketone is vaporized and recovered from the resin. The resins have melting points ranging from about 70° to 150°C (158° to 302°F), if a solid resin is being made, and the final temperature is such that the resin is molten. It is then drained and cooled to form a solid mass which is crushed to provide the final granular solid product, or blended with a solvent to make a solid resin solution product.

If a liquid resin is being made, after recovery of the solvent, the resin is either packaged directly as a liquid resin or a solvent is introduced to provide a liquid resin solution.

Batch Fusion and Solid Resin: A third process used by both the basic epoxy resin producer and by those customers tailoring the resin to various end uses is a nonaqueous fusion reaction involving the thermoplastic epoxy resin, a reactive diluent and a catalyst as shown in Figure 19. Additional water of reaction is produced, and the wastewater load is primarily due to the barometric condenser water (used to develop the vacuum in the reactor), vent scrubbers (when not reacting under vacuum), and housekeeping and maintenance water. The result is a low wastewater volume that contains a significant quantity of pollutants.

FIGURE 19: BATCH FUSION SOLID EPOXY RESIN PRODUCTION

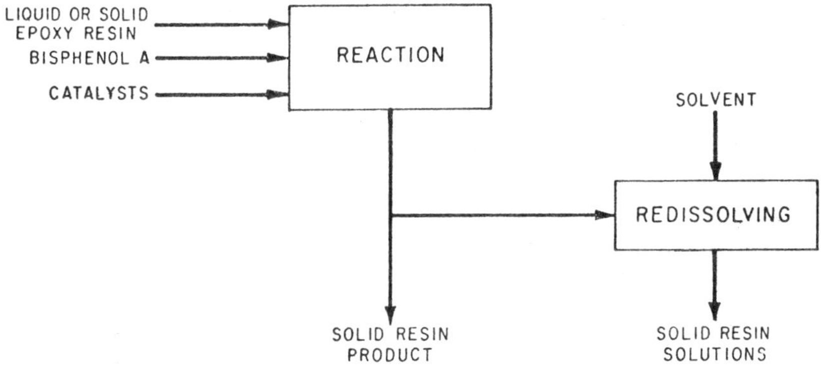

Source: EPA (8)

Ethylene-Vinyl Acetate Copolymer Manufacture

Ethylene-vinyl acetate (EVA) copolymers with vinyl acetate contents in the range of about 7 to 40% by weight are manufactured in the same facilities as low density polyethylene (LDPE) and often with high density polyethylene (HDPE). A process flow diagram is shown in Figure 20. The same equipment is used for EVA copolymers as for LDPE except for additional facilities needed for recovery of unreacted vinyl acetate and ethylene. This equipment consists of a separator downstream from the polymerization autoclave, where the solid EVA particles are sent to the pelletizing operation and the liquid phase is distilled to recover ethylene and vinyl acetate. The distillate wastes consist of a waxy residue that is incinerated or used as fuel.

FIGURE 20: ETHYLENE-VINYL ACETATE COPOLYMER PRODUCTION

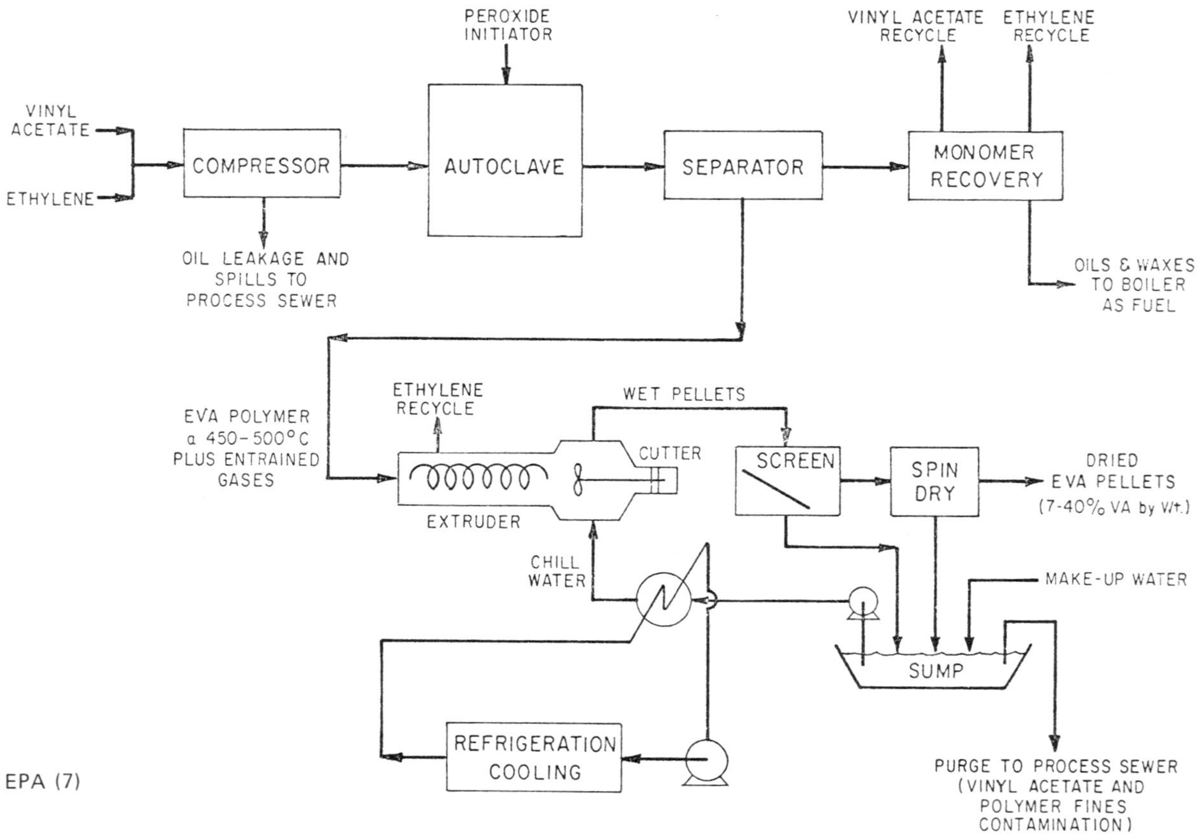

Source: EPA (7)

In the overall process, shown in Figure 20, vinyl acetate and ethylene monomers are fed to a compressor to build up the pressure necessary for polymerization. Polymerization is carried out in an autoclave using a peroxide type initiator. Following polymerization, the pressure is reduced and the mixture of unreacted vinyl acetate and ethylene, together with EVA copolymer, is sent to a separator. The separated EVA copolymer is fed to an extruder (where residual ethylene gas is removed and returned to the compressor) which extrudes continuous strands into a water chill bath where they are mechanically cut into pellets. The polymer pellets are screened from the water and spin dried.

The liquid phase from the separator is distilled for recovery of monomers, and the final residue incinerated as described above.

As the final process step, the EVA pellets are remelted, combined with additives and re-pelletized. Examples of additives are diatomaceous earth, amides, butylated hydroxy toluene, and various cyclic organic compounds. Since many of the end uses for the product involve direct contact with foods, it is produced to meet FDA requirements.

An EVA copolymer with distinctly different characteristics from those described above is produced by an emulsion polymerization process. The emulsion copolymer has a very high vinyl acetate content. It is made at only one plant, which is unrelated to those plants using the LDPE process. The emulsion polymerization process and associated wastewater loads are essentially the same as those reported for polyvinyl acetate homopolymer emulsion.

In plants using LDPE equipment for EVA production, the pelletizer cooling water is generally recirculated through the refrigeration cooling system. A continuous purge is maintained to control vinyl acetate and polymer fines contamination in the recirculated water. Vinyl acetate which enters the wastewaters from the purge stream is biodegradable both as the intact compound and in the form of its hydrolysis products, acetaldehyde, and acetic acid.

Other waste sources are oil leakage and spills from compressors and pumps which enter area surface water drainage ditches. Washdown water from processing and loading areas also flows to the drainage ditches and is another source of vinyl acetate contamination. The waste stream from the ditches is skimmed to recover oil and EVA particles. The oil is incinerated and the EVA is either land-filled or sold to scrap reprocessors.

Fluorocarbon Polymer Manufacture

The term "fluorocarbon polymers" as used in this report refers to addition type polymers in which all, or a significant portion, of the substituent groups on the carbon atoms in the polymer chain are fluorine. Typically, the balance of the substituents are chlorine and/or hydrogen. The fluorocarbon polymer family encompasses a range of homopolymer, copolymer, and terpolymer compositions.

Polytetrafluoroethylene (PTFE) is by far the most important commercial polymer in this group and accounts for an estimated 75% or more of the total production. PTFE is produced in two dry product forms (granular and fine powder) and as an aqueous dispersion. It is the only fluorocarbon polymer produced in three different plants. In view of the relative significance of PTFE, the process descriptions below have been divided to separate PTFE from the other fluorocarbon polymers.

Practice varies widely from plant to plant in this industry. Some plants produce only one type; more commonly several types of fluorocarbon polymers are made at the same plant. However, since most of the polymers are proprietary, the product mix differs from plant to plant. All existing plants are located within larger chemical complexes. The practice with respect to production of monomer and monomer feedstock varies. TFE monomer is produced on-site in all cases. Production of other monomers and monomer feedstocks (chlorodifluoromethane in the case of TFE) may or may not be carried out at the same plant. From examination of available waste load data related to production of the various polymers and in view of the widely varying practices from plant to plant, it has been concluded that wastewater guidelines should be limited to the dominant products – granular and fine powder grades of PTFE – and that aqueous dispersion grade and all other fluorocarbon polymers must be considered as unique products. It is also characteristic of this industry that process technology is considered highly confidential. The process descriptions that follow, therefore, are necessarily general in nature.

Polymerization of TFE to the homopolymer, PTFE, proceeds by free-radical addition polymerization typical of olefins. The polymerization is carried out under pressure in aqueous media in batch reactors. The literature suggests initiators such as sodium or potassium peroxydisulfates may be employed. A generalized flowsheet of the process is shown in Figure 21.

PTFE is produced for sale in several forms: a granular or pellet form, a fine powder, and aqueous dispersion. The dry product forms, granular and fine powder, account for the major portion of PTFE production. Guidelines have been proposed for these dry product forms only. Dispersion grade is made by only two of the three plants presently producing PTFE. Water is used as the polymerization medium in producing all forms of PTFE, but subsequent process water use and wastewater discharge varies with the form being produced. High purity, demineralized process water is required in all cases.

For granular or pellet grades, process water use in addition to polymerization medium includes polymer wash water and, for pellets, chill water for extrusion/pelletizing operations. The process water discharged from these operations is very clean. In one plant, all the process water is collected and recycled through a purification system. In the production of fine powder grade PTFE, wash water is required to purify the polymer particles. The conditions required to produce fine powder grade are such that the water discharged from the polymerization and washing steps may contain a higher level of suspended and dissolved solids than in the case of granular product.

In the production of the dispersion form of PTFE, the polymerization batch is concentrated after addition of a surfactant to stabilize the dispersion. The supernate liquor from the concentrating step is discharged as a waste stream. This stream has a BOD load due to presence of the surfactant.

FIGURE 21: PTFE PRODUCTION – PTFE POLYMER PROCESS

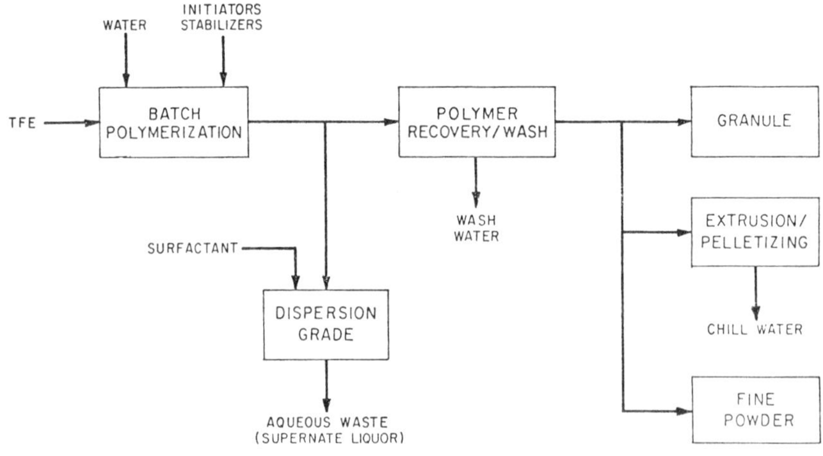

Source: EPA (7)

Nitrile Barrier Resin Manufacture

This class of resins has assumed importance primarily because nitrile barrier resins are transparent polymers with good resistance to passage of gases and solvents. They are in the early stages of being utilized for beverage containers. The nitrile group is the source of these good barrier properties. The nitrile group originates from the presence of either acrylonitrile or methacrylonitrile in the final polymeric structure. The nitrile content was previously restricted to below 30% in order to produce resins with acceptable processibility. At this level, the barrier properties were not exceptional. Recent developments have been in the direction of producing resins which have higher nitrile content while still retaining adequate processibility by conventional thermoplastic methods (i.e., extrusion, injection molding, blow molding, thermoforming).

The exact details of resin composition are considered proprietary by the resin manufacturers. The general structure however may be viewed as a butadiene backbone to which acrylonitrile/methacrylate or acrylonitrile/styrene copolymers are attached by grafting.

The exact nature of the technical developments which have resulted in the breakthrough is closely held proprietary knowledge by the three U.S. resin suppliers competing in this field.

Any of the generally known polymerization methods (such as bulk, solution, or emulsion) could be used to prepare these resins. Emulsion polymerization is undoubtedly the preferred method. The final composition may result from a two-step polymerization scheme in which a copolymer (such as acrylonitrile/acrylate) is polymerized by emulsion techniques in the presence of a previously formed graft copolymer (such as acrylonitrile/butadiene).

The polymerization scheme described below is speculative and is based on a review of in-house information and published literature. It is believed, however, that it is a reasonable representation of a typical polymerization process. A typical polymerization procedure would involve a two-step process in which the acrylonitrile/butadiene graft copolymer is made by batchwise emulsion polymerization of a recipe such as that listed below.

Typical Latex Recipe	Parts
Acrylonitrile	40
1,3-butadiene	60
Emulsifier	2.4
Azodiisobutyronitrile	0.3
tert-dodecyl mercaptan	0.5
Water	200

Before starting the reaction, pH is adjusted to about 8 using potassium hydroxide. Conversion of 92% can be obtained in 22½ hours at 45°C (113°F) giving a total solids content of 33.1%. The final resin is then prepared by mixing the following:

Latex (from reaction above)	31.9
Acrylonitrile (or methacrylonitrile)	70
Ethyl acrylate	30
Potassium persulfate	0.06
Emulsifier	3.0
n-dodecyl mercaptan	1.0
Ethylenediaminetetraacetic acid	0.5
Water	200

Adjust to pH 7 using potassium hydroxide. Twenty hours polymerization time (absence of oxygen) at 60°C (140°F) results in 97% conversion to 33% solids. The polymer is then coagulated using aluminum sulfate, washed and dried. At this point the dried polymer chips are then probably densified and passed to an extruder for processing into pellets. A batch process, as shown in Figure 22 is commonly used. Typical reactor size is 19 m^3 (5,000 gal). The batch cycle consists of the continuous introduction of a water-monomer emulsion to the stirred reactor. Polymerization occurs at about the rate of monomer addition; the heat of reaction is removed to cooling tower water circulated through the jacket.

The reactor is vented through a condenser for monomer recovery; and the condensate, including any water, is returned directly to the vessel. On completion of the batch, a short "soaking" time is allowed for completion of the reaction, and water is then added to dilute to the desired end composition. The batch is drawn off through a screen to product storage. Oversize screenings (a very small amount) are disposed of to landfill.

Monomers, the principal raw materials, are often protected during shipping and storage by an inhibitor, such as catechol, which may be removed prior to polymerization by washing. This contributes to the wastewater load. Sources of wastewater from a typical emulsion polymerization include the following:

- Reactor cooling water
- Cooling tower and boiler blowdown
- Monomer washing
- Liquid or solid waste from monomer stripping or recovery operations
- Discarded Latex batches
- Coagulant wastes
- Startup, spills, etc.
- Demineralizer wastes
- Possible liquid wastes from monomer scrubbing

As regards other pollutants, organic nitrogen occurs as a result of losses from dissolution and emulsification of reactants. Cyanides will be detectable by analytical methods due to the presence of acrylonitrile.

FIGURE 22: NITRILE BARRIER RESIN PRODUCTION—EMULSION POLYMERIZATION PROCESS

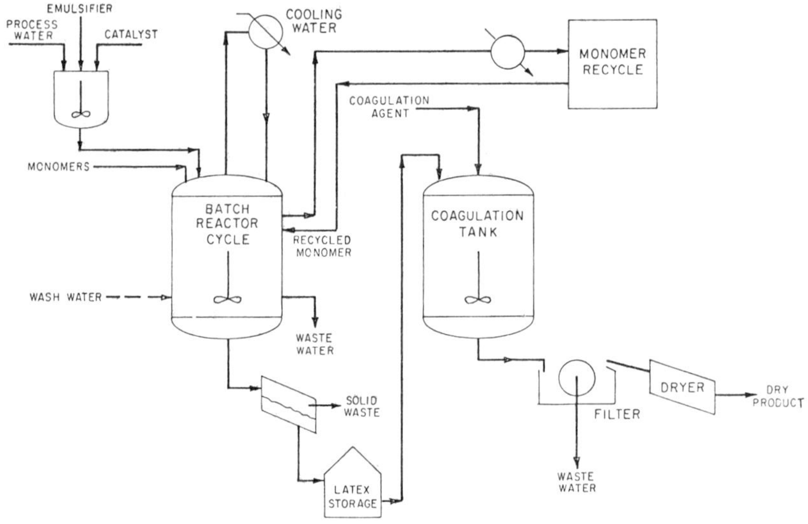

Source: EPA (7)

Phenolic Resin Manufacture

The family of phenolic resins includes the oldest synthetic polymers. The term is used to describe a broad variety of materials, all of which are based upon the reaction between phenol, or a substituted phenol such as creosol or resorcinol, and an aldehyde such as formaldehyde or acetaldehyde. Nearly all industrially significant resins, however, are based upon the reaction of phenol with formaldehyde.

Phenol, commonly known as carbolic acid, is a solid at room temperature but melts at between 42° and 43°C (107° to 109°F). It is usually shipped and handled as a liquid by keeping it above its melting point. Formaldehyde is normally a gas. Its most common commercial form is formalin, a 37% by weight solution of formaldehyde and water. There are two broad types of resins produced by this industry: resols and novolaks.

Resols are formed from a mixture of phenol and formaldehyde which contains an excess of formaldehyde. Often the mol ratio is about 1.5 to 1. An alkali such as sodium hydroxide is used to catalyze the polymerization which takes place at a pH of between 8 and 11. The reaction is shown in Figure 23.

The reacting mixture contains sufficient formaldehyde so that, if allowed to proceed to completion, a cross-linked thermoset resin would be formed. The reaction, however, is stopped short of completion at an average molecular weight of the polymer appropriate for the end use of the material. Three classes of products are produced under the general grouping of resols:

[1] The least degree of reaction produces a water-soluble bonding resin which is either sold "as is" or neutralized and partially dehydrated.

[2] Further reaction produces a water-insoluble resin which is vacuum dehydrated and dissolved in solvents to produce laminating resins and varnishes, using much less basic catalyst than in making bonding resins. Some resols, such as varnishes that are used in rubber cement, require washing of the resin to remove salts; most other resins do not require washing.

[3] The third class of product is similar to [2], but the water is removed and the reaction carried even further to make a "one-stage" solid resin that is then vacuum dehydrated and dropped from the reactor for cooling and solidification. These one-stage resins are then compounded into
 (a) bonding compounds and surface coatings, by adding catalysts and lubricants;
 (b) thermosetting molding powders by adding catalysts, lubricants, pigments and fillers.

The compounding is sometimes performed in the same facilities where the resin is produced, but more often the resin is shipped to custom compounders. Compounding is a dry operation and adds negligible wastes to the resin manufacturing waste generation.

The material already contains sufficient formaldehyde to completely cross-link the ultimate product so that it can be thermally set into an infusable material by the application of heat at the customer's facilities. Since cooling the mixture in its partially polymerized form does not completely stop further polymerization but merely retards it, these materials have a somewhat limited shelf life (in the order of 60 days for many types).

Novolaks are the second category of phenolic resins. These are formed from a reacting mixture which contains a deficiency of formaldehyde. The normal commercial range for this mol ratio is between 0.75 and 0.90. To produce this material, polymerization is carried out in an acid medium using a catalyst such as sulfuric acid. The pH of the reaction usually ranges from 0.5 to 1.5. For special uses where high ortho linkage is desired, the polymerization may be carried out at a pH of from 4 to 7, but this is not typical. The reaction is shown in Figure 24. Since the reacting mixture contains a deficiency of formaldehyde, essentially all of the formaldehyde is consumed during polymerization. Since no further polymerization can take place, the product is a low molecular weight, thermo-

FIGURE 23: TYPICAL REACTION TO FORM ONE-STEP RESINS OR RESOLS

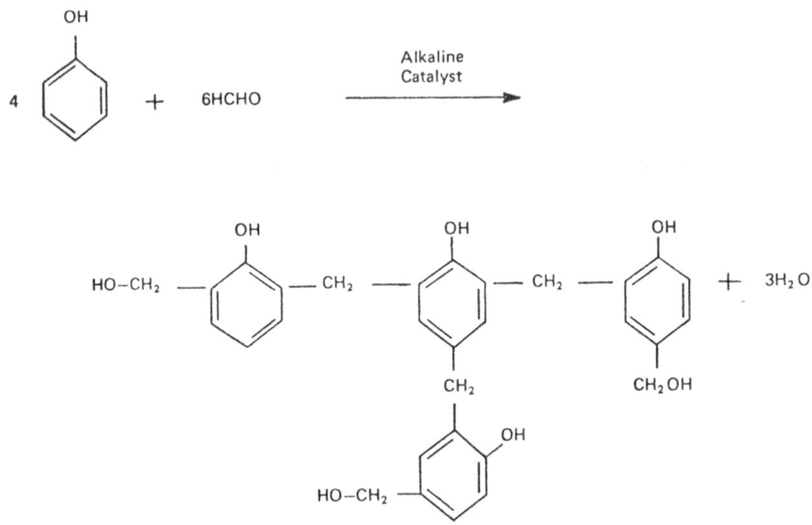

FIGURE 24: TYPICAL REACTION TO FORM NOVOLAK RESIN

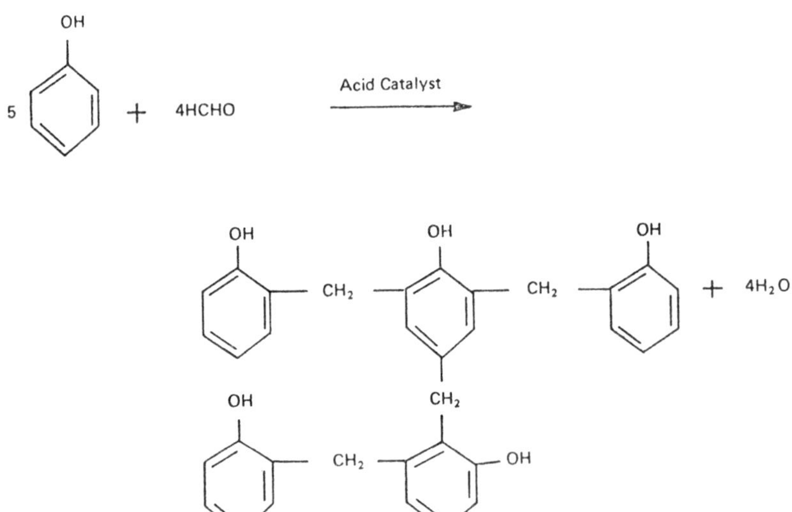

Source: EPA (8)

plastic, stable material. The water which enters with the formaldehyde plus the water of reaction is removed under vacuum at the end of the reaction, and a solid, meltable material results.

In order to complete the polymerization, the user must add additional formaldehyde. Sometimes this is done by using paraformaldehyde, a solid polymer of formaldehyde, but the extremely irritating nature of this material has limited its use. Most users complete the reaction by using hexamethylenetetramine. With this material ammonia is evolved from the reacting mass, leaving the same types of methylene linkages as can be obtained by using additional formaldehyde.

The basic resins described above are sometimes modified by the use of materials such as drying oils or epoxy compounds in the final stages of polymerization. These modified phenolics find many specialty uses but do not affect the basic manufacturing processes to any significant degree.

Although continuous processes for the production of phenolic resins have been developed, they are seldom used. The production of these continuous units must be high, and the industry calls for such a wide variety of materials that it is seldom possible to have a large enough run on a single grade of polymer to justify their use.

The standard producing unit of the industry is typically a batch resin kettle arrangement, such as is shown in Figure 25. The heart of the process, the resin kettle, varies in size from 7.6 to 38 m^3 (2,000 to 10,000 gal). These are jacketed, and in the larger sizes internal cooling coils are used in order to provide sufficient surface-to-volume ratio to remove the considerable amount of heat generated during polymerization. The kettles are agitated and can operate under either pressure or vacuum conditions.

The feed system generally consists of two weigh tanks which weigh in the required amounts of phenol and formaldehyde solution. Commercial formaldehyde solution is available at 37% by weight formaldehyde, but other concentrations are used, such as 44 or 50%. This solution often contains about 5% methanol (from incomplete conversion or separation in formaldehyde manufacture) which acts as a stabilizer. Other stabilizers are sometimes used. The kettle is equipped with a water-cooled condenser, which is also joined to a vacuum system.

Figure 26 shows the basic manufacturing steps and the resulting products for resol and novolak production. In a typical production cycle for a resol resin, the phenol is charged in a molten form to the kettle followed by formaldehyde, which washes any residual phenol out of the lines leading to the kettle. A sodium hydroxide catalyst solution is then added, and the kettle is heated to bring the mixture to a temperature of about 60°C (140°F).

During this period the condensation reaction starts and the reaction becomes highly exothermic so that a change is made from supplying steam to the coils to supplying cooling water. The mixture is held at a temperature ranging from 60° to 80°C (140° to 176°F) for a period of 3 to 5 hours. During this period temperature is controlled by circulating cooling water through the coils as well as by using total reflux returning from the water-cooled condenser mounted above the kettle. When the polymerization has reached the desired degree, as shown by laboratory tests, the mixture is cooled to about 35°C (95°F) to essentially stop further reaction. At this point the caustic may be neutralized by the addition of sulfuric acid, which brings the mixture to a pH of about 7.

The mixture is then heated by admitting steam to the coil, and the resin is purified by distillation. The water from this distillation is a concentrated waste which contains unreacted formaldehyde and phenol and low molecular weight resin, and may be segregated for disposal by incineration. The batch is then dumped. A few resins, such as varnish type resols used as tackifiers for rubber cement, are washed two or three times, thereby resulting in a considerable increase in wastewater and contaminants. If a resin is required which contains a very small amount of water such that it cannot be dehydrated at a

FIGURE 25: PHENOLIC RESIN PRODUCTION

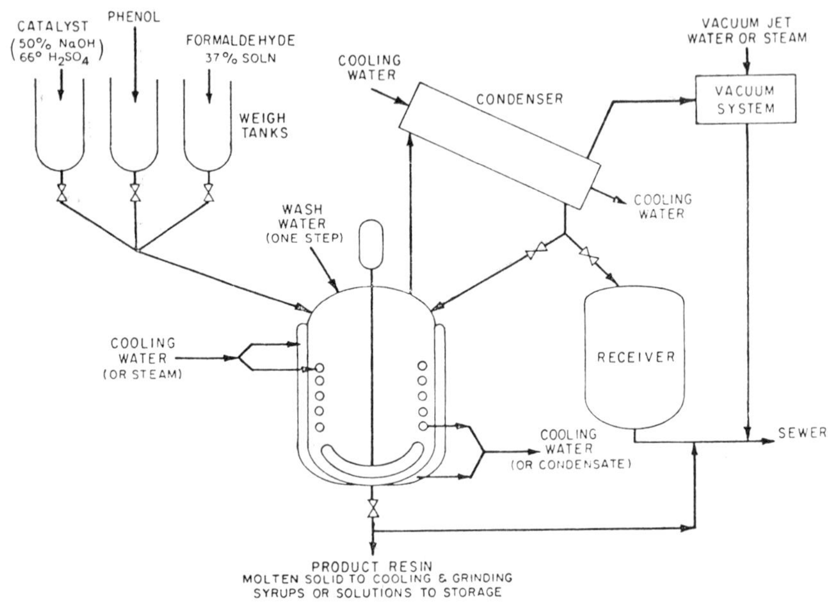

FIGURE 26: PHENOLIC RESIN PROCESSING

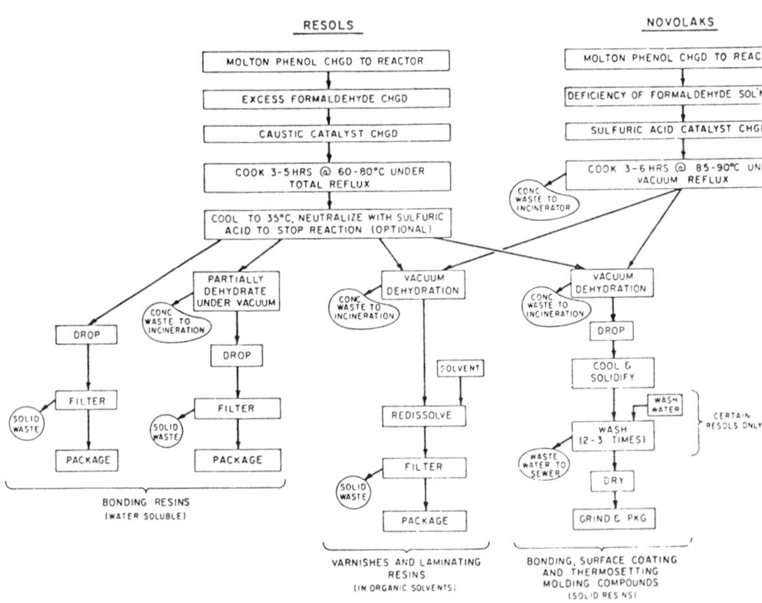

Source: EPA (8)

temperature low enough to prevent further polymerization, a vacuum is applied during the latter part of the dehydration cycle. This technique can be used to produce an essentially anhydrous melt of a single-step resin.

The molten resin must be quickly discharged from the bottom of the kettle through cooling plates for a quick quench in order to prevent the mass from setting up into an insoluble, infusible material. The cast material, when solidified, can be broken up and crushed for shipment as a powder.

The manufacture of novolak resins is entirely analogous except that an acid catalyst, such as sulfuric acid, is added at the start of the batch. With strongly acid catalysts, it is necessary to utilize a vacuum reflux in order to maintain temperatures at 85° to 90°C (185° to 194°F), a slightly higher temperature range than that used for the one-step reaction. Under milder reaction conditions, atmospheric reflux is adequate to control the temperature.

At the end of the reflux period, three to six hours after initiating the reaction, the condensate is switched to the receiver and water is removed from the batch. When the temperature reaches the order of 120° to 150°C (248° to 302°F) the vacuum is applied to aid in removing the final traces of water and part of any unreacted phenol. Final temperatures may rise to about 160°C (320°F) under a vacuum of 63.5 to 68.5 cm (25 to 27 inches) of mercury. These higher temperatures are possible since the reaction has proceeded to completion and, therefore, no further polymerization can be carried out until additional formaldehyde is added. The completed batch is dumped in the molten form onto cooling pans where it solidifies, or onto a flaker. If the product is needed in solution form, solvent is added at the end of the batch as it cools in the kettle and the solution discharged from the kettle to storage tanks for drumming.

The finished products may be shipped to customers as such or may be compounded with additives at the resin-producing point. The solid resins may be ground, and wood fillers, pigmenting materials and hexamethylenetetramine added to form a finished molding compound. These processes all involve solids-handling and do not normally give rise to wastewater generation.

A variety of processes have been described in the periodical and patent literature for phenol removal from wastewaters such as those from phenol-formaldehyde resin manufacture. These include

- ozone treatment
- extraction with trialkyl or triaryl phosphates
- extraction with esters
- extraction with cumene
- chlorine treatment followed by extraction of the chlorophenols produced

All these processes, however, relate to the removal of phenol only, whereas, when preparing phenol-formaldehyde resins, the wastewater obtained contains phenols and also formaldehyde, and consists partly of (a) water that is formed during the condensation step, (b) the formaldehyde solution, (c) the phenol, and (d) the catalyst diluent. To recover the phenols and the formaldehyde, the water has to be removed by decanting, for instance, or by evaporating under reduced or normal pressure and recondensing by means of a condenser.

If the wastewater should be simply evaporated by heating the bottom and walls of a vessel, an unusable resin would be formed, which rapidly reaches the C-stage or resite phase, i.e., becomes an infusible and insoluble resin. In order to avoid this, and to develop a process that is both acceptable and economical, it is necessary for the wastewater to be purified in such a way that the phenols and the formaldehyde are separated off by converting the products into low molecular resins, while the water is evaporated.

According to a process developed by *A. Adegeest; U.S. Patent 3,741,392; June 26, 1973; assigned to Corodex NV, Netherlands,* wastewater in which one or more phenols and formaldehyde have been dissolved may be purified by adding phenol(s) thereto in such quantities that the phenol:formaldehyde molar ratio will become from 1:1.02 to 1:1.12, and by adjusting the normality of the total acid to from 0.015 to 0.10 N; heating the solution at a temperature of 80° to 85°C, and subsequently removing the water by heating at a temperature that is at or above 5°C below the boiling point of the liquid mixture.

The composition of the wastewater is determined by the type of resin that is prepared. It will generally vary between the following limits:

3 to 7% by weight of phenols (normally 4%)
5 to 8% by weight of formaldehyde (normally 6%)
5 to 7% by weight of methanol (summer or winter formula)

and small percentages of: (a) esters formed during the reaction (BP 40°C), (b) organic chlorides, (c) methylal (BP 44°C) formed during the reaction, and (d) formic acid. The HCl normality (if HCl is used as a catalyst) of wastewater to be purified is 0.015.

Figure 27 shows a suitable form of apparatus for the conduct of this process. A is the reaction vessel in which the resin is prepared. The apparatus comprises a mixing tank **1** which is supplied through a pipe **1'** with wastewater to be purified. Vacuum created in the tank through vacuum line **22** draws in phenol through the suction line **8**, and acid, preferably mixed with water, through the suction line **21**. Appropriate valving is supplied for closing these suction lines when the mixing tank is to be emptied by means of pressure applied through pipe **2**, which forces the water, phenol and acid through the pipe **3** to the reservoir **B**. **5** is a gauge-glass; **S** and **C** are, respectively, steam feed and steam discharge lines.

FIGURE 27: APPARATUS FOR PURIFICATION OF WASTEWATER FROM PHENOLIC RESIN MANUFACTURE

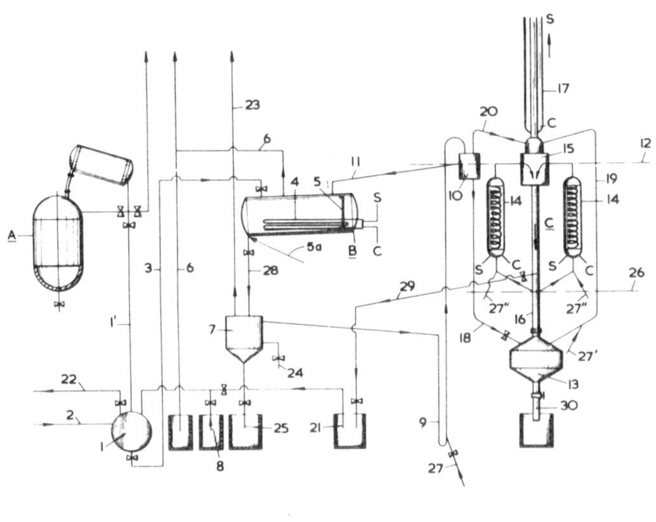

Source: M. Adegeest; U.S. Patent 3,741,392; June 26, 1973

The reservoir **B** is installed at an inclination. It is equipped with a heating member **4** and it is connected at an upper point to the discharge pipe **6** for the removal of methanol vapor and other volatile compounds. Finely divided air may be forced into the tank through an air injector **5a**. The pipe **28** carries fluid from the lower end of the reservoir **B** to a resin settling tank **7**, which is connected to a pressure safety valve by way of line **23**, and also has a gravity drain line **25** for removing the resin which has settled out in the lower levels of the tank. In line **28**, near vessel **B**, means may be provided to introduce finely divided air into **B**. A drain **24** may be provided for removing fluid from the tank **7**. A pipe **9** is connected to an upper portion of the settling tank **7** for removing lighter material from that tank.

An air injector **27** is connected into the pipe **9**, and carries the fluid from the tank **7** to an overflow tank **10**, from which fluid may be led back to the tank **B** by way of the overflow line **11**. Heavier constituents of the material in the tank **10** may flow through the line **18** into the tank **13**, where separated resin is collected. Lighter constituents in the tank **10** are carried to an expansion boiler or evaporator **15** by way of a line **20** equipped with a pressure equalizer. **C** is the after-reaction unit.

Between the expansion tank **15** and the separating tank **13** there is a primary circulating pipe **16** which interconnects these two vessels. At or about the level indicated by the broken line **26**, indicating the interface of hot and cold zones, connections are provided from the line **16** to heat exchangers **14,14**, these being supplied with fluid from the line **16** by the aid of air injectors **27″,27″**. The point at which the connections to the heat exchangers come off is just slightly below the zone at which the hot and the lukewarm water meet in the pipe **16**. Slightly above this level is connected the drain pipe **29**, which withdraws water mixed with acid, and conveys it back to the tank from which the line **21** extends.

A heated discharge pipe **17** leads from the top of the expansion tank **15** for the removal of volatile compounds. There is a gauge **19** mounted between the tank **13** and the tank **15** for the purpose of checking on the fluid level. This gauge may be provided with an air injector **27′**. At the bottom of the tank **13** there is a drain **30** for the removal and recovery of the liquid resin. The following is a typical example of the conduct of this process.

Example: 624 liters (683 kg) of wastewater, remaining after the preparation of a phenol-formaldehyde resin which had been condensed in a basic medium, was admitted to the mixing tank together with 312 liters (341 kg) of wastewater resulting from the condensation of phenol-formaldehyde resin in an acidic medium. 84 liters (90 kg) of water containing acid was mixed with these wastewater portions, this water-containing acid having been drawn through the continuous drain **29** from the after-reactor of the evaporating plant. 109 kg of phenol was added to this mixture.

Before the addition of phenol, the phenol:formaldehyde molar ratio was 0.43 gram mol per liter to 1.8 gram mol per liter, i.e., 1:4.18. This is 40.46 g of phenol and 54 g of formaldehyde per liter. After the addition of phenol the phenol:formaldehyde ratio was 0.57 gram mol per liter to 0.61 gram mol per liter, i.e., 1:1.07. This is 53.64 g of phenol and 18.38 g of formaldehyde per liter.

The normality of the total acid amounted to 0.015. After the addition of phenol the mixture was held at a temperature of 80° to 85°C for 48 hours. The water was subsequently removed by being heated to boiling temperature. There was thus obtained 200 kg of a thin, liquid resin mixture which was separated off by settling.

Inasmuch as 109 kg of phenol was added, there have thus been recovered 91 kg of resin product, i.e., 8.8% by weight from the acid wastewater. Subsequently, another 84 liters of water-containing acid was obtained from the continuous drain **29** from the reaction-evaporation zone **C**. This water is required for maintaining the acidity in this zone at a constant level, while it also serves to adjust the acidity in the mixing tank.

According to a process developed by *J.H. Thayer and E.C.Y. Fan; U.S. Patent 3,736,292; May 29, 1973; assigned to General Electric Company* the phenol content of wastewater from phenol-formaldehyde resin manufacture may be markedly reduced.

Aqueous solutions produced by the reaction between the phenol and aldehyde at different stages of the reaction and containing different proportions of phenol are mixed together at a temperature in the range of about 30° to about 45°C to provide rapid separation into two solution layers, of which one contains low phenol concentration and may be discharged as waste, and the other contains high phenol concentration which is returned to the reaction vessel for further use in resin production. The low phenol waste solution thus discharged enables a reduction in pollution of streams into which products of the resin process are usually discharged.

Figure 28 shows the apparatus for producing phenol-formaldehyde as a means of discussing waste generation by this manufacturing process. The reactants are introduced through feed pipe **1** into closed reaction vessel **2** which has a steam jacket **3** for heating the contents of the reaction vessel.

In a usual process, the phenol in liquid form and a suitable acid catalyst are first introduced into reaction vessel **2**, and the aldehyde compound, typically paraformaldehyde in granular form, is then added gradually through feed pipe **1**. The reactants, as well understood in the art, are added in stoichiometric proportions. Initially, the reaction vessel is heated to a temperature of about 80°C by means of steam jacket **3**. With the continuing addition of the aldehyde, the temperature of the vessel contents rises to about 150°C due to the exothermic nature of the reaction between the aldehyde and phenol.

As a result of the reaction, phenol-aldehyde resin **10** is formed and settles to the bottom of reaction vessel **2** where it remains in liquid form below the supernatant water solution **11** of unreacted phenol and aldehyde. During the reaction, gaseous reaction products including vaporized water of formation are given off, and such gases, including vaporized phenol, pass through vapor line **4** to condenser **5** where the vapors are condensed to a liquid which passes through line **6** to a condensate receiver **7**. By means of return line **8**, the condensate in receiver **7** is returned to reactor vessel **2** for further reaction.

After all of the desired amount of aldehyde has been added to the reaction vessel **2**, and while reaction is still continuing, it has been the practice in the past to remove from receiver **7** the condensate formed during the reaction, which is known as the "reaction condensate" and which contains normally about 10 to 12% by weight of phenol in water solution, and to dispose of this solution through discharge lines to a stream or waste treatment facilities.

After the reaction in vessel **2** has proceeded to substantial completion, the supernatant solution of unreacted phenol and aldehyde is vaporized due to the heat in the vessel and the reduced pressure therein obtained by use of vacuum pump **12**, so that the vapor passes into line **4** for passage through condenser **5** and to receiver **7**. The solution, which is known as "strip condensate" and contains about 45 to 50% by weight phenol, has also in the past been disposed of as waste. Although the strip condensate has a relatively large unreacted phenol content, this liquid has not been found suitable for recycling and reuse in the described process because the phenol is largely in emulsified condition therein, and to recover the phenol would require a prolonged period of detention, e.g., over 12 hours, of the strip condensate in an undisturbed settling tank to allow the phenol to separate out into a discrete layer which could be removed for reuse.

This process which has both phenol recovery and the prevention of phenolic water pollution as its goals is shown schematically in Figure 29. The reaction condensate **15** and the strip condensate **16** are mixed together in a suitable vessel as indicated in step **17**. In a usual process, the proportion of phenol in the mixture is in the range of about 25 to 40% and the temperature of the mixture is held in the range of about 30° to 45°C, and preferably at about 35°C. As a result, the mixture in a short time, usually less than 1 minute, separates

FIGURE 28: APPARATUS FOR PHENOLIC RESIN MANUFACTURE

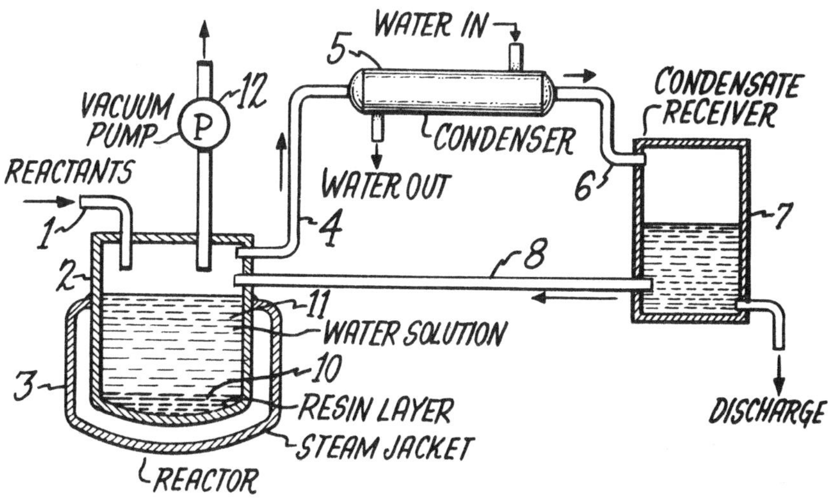

FIGURE 29: BLOCK FLOW DIAGRAM OF PHENOL RECOVERY PROCESS

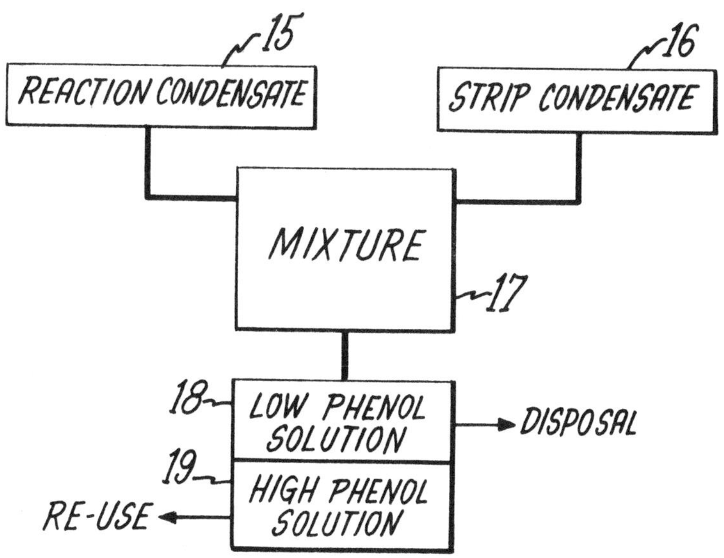

Source: J.H. Thayer and E.C.Y. Fan; U.S. Patent 3,736,292; May 29, 1973

out into discrete layers **18** and **19**, with upper layer **18** having about 8 to 12% phenol in solution and lower layer **19** having about 70 to 75% phenol in solution. The high phenol solution layer **19** may then be readily drawn off and returned to reaction vessel **2** for further use in the manufacture of phenolic resins. Solution **18** is sufficiently low in phenol content to be disposed of through usual sewage treatment facilities, where the remaining phenol is readily removed by suitable treatment, such as biological oxidation reactions commonly employed in such facilities.

Various methods may be employed for mixing the strip condensate with a suitable liquid for obtaining the desired discrete layers of phenol solution. In one method which has proved particularly practical, the strip condensate from a previous batch process of making phenolic resin as described above is placed in a thermally insulated mixing tank, and the reaction condensate from the following batch process is pumped into the mixing tank through a pipe extending into the strip condensate solution, so that sufficient mixing action takes place due to the force of the flow of the reaction condensate below the surface of the strip condensate.

In such a process, in which the volume of reaction condensate added is usually greater than that of the strip condensate, the strip condensate typically is at a temperature of about 25°C and the reaction condensate is about 35° to 40°C, so that the mixture attains an intermediate temperature within the desired range as described above.

As a result, separation of the mixture into two layers takes place very rapidly without the necessity for mechanical mixing means or the use of temperature control means. If desired, however, the mixing tank may be equipped with heating means or mechanical mixing means if these should be required for particular circumstances.

If in certain situations the reaction condensate is, for any reason, not available or not desired for use in mixing with the strip condensate, plain water without any phenol may be mixed with the strip condensate at the specified temperature to provide the layer separation desired, in accordance with the process. While in general the reaction condensate used will have a phenol content of around 10 to 12%, it may in some cases contain up to about 15% phenol. Thus the solution mixed with the strip condensate may, in the broad aspects of the process, contain a range of 0 to about 15% phenol.

While the range of phenol content of the mixture is preferably about 25 to 40%, this range may be about 20 to 45% phenol. With an amount of phenol below this range, insufficient phenol is present in lower layer **19** to make the process practical. With amounts of phenol above this range, it is necessary to use temperatures above 45°C for the mixture to provide layer separation, and under these conditions the volume of lower layer **19** is reduced to an impractical amount.

Further, it has been found that the use of mixture temperatures substantially below the 30° to 45°C range mentioned above results in unduly prolonging the time necessary for the described discrete layers to separate out.

A process developed by *J. Ackerman and P. Radici; U.S. Patent 3,843,643; October 22, 1974; assigned to Società Italiana Resine SpA, Italy* is one in which formaldehyde and phenol are recovered from waste liquors, with purification of the liquors, by addition of ammonia to the formaldehyde liquor to form hexamethylene tetramine, evaporation of water, addition of the concentrate to the phenol liquor, and recovery of the hexamethylene tetramine/phenol adduct formed.

Polycarbonate Manufacture

Polycarbonates are a special variety of linear thermoplastic polyesters in which a derivative of carbonic acid is substituted for the acid and a diphenol is substituted for the glycol. Polycarbonates are end products in themselves, undergoing thermoplastic processing utilizing the conventional equipment of the plastic industry. By far the most important polycarbonates from the commercial point of view are aromatic polycarbonates derived exclusively from the reaction of bisphenol A with phosgene. In addition, aliphatic or aliphatic-aromatic polycarbonates may be derived, but indications are that they find little industrial usage. Thus they will not be discussed.

Phosgene is a liquefied gas, boiling at 7.6°C (45.7°F), and is only very slightly soluble in water. Bisphenol A is a solid melting at 152°C (305°F) and is soluble in water to the extent of 3,000 mg/l at 85°C (185°F). The polymerization may be carried by means of the following mechanisms: condensation, interfacial, and transesterification. Of the three, condensation is preferred. Interfacial polymerization is seldom used for it is inconvenient and slow, and in order to achieve sufficiently high rates of transesterification, the reaction conditions must be so drastic that a portion of the polycarbonate is decomposed.

The reaction between the two raw materials takes place under alkaline conditions in the presence of catalyst and pyridine. In contrast to base-free condensation, which proceeds only at high temperatures in the presence of special catalysts and yields polymers with insufficient molecular weights, condensation in the presence of basic substances proceeds at high rates at room temperature to give high molecular weight polycarbonates.

As shown in the above equation, two mols of hydrochloric acid are formed per mol of phosgene consumed. To maintain basic conditions in the reactant mix, the hydrochloric must be either neutralized or consumed. Pyridine is added, usually in excess, to act as an acid acceptor and so that the resulting polycarbonate forms a more or less viscous solution. Alternatively, a good portion of the pyridine may be replaced by an organic solvent in which the polycarbonate is soluble. The reactant is strongly catalyzed by Lewis acids such as aluminum chloride, aluminum isopropoxide, stannic chloride and titanium tetrachloride. Suitable reaction media solvents are the chlorinated aromatic hydrocarbons, such as chlorobenzene, methylene chloride, or o-dichlorobenzene. A typical catalyst is benzyl triethyl ammonium chloride.

It is important to use equimolar quantities of phosgene and bisphenol A, since the molecular weight of the product will depend on the ratio of the starting materials. It is also important to avoid the presence of substances such as monofunctional alcohols or phenols which act as chain terminators. The molecular weight is also indirectly dependent on possible concurrent side reactions, which in turn depend on the usual parameters such as temperature, time, and reactants. In choosing the solvent, chlorobenzene is preferable since the pyridine hydrochloride formed is insoluble in this medium and may be readily separated by filtration. Traces of the hydrochloride that are in solution may be removed by distillation.

The process scheme is shown in Figure 30. The bisphenol A is charged with excess pyridine and a solvent such as methylene chloride into the vessel. Phosgene is vaporized in a still, and then bubbled through the reaction mixture. Total mols of phosgene fed are in slight excess of the mols of bisphenol A charged. Phosgene and bisphenol react to form the carbonate monomer, which then polymerizes. The reacting mixture is kept under 40°C (104°F); residence time in the vessel is 1 to 3 hours. The reaction can be controlled through any of several variables, including residence time, temperature, and proportions of the components introduced. Component purity also has great effect on the reaction.

Reacted mixture, consisting of the polymer, pyridine hydrochloride, and unreacted pyridine in solvent, is fed to water-wash tanks. Washwater and hydrochloric acid are added here, the acid reacting with residual pyridine. Additional solvent may also be introduced at this stage to lower the viscosity of the mixture.

The next step is the removal of water and pyridine hydrochloride by decantation or the equivalent. The aqueous phase goes to pyridine recovery; the solvent phase, containing the dissolved polymer, goes to an agitated precipitation tank. Here the polymer is precipitated by addition of an organic "antisolvent" such as an aliphatic hydrocarbon. This forms a solution with the carrier solvent and causes the polymer to precipitate. The

FIGURE 30: POLYCARBONATE PRODUCTION–SEMICONTINUOUS PROCESS

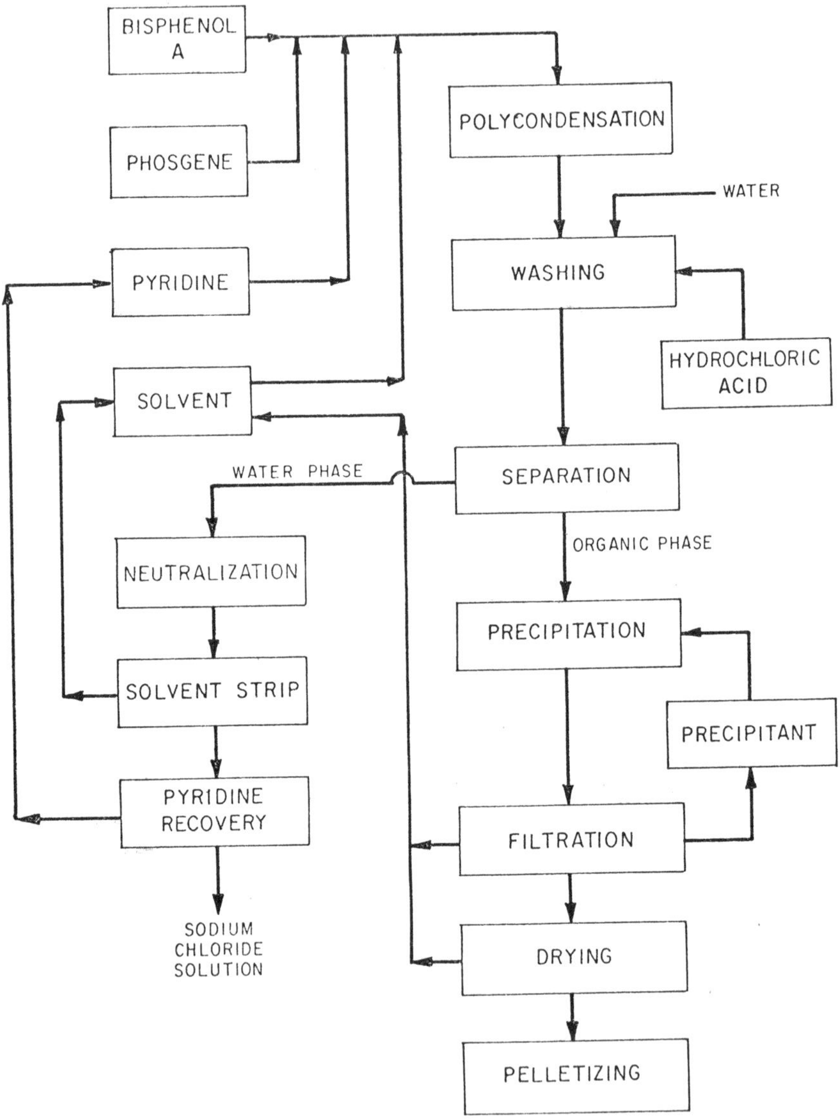

Source: EPA (7)

resulting slurry goes to a rotary filter, and the separated polymer goes to a hot air dryer where the remaining solvent is removed. The polymer leaves the dryer as a powder and is sent to blending, extrusion, and pelletizing. Aqueous pyridine hydrochloride solution is combined with caustic solution in a mix tank for removal of chloride ions as sodium chloride. The stream then goes from the mix tank to a fractionating column where steam strips out any solvent present.

The stripped solution goes to an azeotropic distillation column, and the sodium chloride solution is removed as bottoms and discarded. The overhead stream is an azeotrope consisting of about 43% water, 57% pyridine.

The azeotrope is condensed and then combined with a breaking agent. The mixture goes to a pyridine distillation column where water-free pyridine is removed overhead. The bottoms are treated to recover the breaking agent.

The wastewater originating from the process is principally due to polymer washing. The major substance present in the wastewater stream will be sodium chloride. The amount and concentration of sodium chloride will depend upon the excess of pyridine required and the dilution necessary to effect adequate washing. It is expected that the wastewaters will be alkaline from excess sodium hydroxide used in the recovery of pyridine. Also, the wastewater stream would be expected to contain traces of pyridine, solvents, breaking agents, bisphenol, polycarbonate, and side reaction products.

Polyester Resin (Unsaturated) and Alkyd Resin Manufacture

Unsaturated polyester resins are made by an esterification reaction involving a glycol and both an aromatic dibasic acid and an unsaturated dibasic acid. The unsaturated dibasic acid is used to incorporate an ethylenic linkage into the polymer and is a compound such as maleic acid or fumaric acid. The aromatic dibasic acid can be phthalic acid, isophthalic acid or the like. The glycol is commonly propylene glycol. The basic polyester resin is made typically by a batchwise reaction process in stirred, glass-lined, or stainless steel vessels and is later dissolved in a reactive monomer such as styrene which can cross-link with the ethylenic bonds in the main polymer. The resultant viscous liquid, diluted with styrene, is the current item of commerce known as polyester resin.

The chemical structures of the various materials involved in polyester resin fabrication are shown in Figure 31, along with a representation of the basic reactions involved. All of the starting materials for polyester resin manufacture are derived from petroleum fractions. The aromatic acids are made from xylenes, and the unsaturated acids are made from benzene by oxidation. The most common glycol, propylene glycol, is made from propylene via oxidation. Styrene is made from benzene and ethylene.

Phthalic acid and maleic acid are both easily dehydrated and are therefore used in the form of anhydrides rather than as acids in order to avoid the costs associated with shipping water.

The major use for unsaturated polyester resins is in the manufacture of reinforced plastics. They constitute about 80% of all the materials used for reinforced plastic applications. Glass fiber is the most common reinforcing agent, although other reinforcements such as metallic fibers and natural fibers are occasionally used. Typically, glass fiber averages about 35% of the weight of the reinforced polyester. Nonreinforced applications for unsaturated polyester resin include molded plastic and resins used for castings, surface coatings, and putty-like compounds used as body solder on automobiles.

There are two somewhat different procedures for carrying out the polyester polymerization reaction. The differences are related to the manner in which the water of reaction is removed. The fusion process removes the water by passing an inert gas, usually nitrogen, through the reaction mixture. In the second technique, which is referred to as the solvent or azeotropic process, a solvent (usually xylene or toluene) is added to the reaction mixture

FIGURE 31: TYPICAL REACTION AND RAW MATERIALS USED TO FORM UNSATURATED POLYESTER RESIN

HYDROCARBON STARTING MATERIALS

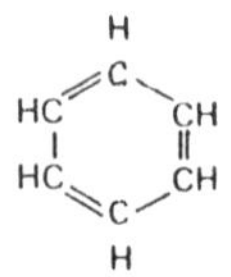

Benzene

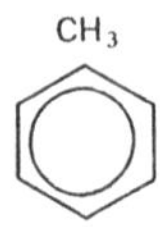

Toluene

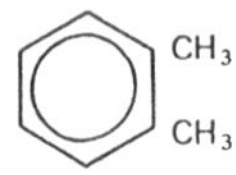

Ortho-xylene

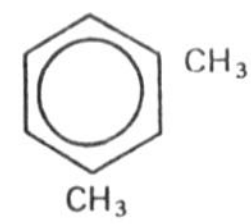

Meta-xylene

ACIDS

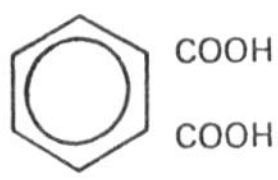

Ortho-phthalic Acid

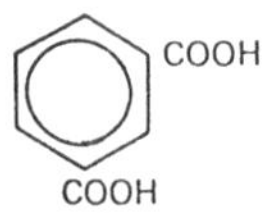

Iso Phthalic Acid (IPA)

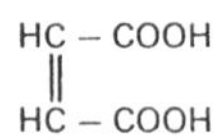

Maleic Acid

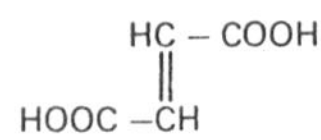

Fumaric Acid (FA)

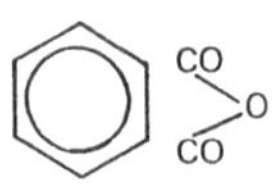

Phthalic Anhydride (PA)

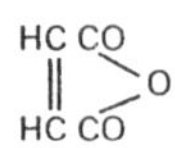

Maleic Anhydride (MA)

GLYCOLS

$HO\ CH\ (CH_3)\ CH_2\ OH$

Propylene Glycol (PG)

$HO\ CH_2\ C\ (CH_3)_2\ CH_2\ OH$

Neopentyl Glycol (NPG)

REACTIVE SOLVENT

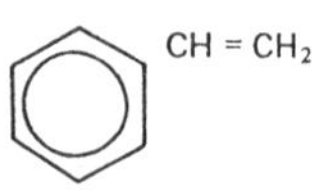

Styrene (S)

POLYESTERS

$$HOOC - R\text{-}COOH + HO - R' - OH \longrightarrow (- OOC - R - COOR' -)_n + H_2O$$

Note: Some of the R groups contain the reactive ethylenic linkage.

RESIN

PA + MA + PG = Base Resin (Solid)

Base Resin + Styrene = Polyester Resin (liquid)

Source: EPA (7)

and forms a constant boiling azeotrope with the water of reaction. This azeotrope is distilled off during the esterification reaction and the solvent is recovered.

After carrying the polymerization reaction to the desired molecular weight and removing the water of reaction, the polymer is transferred to another vessel containing the reactive monomer which is typically styrene, although methyl methacrylate and vinyl toluene are sometimes used. After mixing in the thinning tank, the final composition containing reactive monomer is either discharged to a filtration press prior to being loaded into 208.2 liter (55 gal) drums or bulk tanks. Discharge from the thinning tank is sometimes carried directly to drums or bulk without filtration. The concentration of reactive monomer can vary considerably (from 20 to about 55% by weight). A typical formulation contains about 35% styrene.

Although the vast majority of polyester reactions are carried out in batch reactors, there are several plants in the U.S. which have continuous esterification reactors. This mode of operation can usually only be justified when a large quantity of a specific type of polyester resin is desired. The continuous reactor undoubtedly generates less waste per pound of product as compared to batchwise production due to more infrequent cleanout, more efficient operation, and more careful control of operating parameters. In addition to boiler and cooling tower blowdown, the sources of effluent are as follows.

Water of Reaction: The water of reaction which is passed out overhead is condensed by some means and is either removed directly from the condenser, or more typically from the decanter following the condenser. This condenser also serves as a means for separating the solvents used in azeotropic distillation and sending them back to the reactor. The water of reaction may contain a variety of contaminants, including glycols, acids, and minor quantities of dissolved solvent.

Scrubber Waste: Scrubbers are used on the overheads leaving the reactor in order to reduce the concentration of entrained liquids and solids. Both recirculating and once-through scrubbers are used. The scrubber operation usually consists of passing the gaseous stream leaving the top of the reactor through a column into which water is sprayed. The scrubber may either use a recirculating water stream or once-through water.

In the latter case, although the water use is quite high, the effluent concentration in the water stream leaving the scrubber is not high and the stream can therefore be passed to a treatment system. In the case of a recirculating scrubber, however, the BOD and COD concentration can often be 200,000 to 400,000 ppm. Such high concentrations could upset the operation of conventional biological or municipal waste treatment plants, and therefore it is common practice in the industry to either discharge such concentrated recirculating scrubber wastes to a landfill or to incinerate the wastes.

Caustic Cleanout of Reactors: Concentrated caustic solution is typically used to clean out the polymerization reactors. The frequency of reactor cleanout is highly variable depending on the grades and types of resin produced by the manufacturer. Cleanout periods ranging from once every three weeks to once a year were encountered during interviews. The caustic solution is also used to clean out tank cars and tank trucks. Generally, some form of recycling is practiced on this caustic cleanout solution.

Although there are some notable exceptions, most polyester resin plants also manufacture the alkyd resins which are used in paint manufacture. These resins are quite similar to polyester resins with the exception that the acid portion of the polymerization recipe contains a significant quantity of a long chain of unsaturated fatty acid, and the polymerized resin instead of being diluted with a reactive monomer such as styrene is diluted with a solvent such as xylene or naptha.

With these two exceptions, the processes used to manufacture alkyd resins are quite similar to those used for manufacturing unsaturated polyester resins. Both azeotropic and fusion cook processes are used for manufacturing alkyd resins and, in many cases, identical reactors are used. The nature of the effluents in alkyd resin and unsaturated polyester resin manufacture is also quite similar, with the notable exception of an increased amount of an oily material which originates from the long chain, unsaturated fatty acids used in the alkyd recipe. The presence of this material often appears as a high "oil and grease" analysis in the discharge. It is also worthy of note that although a typical polyester reaction gives off about 12% of the original weight of the reactants as water of reaction, the alkyds, because of the high molecular weight of the reactant, typically give off 5% of the weight of the original reactants as water of reaction.

Litharge (lead oxide) and other catalysts such as lithium compounds are often used as alcoholysis catalysts in the reaction vessel in concentrations of a few ppm. Occasionally benzene sulfonic acid is used as an esterification catalyst. Most of these catalysts go out with the polymer or are trapped in the filtration step.

Various additives are often mixed with the resin in the thinning tank. These are color stabilizers such as triphenyl phosphate, amines, fire retardants such as chlorophthalic anhydride, curing accelerators such as cobalt naphthenate, and thixotropic agents such as Cab-O-Sil. All these components appear to go out with the resin.

Discharge to municipal waste is the waste treatment method utilized by 90% of the polyester/alkyd resin industry. There are some notable exceptions, however, who carry out their own biological treatment or who truck all of the wastes to a large, centralized municipal treatment plant. The exact nature of the wastes can be highly variable depending on such factors as:

The use of once-through vs recirculating scrubbers.
The extent to which cooling water is recycled.
The frequency of reactor washout.
The product mix (polyesters vs alkyds).
Whether or not polymerization catalysts are used.
Whether or not wastes are incinerated.

Polyolefin Manufacture

Some of the characteristics of four different variants of polyolefin manufacture will be reviewed here as a basis for consideration of pollution problems presented by these processes. These variants are:

[1] ICI high pressure mass polymerization for production of low density polyethylene.
[2] Phillips solution polymerization for high density polyethylene and other polyolefins.
[3] Ziegler solution polymerization for high density polyethylene and other polyolefins.
[4] Phillips slurry polymerization for high density polyethylene and other polyolefins.

The high pressure process for low density polyethylene is a very simple one, as illustrated in Figure 32. Ethylene gas is mixed with a very small quantity of air or oxygenated organic compounds as a catalyst, and with recycled ethylene, and raised to high pressure in reciprocating compressors. The operating pressure is considered to be confidential information, but the trend in the industry has been to the highest practical pressures, and literature references to design ratings of 40,000 psi (2,722 atm) and up are common.

At the operating pressure and at an appropriate temperature, polymerization is carried out in jacketed tubular reactors. The heat of reaction is removed to hot water in the jacket, which circulates through a waste heat boiler for the generation of steam. On completion of the reaction, the pressure is reduced and specification polymer separated in flash drums. This molten material is pumped through a multiple orifice extruder to an underwater

FIGURE 32: LOW-DENSITY POLYETHYLENE PRODUCTION—HIGH PRESSURE PROCESS

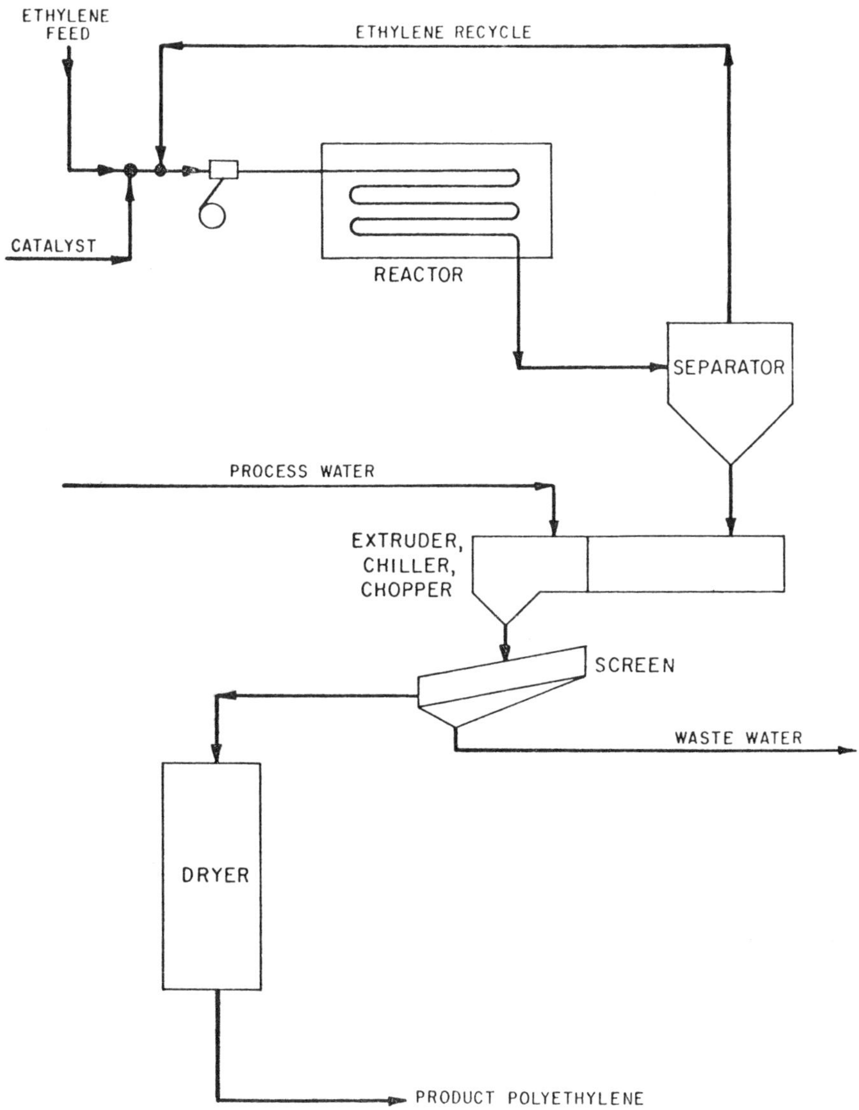

Source: EPA (9)

chiller and chopper to produce polyethylene pellets. The water is separated on a screen and pumped through a cooler for recycle. A purge stream of this water is removed and replaced with high-quality, clean water. The purge is at a rate sufficient to remove polymer fines generated in chipping. The quantity of fines depends on the grade of polymer produced and with some grades is negligible. Wet polymer from the screen is dried and stored in silos.

In the solution process for polyolefins shown in Figure 33, the polymer is dissolved in the reaction solvent as it is formed, and the catalyst is present as a separate solid phase. The catalyst system is activated chromium oxide deposited on a carrier such as alumina. This process is one of two for polyolefins which first came into prominence in the late 1950s; the other is the Ziegler process, in which the polymer precipitates as it is formed. Products of the solution system include high density polyethylene and a limited number of copolymers.

As the concentration of polymer, or the molecular weight of the polymer in solution, increases, the viscosity of the solution also increases markedly. This phenomenon places severe limitations on the processability of the reaction mass. Temperature control is accomplished by indirect cooling with refrigerated water, and the viscosity must not be allowed to exceed a reasonable limit for efficient heat transfer.

Viscosity is also an important limitation in the next step, which is the removal of the catalyst by filtration or centrifugation. From the filter, the catalyst, wet with solvent, is mixed with hot water and the solvent removed by steam stripping. Solvent-free catalyst slurry is processed in a skimmer and solid catalyst removed to land fill.

The aqueous phase is recycled to the steam stripper. Vapor from the steam stripper is combined with other recovered solvent for purification. The catalyst-free polymer solution is processed in a system which concentrates and precipitates the polymer, and then removes the last traces of solvent by steam stripping, leaving the polymer as a slurry in water. The slurry is filtered or centrifuged, and the filtrate recycled to the stripper.

FIGURE 33: POLYOLEFIN PRODUCTION – SOLUTION PROCESS

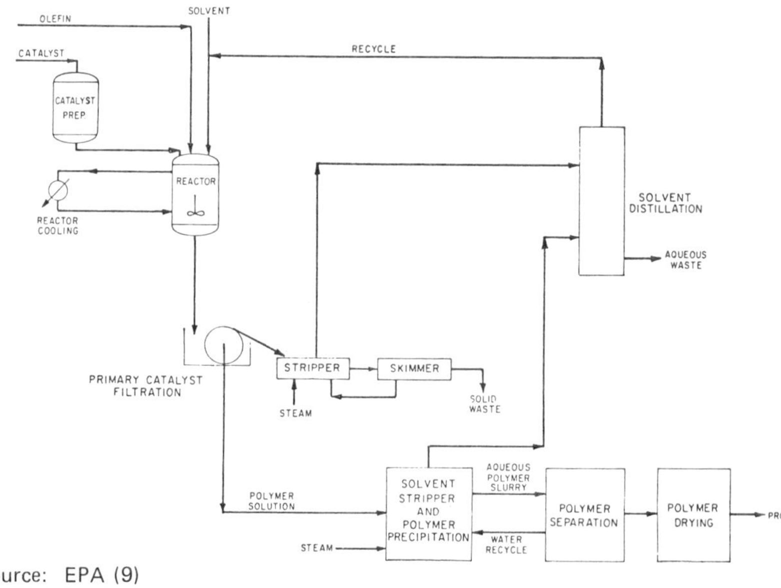

Source: EPA (9)

Solvent recovered in the concentrator and vapors from the steam strippers are processed by distillation in the solvent recovery section. All process water used in the catalyst and polymer separation area appears as an aqueous waste stream from this distillation unit. It contains small quantities of dissolved hydrocarbons, but in at least one plant it is used as boiler feed water.

Dry polymer crumb or flake is blended, melted, extruded and pelletized. This pelletizing operation is carried out under water, with cooling and transport accomplished with recirculated, clean, softened water. A purge stream amounting to a few percent of the circulation rate is withdrawn to waste. This sytem is the same as already described for the low-density polyethylene process.

The Ziegler process depends on a catalyst system discovered and patented by Dr. K. Ziegler. There have been a number of improvements by companies using the basic principle, and the name in fact applies to the catalyst system. Each user has had to design his own plant. It is convenient, however, to group under this name all polyolefin processes which employ a reaction solvent in which the polymer precipitates as it is formed. Figure 34 details this type of polyolefin production. The catalyst is a relatively complex alkyl, or alkyl halide, of metals such as titanium and aluminum. Products of this process include:

High density polyethylene
Polypropylene
Polybutene
Copolymers

Catalyst preparation, monomer addition, and reaction proceed as already described for the solution process. Temperatures and pressures are lower; and, because the polymer does not dissolve, problems caused by excessive viscosity do not arise.

The next step is the removal of the catalyst, which historically has been the most troublesome part of the system. The residual catalyst content of the final polymer must be very low, and for this reason a system is employed which allows transfer of catalyst to a separate liquid phase. Aqueous alcohol is used for this purpose and the catalyst is removed in solution, leaving the polymer slurried in the hydrocarbon solvent.

The aqueous alcohol phase is treated to precipitate the catalyst as the oxides (e.g., titanium, aluminum), and these materials eventually appear as finely-divided suspended solids in the aqueous waste. They will settle sufficiently to permit discharge of a clarified effluent, but consolidation of the sludge left behind has been a problem. Alcohol is recovered for reuse by distillation. The aqueous phase remaining is the principal waste product of the plant. This water contains a finite amount of dissolved alcohol, and this chemical constitutes the largest raw waste load on the treatment facilities.

The polymer slurry is processed by steam stripping, filtration, drying, extruding and pelletizing as is done for the solution process, and the hydrocarbon solvent is purified by distillation. A small quantity of aqueous waste is recycled to the alcohol unit.

The problems of the solution process for polyolefins described above have to a large degree been overcome in a newer version called the particle form process, and the method has a growing commercial acceptance. Figure 35 details this method of production. There have been three major changes:

[1] The catalyst system has been modified and its activity increased to the point that special measures for catalyst removal are unnecessary for many grades of polymer.

[2] The solvent system has been modified so that the polymer is obtained as a slurry rather than a solution in the diluent.

[3] Special design loop-reactors have been developed which allow the polymerization system to operate under good control of reaction conditions and at satisfactory rate.

FIGURE 34: POLYOLEFIN PRODUCTION – ZIEGLER PROCESS

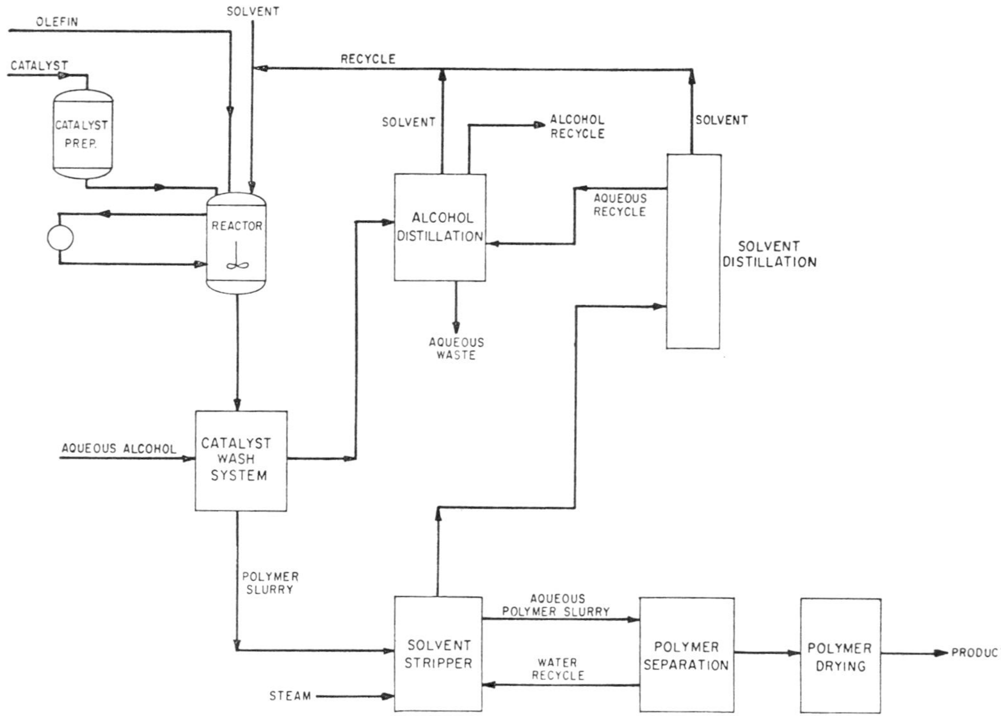

Source: EPA (9)

FIGURE 35: POLYOLEFIN PRODUCTION – PARTICLE FORM PROCESS

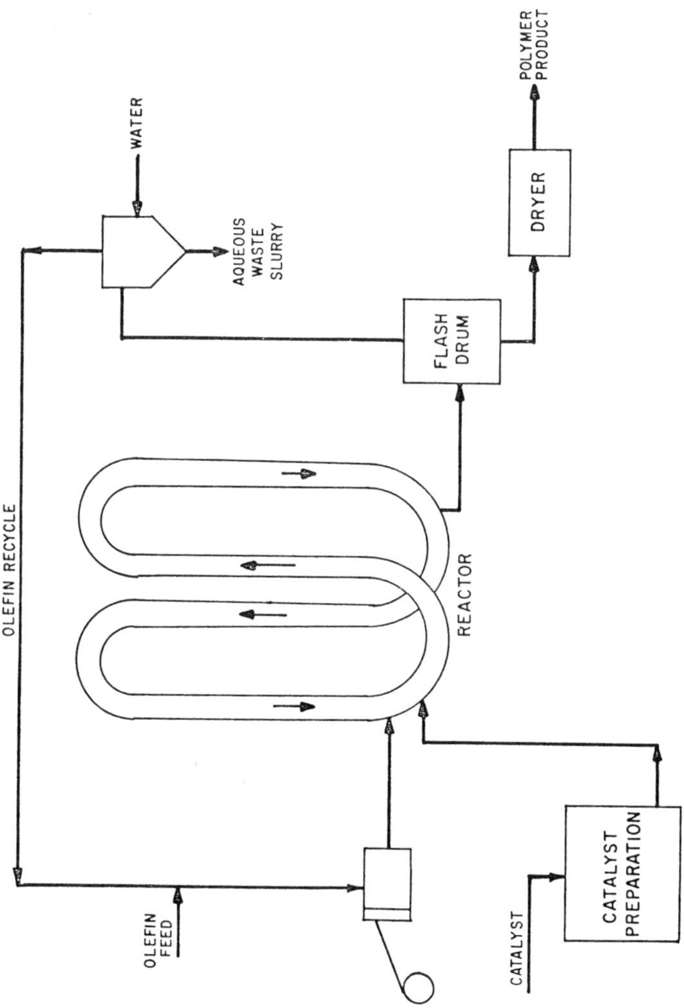

Source: EPA (9)

In practice of this method, catalyst and olefin feed are added to the reaction mass which is circulated continuously through the loop reactors. A stream is also withdrawn continuously from the reactor to a flash drum. Polymer is removed from the bottom of the flash drum, dried, and processed through an extruder-pelletizer as with the other methods.

The vapor stream from the flash drum is scrubbed to remove polymer fines, this step produces a small quantity of wastewater. Both unreacted olefin, and recovered diluent are then separated from the overhead stream and recycled to the reaction step.

Polyphenylene Sulfide Manufacture

Polyphenylene sulfide polymers possess recurring units of sulfur which provide linkage for aromatic compounds. Polyphenylene sulfide (PPS) is a finely divided light-colored powder and is insoluble in any known solvent below 190°C (375°F). Above this temperature it has limited solubility in some aromatic and chlorinated aromatic solvents and certain heterocyclic compounds. It is highly crystalline with a melting point near 285°C (550°F). Although curing in air at elevated temperatures is required to effect chain extension and cross-linking, the resin remains thermoplastic in nature. It can be processed through conventional equipment for compression and injection molding.

Polyphenylene sulfide is a recent thermoplastic on the market; therefore only a small amount of information on its synthesis is available in the literature. It is believed that the process employs p-dichlorobenzene, sodium sulfide, and polar organic material such as n-methyl pyrrolidone to yield PPS which may or may not have cross-linking agents added to the reaction medium.

p-Dichlorobenzene is solid with a melting point of 53.1°C (127.5°F), a specific gravity of 1.46, and is insoluble in water. Sodium sulfide in the hydrated form contains 9 mols of water and is very soluble in water. N-methyl pyrrolidone is a liquid which boils at 197°C (387°F) and is soluble in water.

The reaction between the two raw materials takes place in the presence of a polar organic solvent. The reaction is carried out at a temperature in the range of 130° to 175°C (266° to 347°F). The mol ratio of p-dichlorobenzene to sodium sulfide should be in the range of 0.9:1 to 1.3:1. If ratios above this range are employed, the amount of unreacted dichlorobenzene will be increased, requiring separation and recycle. Larger excess of either reagent leads to lower molecular weight polymers, and still shorter polymers are produced by an increase in the reaction temperature.

In general, the synthesis reaction is carried out by reacting a polyhalo-substituted compound with an alkali metal sulfide which has been partially dehydrated by the polar organic compound that also serves as the solvent for the reactants. The solvent should be stable at the elevated temperatures for the reaction. The polymer formed may be heat treated in the absence of oxygen to yield a higher molecular weight polymer. Molecular weight control may be achieved by introducing a monohalo-substituted aromatic compound to the reaction medium, which causes termination of the chain growth. Yields of finished polymer are higher if a catalyst such as copper or a copper compound is added to the reaction medium.

It is possible to obtain a cross-linked polymer through addition of a polyhalo-substituted aromatic compound which contains substituents through which cross-linking can be effected by further reaction. A typical process scheme is shown in Figure 36.

The hydrated sodium sulfide in N-methyl pyrrolidone is charged to a stainless steel autoclave. The temperature is brought up to 190°C (374°F) while flushing with nitrogen to remove the water of hydration. Upon removal of water, the p-dichlorobenzene is charged to the autoclave and the temperature raised to approximately 250°C (482°F) for the desired reaction time after which the polymer and reaction medium is dropped to a stainless steel tank. The contents are washed with water and then with acetone, which carries along unreacted reaction media. The finished polymer is dried and packaged.

FIGURE 36: POLYPHENYLENE SULFIDE PRODUCTION

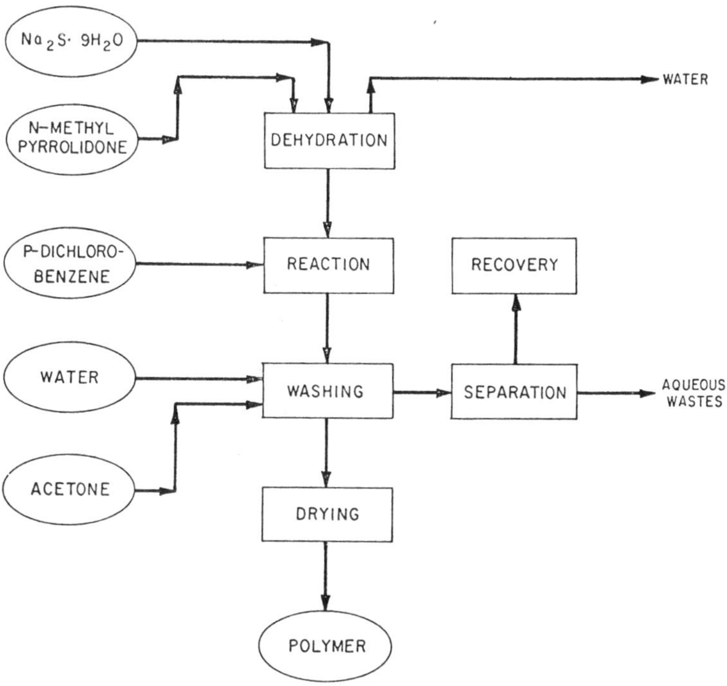

Source: EPA (7)

Aqueous wastes would be generated from two sources:

[1] Water of hydration.
[2] Process washwater. The primary contaminant would be sodium chloride, of which two mols would be generated per mol of feedstock.

The water of hydration is associated with the water bound to the sodium sulfide which is removed in a prereaction step by heating. It is likely that the water removal would carry along traces of sodium sulfide and N-methyl pyrrolidone which may or may not be recovered. Process washwater is used to wash sodium chloride from the reaction mass.

Due to the high solubility of N-methyl pyrrolidone and sodium sulfide in water, the waste generated from this source is most apt to contain a large waste loading. It is likely that solvent extraction and subsequent further recovery operations will be applied to this stream to recover the raw materials for recycle.

If it is assumed that the sodium chloride is removed in a water solution at a concentration of about 10% by weight, then about 10,000 mass units of water per 1,000 mass units of product would be required.

Polysulfone Resin Manufacture

Polysulfones are high molecular weight polymers containing sulfone groups and aromatic nuclei in the main polymer chain. The term "polysulfone" is also used to denote the class of polymers prepared by radical-induced copolymerization of olefins and sulfur dioxide. The latter do not have commercial significance, and will not be discussed.

Among the aromatic polysulfones synthesized, the polysulfone derived from dihydric phenols (bisphenol A) and 4,4'-dichlorodiphenyl sulfone have achieved commercial application under the name Bakelite polysulfone. This is a rigid, strong thermoplastic which can be molded, extruded, or thermoformed into a variety of shapes. It is both stable and self-extinguishing in its natural form. Bakelite polysulfone is prepared from the two raw materials under alkaline conditions according to the equations listed in Figure 37, Equations (1) and (2).

The disodium salt is prepared in situ by reaction of bisphenol A with exactly two mols of aqueous sodium hydroxide. A solvent is required for this polymerization, dimethyl sulfoxide being the most suitable. Very few others are effective. The reaction must be carried out at 130° to 160°C (266° to 320°F), primarily because of the poor solubility of the disodium salt at lower temperatures. Polymerization is, however, very rapid at these temperatures leading to molecular weights as high as 250,000 in an hour's time. As these molecular weights are too high for commercial processing, chain growth must be regulated by the addition of terminators. A variety of monohydric phenolic salts or monohalogen compounds have been found to be effective.

In the polymerization, the highest molecular weights will be obtained when the mol ratio of comonomers approaches unity. Since the comonomers contain two functional groups per molecule, as is required for this type of polycondensation, addition of a compound containing one functional group per molecule will result in an overall imbalance in functionality, terminating chain growth.

All but traces of water must be removed from the reaction mixture before polymerization. Hydrolysis of the dihydric phenol salt occurs otherwise, resulting in the formation of sodium hydroxide, which reacts very rapidly with the dichlorodiphenyl sulfone, forming the monosodium salt of 4-chloro-4'-hydroxydiphenyl sulfone [Figure 37, Equation (3)].

Two mols of caustic soda (from 2 mols of disodium salt) are used per mol of dichlorodiphenyl sulfone, which creates an imbalance in functionality between the comonomers. Consequently, it becomes impossible to obtain high molecular weight.

Another, somewhat less important, side reaction may occur if caustic soda is present during polymerization. Cleavage of polymer chains para to the sulfone groups results in the formation of two phenoxides, as shown in Figure 37, Equation (4), for polysulfone.

Other bisphenol A derived polysulfones are prepared by using various combinations of dihydric phenol sodium or potassium salts and dichlorodiphenyl sulfone. Use of certain other aromatic dihalides besides dichlorodiphenyl sulfone expands the list considerably.

A typical process scheme is shown in Figure 38. Polymerization of bisphenol A and 4,4'-dichlorodiphenyl sulfone (DCDPS) is carried out batchwise by first forming the disodium salt of bisphenol A. This is done by charging bisphenol A and an excess of dimethyl sulfoxide (DMSO) to a reactor where the temperature is brought up to 60° to 80°C (140° to 176°F). Sodium hydroxide as a 50% solution is added stoichiometrically to the mixture over a period of about 10 minutes. Water is removed from the system, the DMSO that codistills being returned continuously.

In so doing, the temperature of the contents rises from about 120°C (248°F) to 140°C (284°F) at the conclusion of this step. When this point is reached, most of the water originally present has distilled, and the disodium salt of bisphenol A appears as a precipitate.

FIGURE 37: TYPICAL REACTIONS TO FORM POLYSULFONE RESINS

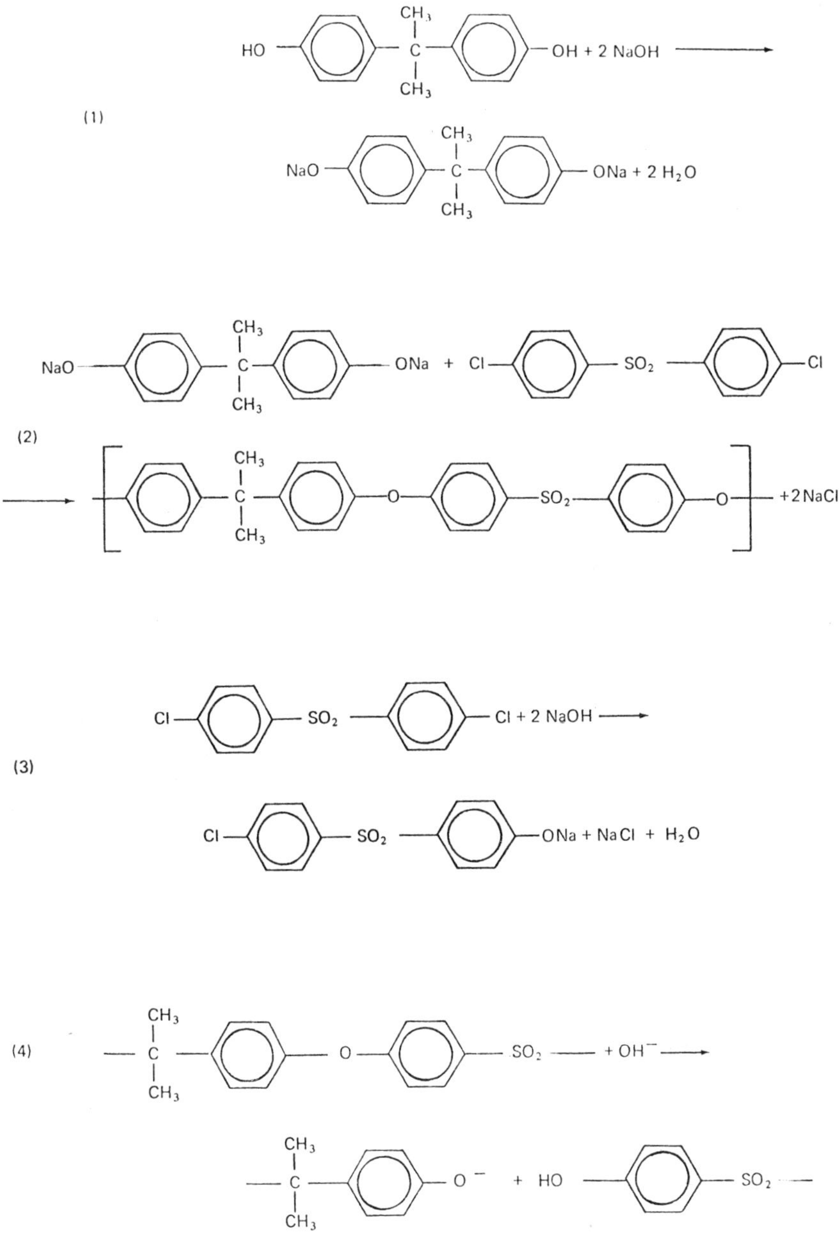

Source: EPA (7)

FIGURE 38: POLYSULFONE RESINS PRODUCTION

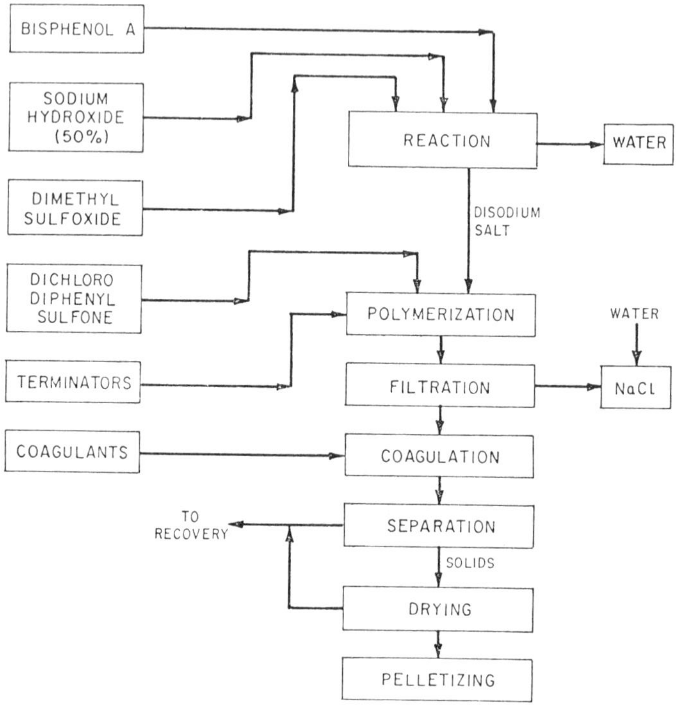

Source: EPA (7)

Excess azeotrope solvent is distilled from the system until the temperature of the contents reaches 155° to 160°C (311° to 320°F). At this point the precipitate will redissolve with the formation of a very viscous solution. It is assumed that at this point only traces of water remain.

A 50% solution of 4,4'-dichlorodiphenyl sulfone (DCDPS) in DMSO maintained at 110°C (230°F) is fed stoichiometrically to the polymerizer. The temperature must not drop below about 150°C (302°F) until the polymerization is well along, since sodium-ended low polymer may precipitate on the walls of the reactor. Too high a temperature during addition of the sulfone and subsequent polymerization is to be avoided, as the reaction is mildly exothermic, extremely rapid over 160°C (320°F), and excessive solvent decomposition and/or discoloration or even gelation of the reaction mass may occur.

The polymerization may be terminated in a variety of ways, one of which is to pass methylene chloride into the polymerizaing mixture when the desired degree of polymerization is reached.

Additional DMSO is added to the reactant mixture to reduce the viscosity to a workable level. The mixture is then passed through a rotary drum filter to remove sodium chloride. The process stream is fed to a coagulation vessel where an alcohol such as ethanol is added to coagulate the polysulfone. The solvent and unreacted feed materials are separated from the polymer and sent to recovery. The polymer is then dried and pelletized while the vented solvent is recovered and recycled.

As regards wastewater generation, the reaction products include 810 mass units of water per 1,000 mass units of the disodium slat, and an estimate 1,800 mass units of water is added with the sodium hydroxide for a total of 2,610. The wastewaters from the azeotropic distillation are likely to contain DMSO and traces of bisphenol. In addition, appreciable concentrations of sodium chloride are expected to occur. The wastewaters are expected to be alkaline because of the excess sodium hydroxide required for pH control. The volume of wastewater generated is highly dependent upon operating conditions, especially those associated with the washing operations.

Polyvinyl Acetate, Polystyrene and Related Vinyl Polymers

A number of related polymers are made by processes in which the monomer is dispersed in an aqueous, continuous phase during the course of the reaction. There are technical differences between emulsion and suspension systems which pertain to the polymerization reaction itself, but these do not have a bearing on the potential aqueous pollution problem. Therefore both methods will be covered by this discussion. Products of this process include:

Polystyrene (PS)
Acrylonitrile-butadiene-styrene (ABS)
Styrene-acrylonitrile (SAN)
Polyvinyl chloride (PVC)
Polyvinyl acetate (PVA)

A batch process, as shown in Figure 39, is commonly used in emulsion or suspension manufacture. Typical reactor size is 5,000 to 30,000 gal (18.9 to 113.5 m^3). The batch cycle consists of the continuous introduction of a water-monomer emulsion to the stirred reactor. Polymerization occurs at about the rate of monomer addition; the heat of reaction is removed to cooling tower water circulated through the jacket. The reactor is vented through a condenser for monomer recovery; and the condensate, including any water, is returned directly to the vessel. On completion of the batch, a short "soaking" time is allowed for completion of the reaction, and water is then added to dilute to the desired end composition. The batch is drawn off through a screen to product storage. Oversize screenings (a very small amount) are disposed of to landfill.

Monomers, the principal raw materials, are often protected during shipping and storage by an inhibitor, such as catechol, which may be removed prior to polymerization by washing. This contributes to the wastewater load.

A number of products, polyvinyl acetate for example, are marketed in this latex form with no further processing required. Thus, although water is a process material, there is no aqueous waste inherent. When the product is isolated and sold in solid form, the screened latex is pumped to another reactor. A small quantity of a flocculating agent is added which destroys the emulsion and permits subsequent separation of the polymer.

A number of important plastics are manufactured by mass polymerization, a system in which the purified monomer is allowed to polymerize under controlled conditions of temperature and reaction rate. This process is shown in Figure 40. Catalysts and modifiers are used to initiate the reaction, control its rate, and influence the final molecular weight. These materials are used in very small amounts, and their residue remains in the product. Removal of the heat of reaction is a difficult problem in this process and limits the type of equipment which can be used.

FIGURE 39: EMULSION POLYMERIZATION

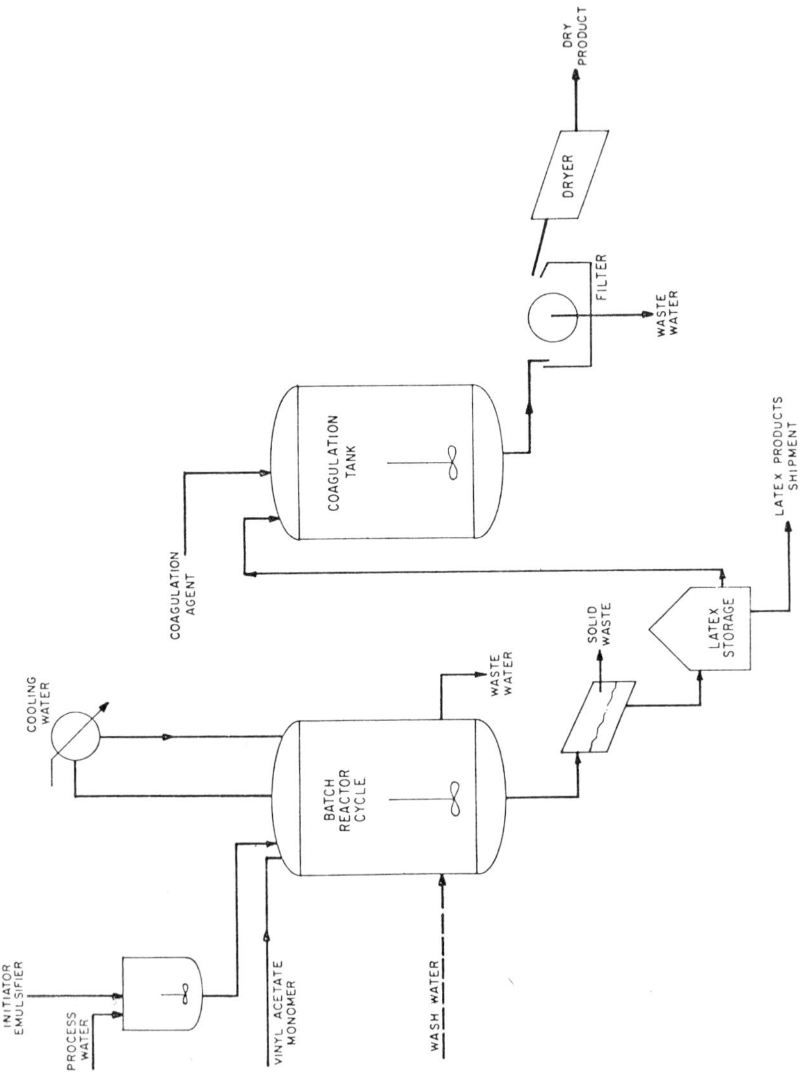

Source: EPA (9)

FIGURE 40: MASS POLYMERIZATION

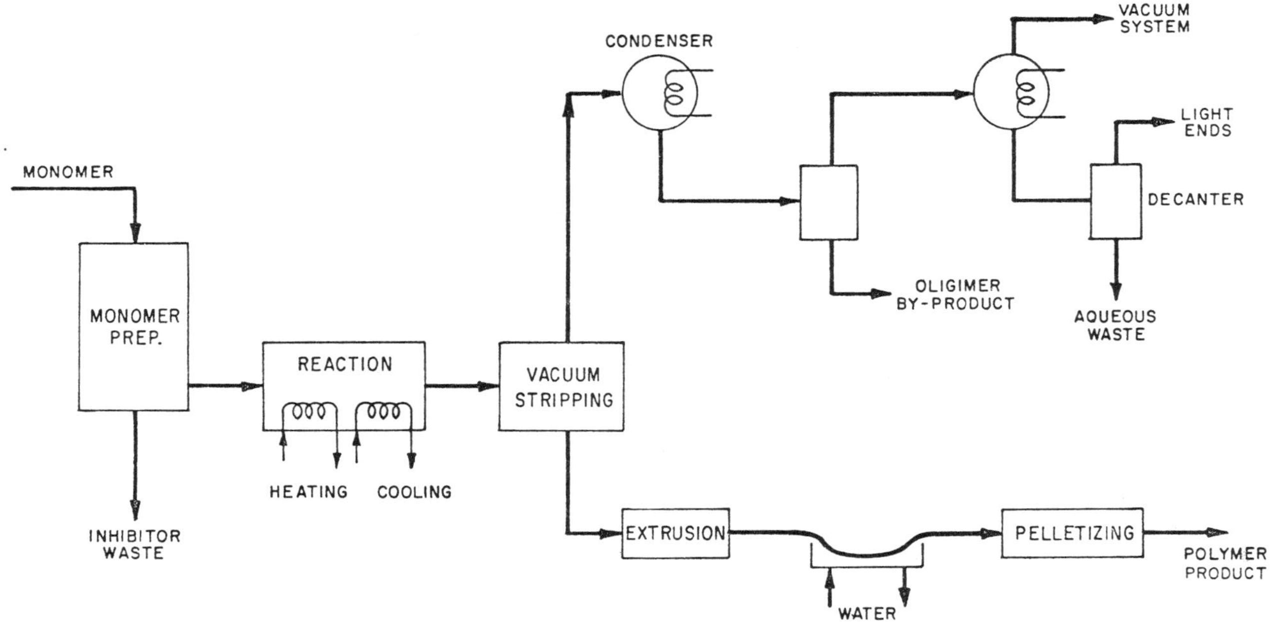

Source: EPA (9)

Products of this process include:

Polystyrene (PS)
Acrylonitrile–butadiene–styrene (ABS)
Styrene–acrylonitrile (SAN)
Polyvinyl chloride (PVC)

It is usually necessary to protect the purified monomers from autopolymerization in storage. The inhibitor used for this purpose is removed by distillation or washing. This frequently results in an aqueous waste. The reaction system is usually continuous, or multistage, and the first step is to bring the monomer to reaction temperature by indirect heating. A heat-transfer oil or fluid such as Dowtherm, circulated from a fired heater, is used. Once reaction begins the heat is removed by transfer to a cooling oil circulated through coils or in a jacket. The circulated oil is cooled by water in conventional heat-exchange equipment.

On leaving the reactor, the polymer contains unreacted monomer and small amounts of contaminants and by-products. These materials are removed by vacuum stripping.

Vapors from this unit pass through an oil-cooled tar condenser. The vent from this condenser is connected to a steam jet ejector, and steam and volatile hydrocarbons condense in a water-cooled surface condenser. Insoluble oils are decanted and recovered, and contaminated condensate goes to the process sewer.

Pure polymer from the bottom of the stripper is forced through multiple orifice extruders to make strands of polymer, which are cooled in a water bath before pelletizing for storage and shipment.

A process developed by *H.-G. Keppler, L. Zuern and E. Stahnecker; U.S. Patent 3,829,378; August 13, 1974; assigned to Badische Anilin- & Soda-Fabrik AG, Germany* is a process for purifying wastewater obtained in the production of styrene bead polymers in the presence of vinyl pyrrolidone polymers or vinyl alcohol polymers used as suspension stabilizers. This process comprises treating the wastewater at 70° to 180°C with 0.02 to 5% by weight of the persulfate of sodium, potassium or ammonium in order to flocculate the suspended matter.

Polyvinyl Butyral Manufacture

Polyvinyl butyral is formed by condensation of polyvinyl alcohol with butyraldehyde in the presence of an acid catalyst (Figure 41).

Two processes, shown in Figures 42 and 43 are used for production of polyvinyl butyral. The process shown in Figure 42, which is used by E.I. DuPont de Nemours and Co., Inc., Fayetteville, N.C., starts with powdered polyvinyl alcohol. The alcohol is dissolved in water and reacted with butyraldehyde according to the equation in Figure 41.

FIGURE 41: TYPICAL REACTION TO FORM POLYVINYL BUTYRAL

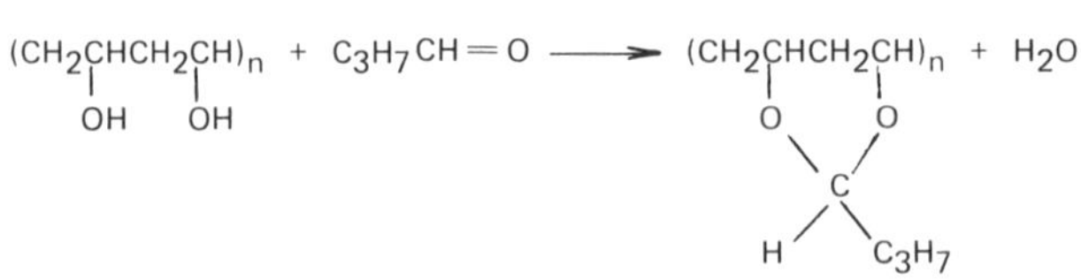

Source: EPA (7)

FIGURE 42: POLYVINYL BUTYRAL PRODUCTION – DU PONT PROCESS

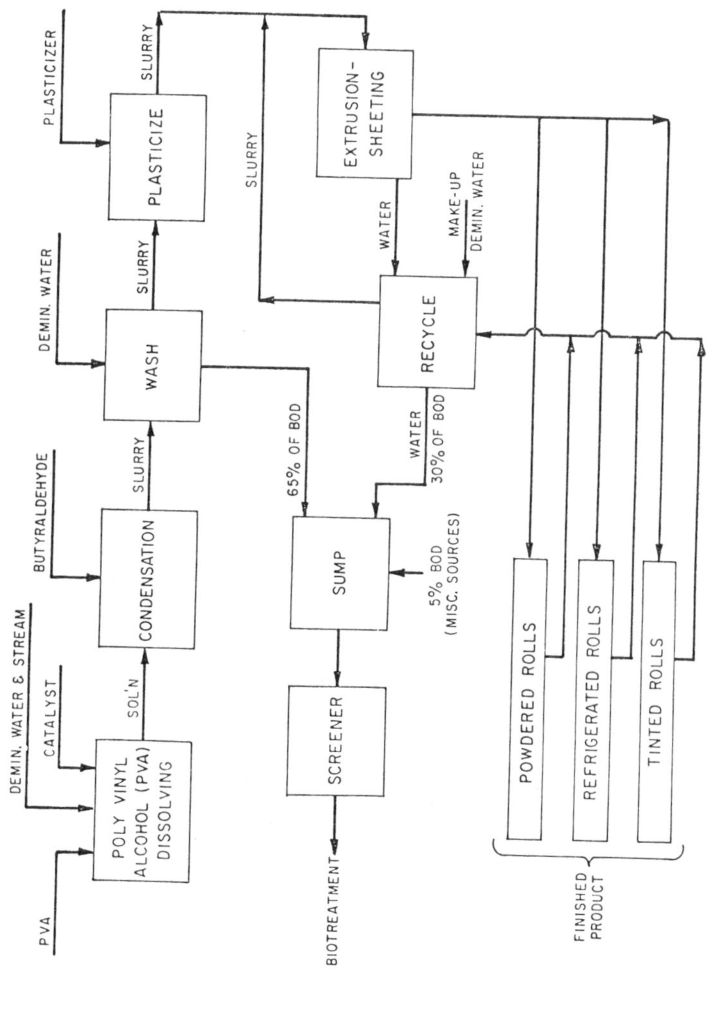

Source: EPA (7)

FIGURE 43: POLYVINYL BUTYRAL PRODUCTION – MONSANTO PROCESS

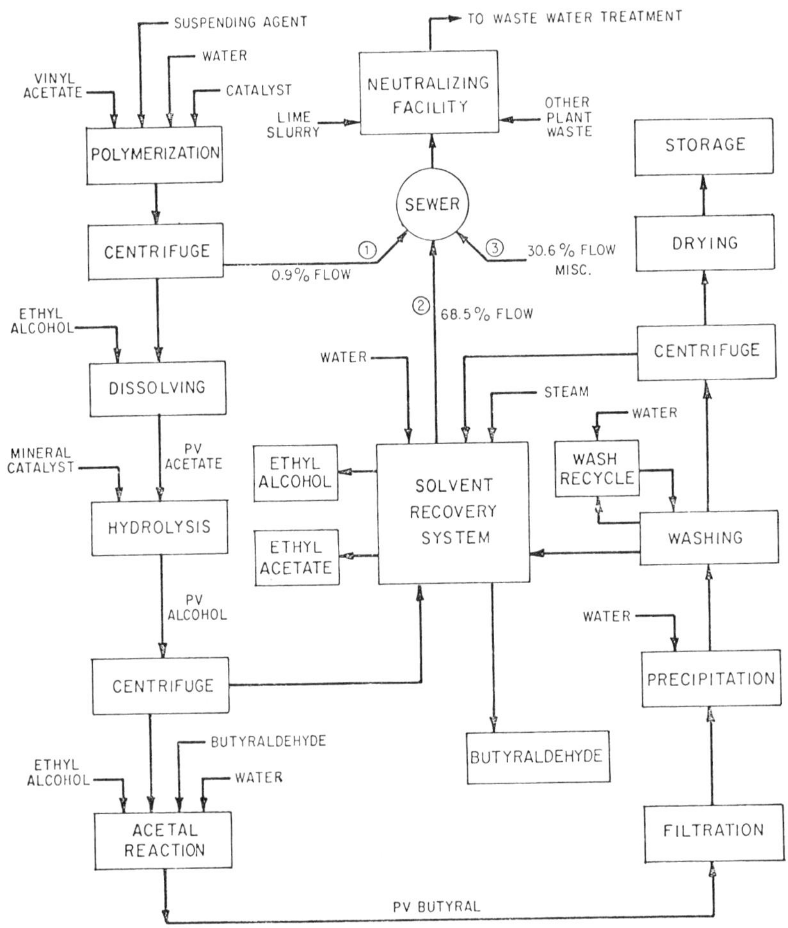

Source: EPA (7)

The butyral is washed and slurried in water. The slurry is treated with a plasticizer, (triethylene glycol di-2-ethyl butyrate is one), and the mixture is sheeted to give the product that is sandwiched and heat sealed between pieces of glass to produce safety glass.

The process shown in Figure 43 is practiced by Monsanto Co., Indian Orchard, Mass., and Trenton, Mich. At these plants, vinyl acetate monomer is polymerized in suspension to give polyvinyl acetate. This polymer is separated, dissolved in ethyl alcohol and hydrolyzed in the presence of a mineral acid. The polyvinyl alcohol is centrifuged and condensed with butyraldehyde in the presence of ethyl alcohol and acid. The butyral solution is filtered, precipitated with water, washed and dried.

The dry product may be sold as such, or transferred to another area of the Monsanto plant for sheeting. In this operation, the polymer is combined with plasticizer and sheeted on rolls. The sheeting process uses very little water, and wastes are negligible.

Wastes generated in the DuPont process are indicated in Figure 42. The acid catalyst, small amounts of the reaction components, and a small amount of the plasticizer are anticipated wastes.

Wastes generated by the Monsanto process, shown in Figure 43, are considerably more complex than those generated by the DuPont process. The Monsanto process requires ethanol, which is not needed in the DuPont process; the ethanol combines with acetic acid which is liberated in the hydrolysis step to give ethyl acetate, which is recovered.

Polyvinyl Chloride Manufacture

A report by the B.F. Goodrich Chemical Co. (10) describes the water pollution aspects of PVC manufacture. In 1966, B.F. Goodrich Chemical Company initiated plans to construct a polyvinyl chloride (PVC) manufacturing facility on the Delaware River in Salem County, New Jersey. This facility would employ emulsion, suspension, and bulk polymerization processes and would generate wastewaters containing monomers, polymers, organic and inorganic salts, organic acids, resin and rubber fine solids, dispersants, wetting agents, and catalyst.

The wastewater discharges from this manufacturing facility were subject to the requirements of the State of New Jersey Department of Health and the Delaware River Basin Commission. A minimum of secondary treatment and a 90% removal of biochemical oxygen demand was required.

In order to competently design an industrial wastewater treatment system to meet these established requirements, extensive laboratory and pilot plant investigations were performed at the company's environmental control laboratory. These investigations were performed under the direction of graduate sanitary engineers from the company and from Roy F. Weston & Associates.

Both primary and secondary treatment methods were investigated. Waste equalization; solids removal by chemical coagulation, dissolved air flotation, and clarification; activated carbon adsorption; contact stabilization; completely mixed activated sludge; and various other methods were considered and studied.

The report cited above discusses the studies performed with a major emphasis placed on those processes found most applicable to the treatment of PVC wastes. A description of laboratory methods and pilot plant equipment is included. Typical data is presented and interpreted in terms of process design. Significant observations, specifically pertaining to the treatment of PVC wastes, are included along with a detailed and comprehensive process design summary. In addition, full-scale plant operation is discussed in relation to individual unit operations and unit processes. Problems encountered during initial start-up operations are presented including recommendations for system improvements. Actual operating data is included.

Lastly, wastewater recycle and reuse is included. The project was supported in part by the Office of Research and Monitoring, Environmental Protection Agency. Since completion of the wastewater treatment system, many visitors from 25 states and 3 foreign countries have toured the facilities. The following conclusions were arrived at as a result of this study.

Waste equalization facilities were found to be a prerequisite to the treatment of this type waste. Variations in organic loading, hydraulic loading, and problems encountered with shock loadings in the pilot plant studies substantiated this need.

Latex solids definitely have an adverse effect on a completely mixed activated sludge system. The solids tend to cling to the biological floc causing stickiness and extreme reduction in organic removals. In addition, these solids should be removed in order to minimize the difficulty in keeping the aeration tank contents in suspension; in order to maximize oxygen transfer; and to reduce the quantity of biological inert solids in the aeration tank.

These conclusions were supported by the initial difficulty encountered in obtaining reproducible data from toxicity studies and by the difficulties encountered in the biological treatment studies.

The method most suitable for removal of this type of solids and for accomplishing primary treatment was determined to be chemical coagulation-flocculation and clarification. The use of coagulants and coagulant aids is recommended. The salt of ferric iron was found to be more effective as a coagulant as compared to alum. Calgon 227 and Nalco 670 were found to be effective coagulant aids. The dosage requirement for coagulants and coagulant aids was found to be approximately 150 mg/l, and 1 mg/l, respectively. Dissolved air flotation was investigated and found to be ineffective for this type of waste. The floc formed was extremely fragile.

The wastes from this type of production plant were found to be nontoxic to a biological treatment system. It was, however, determined necessary to provide waste equalization and primary solids removal prior to such a system.

Of the secondary treatment methods investigated, a completely mixed activated sludge system was found to be most applicable for this type of waste.

Organic loading rates achievable with this system for this particular waste were found to be in a range between those for a conventional activated sludge system and a high rate system treating domestic wastes. A loading rate of 0.70 pounds of 5 day biochemical oxygen demand removed per day per pound of volatile suspended solids under aeration was established. This compares with ranges for conventional and high rate systems of 0.1 to 0.5 and 1.25 to 4.5, respectively.

Energy oxygen requirements were determined to be 0.90 pounds of oxygen per pound of 5 day biochemical oxygen demand removed. This compares with 0.50 for conventional systems treating domestic wastes.

Sludge production was determined to be 0.68 pounds of volatile suspended solids per pound of 5 day biochemical oxygen demand removed. This compares favorably with values of 0.5 to 0.7 for conventional systems treating domestic wastes.

The aeration basin detention time was established at 6 hours. A mixed liquor volatile suspended solids level of 2,500 mg/l was determined optimum for practical design. Aeration basin pH adjustment was provided in order to prevent the growth of large numbers of filamentous organisms and bulking sludge. This type of waste was found deficient in nitrogen and phosphorous levels necessary to support biological life. Supplemental nutrients are essential.

Activated carbon treatment was determined to be impractical for this waste. Poor adsorption capacities were attributed to the presence of water soluble long-chain organic soaps contained in the wastes.

Laboratory data indicated the contact stabilization process was not practical for treatment of this waste. The time requirements for adsorption of the soluble organics was found to be approximately equal to the time required for complete degradation.

Suspended solids removal efficiency has averaged greater than 95% through the flocculator-clarifier system (primary treatment); 99% through the secondary system; and 98% through the complete treatment system.

During initial operations of the plant biochemical oxygen demand average removal efficiency has been in excess of 98% resulting in an average effluent concentration of 2.4 mg/l.

Wastewater recycle has been investigated and appears feasible. Plans are underway for installation of a permanent wastewater recycle and reuse system.

Polyvinyl Ether Manufacture

The various vinyl alkyl ether monomers (the reactants) are normally colorless liquids or low melting solids. All readily add halogens across their double bonds. The lower alkyl vinyl ethers are sparingly soluble in water. These monomer ethers hydrolyze slowly in water at room temperature (and more rapidly in the presence of mineral acids), producing acetaldehyde, as shown in Figure 44, Equation (1).

Hydrolysis of the monomers is avoided by adding alkaline stabilizers (for example, 0.1% of triethanolamine) to the stored vinyl ethers. Stabilizers and impurities such as alcohol, acetaldehyde, and acetals are then removed before polymerization by washing with water or very dilute KOH. Vinyl ethers produce high molecular weight homopolymers when reacted in the presence of Lewis acid catalysts. The catalysts used are related to Ziegler catalysts, for example, diethyl aluminum chloride or Grignard reagents.

The propagation step proceeds as shown in Figure 44, Equation (2), thereby producing a head-to-tail structure. The more highly branched the alkyl groups are, the greater the reactivity of monomer. Long chain alkyl ethers are generally less reactive than the short chain homologs. Aromatic vinyl ethers do not polymerize readily and are susceptible to side reactions such as rearrangements and condensations.

FIGURE 44: TYPICAL REACTIONS TO FORM POLYVINYL ETHERS—INCLUDING MONOMER MANUFACTURE

$$\text{(1)} \quad CH_2{=}CHOR + H_2O \xrightarrow{H+} [CH_2{=}CHOH] \xrightarrow{H+} CH_3CHO$$

(2)

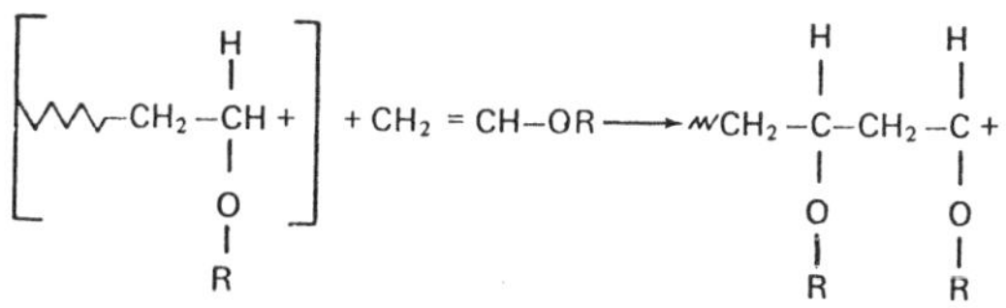

Source: EPA (7)

FIGURE 45: POLYVINYL ETHER PRODUCTION–SOLUTION POLYMERIZATION PROCESS

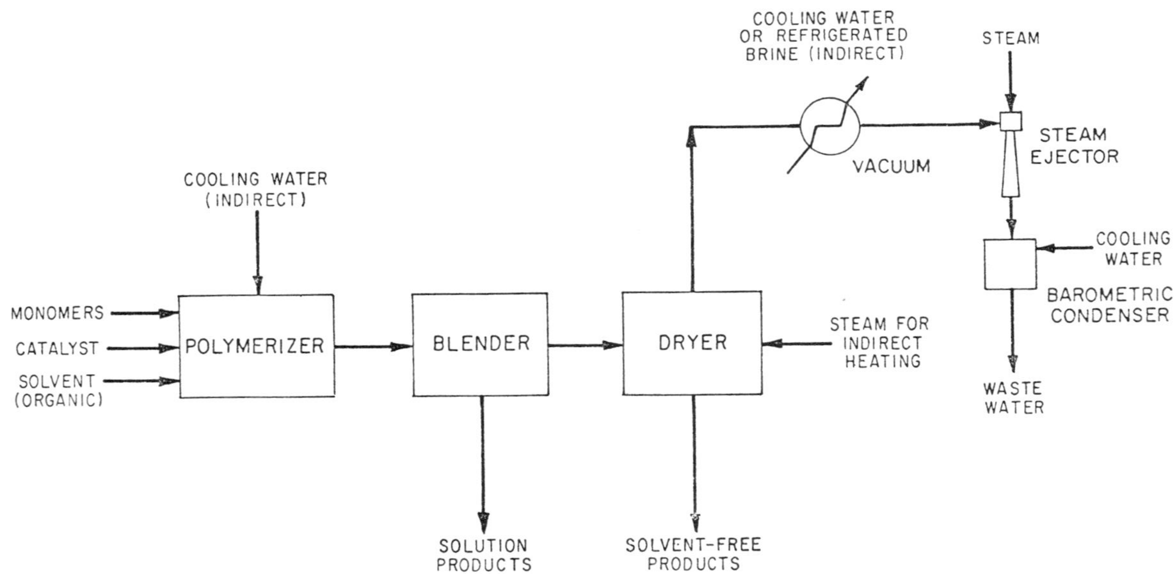

Source: EPA (7)

FIGURE 46: POLYVINYL ETHER PRODUCTION–BULK POLYMERIZATION PROCESS

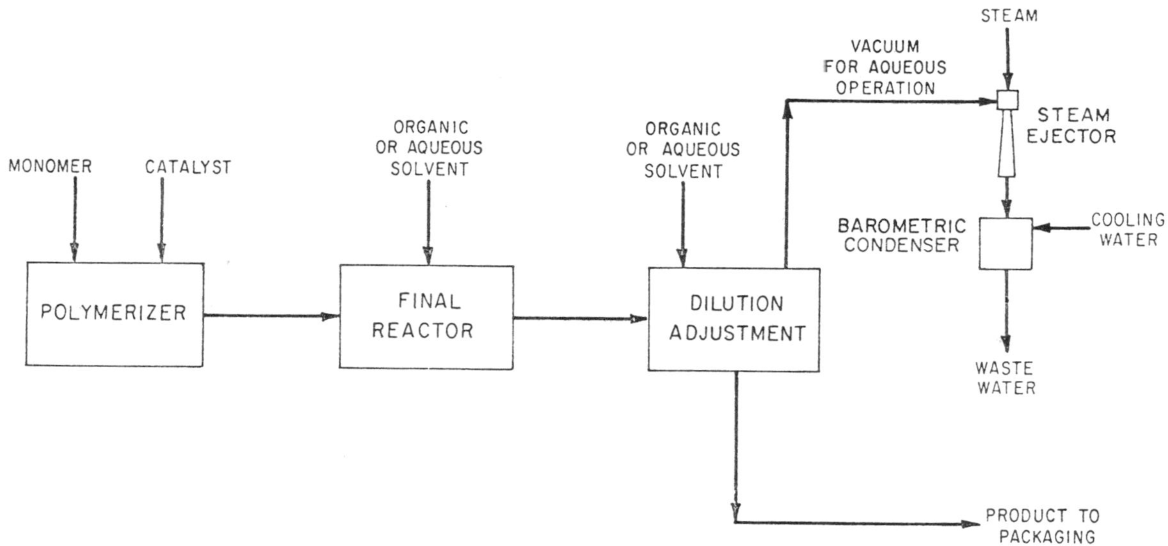

Source: EPA (7)

Vinyl ethers do not copolymerize readily with other vinyl ethers, but they readily form copolymers with a wide variety of ether monomers including dibutyl maleate, maleic anhydride, acrylonitrile, vinylidene chloride, vinyl chloride, vinyl acetate, and methyl acrylate.

Polyalkyl vinyl ethers are utilized primarily for their ability to serve as plasticizers for coatings, or because of their tackiness for use in adhesives. The methyl homopolymer is used as a plasticizer for coatings and is an aqueous adhesive tackifier. The vinyl methyl ether-maleic anhydride copolymer is used as a water thickening agent, suspending agent, and an adhesive.

Commercial processes are typical for the various polymerization techniques. Solution or bulk techniques are presently used in the U.S. Typical flow diagrams for these processes are shown in Figures 45 and 46. In the solution polymerization process, when a solvent-free product is desired, it is dried by heating under vacuum. In the bulk process, aqueous or organic solvent is sometimes added to the product depending on desired properties.

Sources of wastewater will depend upon the polymerization process employed. In solution polymerization, when no drying step is employed, there are no direct contact wastewaters. When drying is employed, some contamination of the water from the steam ejector barometric condenser results. There is no water of reaction in the polymerization of polyalkyl vinyl ethers.

Polyvinylidene Chloride Manufacture

Polyvinylidene chloride latex is commonly used for paper and film coatings. The latex is produced by emulsion polymerization of vinylidene chloride, frequently in the presence of another monomer.

The polymerization is performed in water using an emulsifier, a peroxide, and a reducing agent such as sodium bisulfite. The reaction is conducted in an inert atmosphere, and after several hours it is complete. Following the polymerization, the emulsion may be heated with steam to destroy components in the mixture that might generate odors on standing or in use, and is then ready for sale.

Wastes developed in the process consist of tank washings. These wastes include a low level of all of the ingredients used, and suspended solids corresponding to the polymer produced.

Silicone Manufacture

Plants producing silicones typically produce a wide variety of chemicals incorporating silicone. Silicone chemistry is complex, and the discussion here is limited to indicating the processes conducted and the types and scope of wastes generated. All of the plants examined purchase silicon metal and react it with a wide range of chemicals, used in several steps. The following processes may be conducted at a silicone plant:

[1] Production of methyl chloride, generally by reaction of methanol and hydrogen chloride, Figure 47, Equation (1). Methyl chloride is used in production of methylated chlorosilanes. Other organic chlorides, alkyl or aryl, are also used; e.g., phenyl chloride. Methyl chloride may be purchased by a silicones plant rather than being manufactured there. As far as known, other organic halides are always purchased.

[2] Chlorosilane production. For the methyl fluids, methyl chlorosilanes are produced by the reaction shown in Figure 47, Equation (2). Other organic chlorides (see above) would be used to generate other chlorosilanes. The mixture of products produced in the direct process is separated by fractional distillation to provide each component; dimethyldichlorosilane must be very pure for use in subsequent syntheses (see below). Some of the chlorosilanes (probably methyl trichlorosilane) have

FIGURE 47: TYPICAL REACTIONS TO FORM SILICONES

(1) $$CH_3OH + HCl \longrightarrow CH_3Cl + H_2O$$

(2) $$CH_3Cl + Si + HCl \xrightarrow[300^{\circ}C]{Cu} SiCl_4 + CH_3SiHCl_2 + CH_3SiCl_3 + (CH_3)_2SiCl_2$$

(3) $$CH_3Cl + Mg \longrightarrow CH_3MgCl$$

(4) $$2CH_3MgCl + SiCl_4 \longrightarrow 2MgCl_2 + \begin{array}{c} Cl \\ | \\ CH_3-Si-CH_3 \\ | \\ Cl \end{array}$$

(5) $$(CH_3)_2SiCl_2 + H_2O \longrightarrow (CH_3)_2Si(OH)_2 + 2\,HCl$$

(6) $$(CH_3)_2Si(OH)_2 \longrightarrow$$

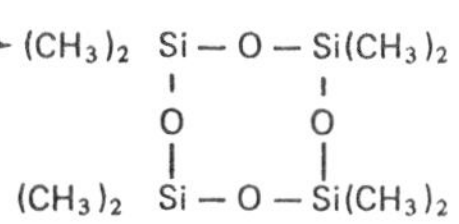

and $$HO[Si(CH_3)_2O]_nH$$

Source: EPA (7)

limited use and are wasted. The above equation represents the direct process for making chlorosilanes; it is widely used for the methyl compounds, the phenyl compounds, and perhaps others. Chlorosilanes may also be made by a Grignard process, represented by Equations (3) and (4) in Figure 47. The Grignard process finds limited use, in part because large amounts of solvent are required, and the metal salts go into a waste stream. It is believed that it is only used for special chlorosilanes. Trichlorosilane ($HSiCl_3$) is produced by reacting directly silicon and hydrogen chloride. For production of still other chlorosilanes, olefins or acetylene may be reacted with appropriate silane monomers.

[3] Hydrolysis. For production of the methyl fluids, dimethyl dichlorosilane is hydrolyzed with water as shown in Figure 47, Equations (5) and (6). The cyclic siloxane may be further processed to the linear polymer. The linear products are manufactured in a broad viscosity range to give the well-known fluids.

Silicone fluids are often sold as emulsions in water; production of these mixtures involves use of emulsifiers and special equipment. In addition, the viscosity of certain fluids may be substantially increased, probably by cross-linking, to provide silicone greases.

[4] Silicone resin production. The resin products are branched and cross-linked siloxane polymers, generally sold as solutions in organic solvents. The resins are manufactured by hydrolysis of mixtures of chlorosilanes in solvents. The mixtures may be complex, including mono-, di- and trichlorosilanes having different organic radicals, alkoxysilanes, and other silanes bearing special functional groups. After hydrolysis, the aqueous layer is separated and the organic phase is neutralized. A catalyst may be added to the organic solution, and the mixture may be heated to polymerize the dissolved chemicals.

[5] Elastomer production. Silicone elastomers are produced from high molecular weight fluids, fillers, and curing agents. The mixtures are often called compounds. Two types of polymer-filler mixtures are produced, those which cure to rubber by application of heat and those which are cured at room temperature. Catalysts used for products cured at room temperature may be tin or organotin salts.

[6] Specialties production. For these purposes, specialties constitute materials which are produced in significant amounts at a single silicone plant but in minor amounts, if at all, at others. Surfactants, coupling agents and fluorosilicones have been included as specialties, but other products may also be classified in this group. It is characteristic of silicone plants that new products are constantly being developed and offered commercially.

Surfactants are produced by reaction of silicone fluids with polyethylene oxide, or polyethylene oxide-polypropylene oxide. The products are water-soluble.

Coupling agents are monomeric silanes which serve as glass surface primers to increase the adhering strength of a resin subsequently applied. The typical final composition is glass-silane-resin. The resin may be epoxy, polyester, melamine, or other. One coupling agent which has been used is aminopropyl triethoxy silane. Production of such chemicals generally involves reaction of a chlorosilane with an appropriate organic compound, followed by exchange of the halogen atoms with alcohol groups.

Resin production generates significant amounts of acid wastes due to the liberation of hydrogen chloride in the hydrolysis step. The acid may be recycled, for example in the production of methyl chloride, but in many cases it is impractical to recover it. Organic solvent wastes, such as naphtha or toluene, are also produced. Most of the solvents are recovered, but trace quantities may appear in the wastewater. Quantitative information on the wastewater generated in silicone polymerization processes is not documented in the literature.

Coupling agents are produced in significant quantities at one manufacturing location. These resins require a vacuum to be applied to the batch reactor kettle during the hydrolysis reaction (5) in Figure 47. Existing plants employ once-through barometric condensers to achieve the vacuum; use of such condensers increases the process water flow from the vicinity of 17,000 to 27,000 gallons per 1,000 pounds of product.

These extra waters cannot be recycled until concentrated, as is practiced in the manufacture of polyesters and alkyds, for instance, since the materials of construction of the process equipment will not withstand the more concentrated hydrochloric acid. New plants, however, could incorporate such a change without undue penalty. On existing plants, this additional condenser water may be used as scrubber water for incineration, thereby not adding to the total hydraulic load by the addition of incineration. Surface condensers of current design are very expensive for this application due to the need for hastalloy type alloys for the metal parts in contact with the process waters, and are also subject to frequent plugging from the products of reaction.

By virtue of being both newer plants and because of the inherently more controllable process from the viewpoint of water recycling and less product variation, the fluid product plants have been able to achieve considerably lower levels of unit water use.

As regards sources of heavy metals, copper catalyst used for the chlorosilane production process enters the wastewater during the hydrolysis step and in subsequent water of reaction produced during polymerization. Fluorides may be present in the wastewater where fluorosilicones are being manufactured.

A process developed by *H. Lapidot; U.S. Patent 3,855,124; December 17, 1974; assigned to General Electric Company* is a process for the purification of an aqueous waste stream from a silicone-polymer producing plant which includes conducting the aqueous waste stream to a flotation and sedimentation zone; adjusting the pH of the wastewater stream to at least about 12; conducting the stream to a clarification zone; conducting the waste stream to an ozonation zone; conducting the waste stream to at least one holding zone prior to the ozonation zone and subsequent to the flotation and sedimentation zone; and obtaining a purified wastewater stream.

A wide variety of different products is usually produced in the same silicone-polymer producing plant, involving the use of enormous quantities of water particularly for the hydrolysis of monomeric chlorosilanes to produce the corresponding siloxane polymers. A major problem facing the silicone industry at this time is the treatment of the aqueous wastewater streams so as to convert materials therein which adversely effect the ecology into harmless compounds in order to safely dispose of such waste streams. The problem of purifying these aqueous waste streams is greatly magnified due to the usual production of a vast number of different products in a silicone-polymer producing plant which in turn results in the contamination of the wastewater stream with a large variety of different types of harmful impurities.

The effect of the many varieties of different types of impurities in the wastewater stream is more than being merely accumulative since the presence of one type of impurity may greatly increase the difficulty of finding a suitable means to remove another type of impurity from the aqueous waste stream.

Another major problem present in attempting to purify aqueous waste streams from silicone plants is that the types and concentrations of impurities vary considerably depending upon which of the many processes in the plant is responsible for the specific aqueous waste stream to be purified at any particular time. Accordingly, the task of designing a relatively inexpensive purification system which can handle all the different waste streams is extremely difficult to accomplish. Moreover, because of the specific types of impurities present in the aqueous waste stream, many conventional types of purification methods are unsuitable or impractical.

For example, the design of a suitable purification process is rather difficult since the aqueous stream contains various inorganic salts and since the concentration of the salts drastically changes from batch to batch. Biological methods are not very practical for purifying aqueous waste streams from silicone producing plants because of the chlorine ion content in the waste stream.

In addition, sorption of impurities by a sorbent such as activated carbon is not a very satisfactory purification step because of the presence of silicones in the aqueous waste stream. The silicones are extremely harmful to the normal regeneration processes for such sorbents as activated carbon. Accordingly, the inability to successfully regenerate the carbon to any appreciable extent would render such a process nonfeasible from an economic viewpoint. In addition, if the sorbent cannot be regenerated, then the additional problem of disposal of the sorbent must be solved. Various oxidizers such as permanganates have been suggested but such materials are not entirely satisfactory for purification purposes. One particular disadvantage of such oxidizers is the necessity of employing extremely high temperatures to effect oxidation. Another suggested means of purification which also has not attained general commercial acceptance is electrolysis. Some of the difficulties associated with electrolysis are high equipment cost, slow reaction rates, and extremely high power requirements.

In addition, it has previously been suggested to treat certain industrial wastewater streams with ozone under certain conditions. Ozone treatments however have been too expensive and too inefficient for the treatment of the large quantities of wastewater which are usually produced in silicone plants. Accordingly, ozone treatment of wastewater streams from silicone plants has not heretofore been used to any appreciable extent in large scale commercial operations. Ozone treatments suggested heretofore require relatively large pieces of equipment as compared to the amount of the material being treated. Also such treatments are extremely expensive due to the poor ozone utilization and high cost of ozone production.

The purification process eliminates or at least greatly minimizes the problems previously encountered with other purification treatments. Figure 48 shows the essential features of this process.

Silicone process wastewater line **1**, connects to a flotation and sedimentation vessel **2**. The sedimentation vessel contains removal means **3** and **4**, and outlet line **5** which connects to holding tank **6**. The holding tank is provided with stirring means **7** and outlet line **8**. The outlet line contains a valve **9** at which point it splits up into lines **10** and **11**. Line **11** connects to a mixing vessel **12**. The mixing vessel contains an inlet line **13**, mixing means **14** and outlet line **15** which connects to mixing vessel **16** which is provided with mixing means **17** and outlet line **18**, which connects to mixing vessel **19** which is provided with mixing means **20** and outlet line **21**.

Outlet line **21** connects to clarifier **22** which has a material introducing means **23**, scraping means **24**, and outlet means **25** and **26**. Outlet means **26** connects to a vacuum degasifier **27** which is provided with outlet lines **28** and **29**. Outlet line **29** connects to storage vessel **33** which has an outlet line **34** which connects to storage vessel **35** and outlet line **31**. Storage vessels **33** and **35** contain stirrers **32**. Outlet line **31** connects to pump **30** which has an outlet line **36**. Outlet line **36** contains a flow measuring device **37**, an organics concentration analyzer **38**, a multiplier element **39**, a valve **41**, and a valve control means **40**. Line **36** connects to ozonation reactor **42** which contains liquid outlet line **46**, gas outlet line **45**, and gas inlet line **43**. Outlet line **45** connects to ozone generator **44**.

Outlet line **46** connects to mixing vessel **47**. Mixing vessel **47** contains inlet line **48** and outlet line **50**. Line 10 from line **8** connects to line **48** before line **48** enters mixing vessel **47**.

An industrial aqueous waste stream from a silicone producing plant is conducted through line 1 through sedimentation tank 2 wherein water-insoluble materials having densities lower than water are skimmed off the top of the aqueous waste material and wherein water-insoluble materials having densities greater than water are removed from the bottom of the sedimentation tank. An aqueous waste stream is removed from this sedimentation tank through line **5** and is conducted to holding tank **6**. The aqueous stream is then removed from the holding tank through line **8**. A portion of the aqueous stream is fed from line **10** to line **48** to mixing vessel **47**, depending upon the requirements discussed above for use of this stream as a back neutralization source of material.

The rest of the aqueous stream is conducted to a mixing vessel 12 through line **11**. In addition, an alkaline material is added to the mixing vessel 12 via line **13**. The stream is removed from vessel 12 via line **15** and conducted to vessel **16** whereafter it is removed

FIGURE 48: SILICONE PLANT WASTEWATER TREATMENT (above) AND OZONATION STEP DETAIL (below)

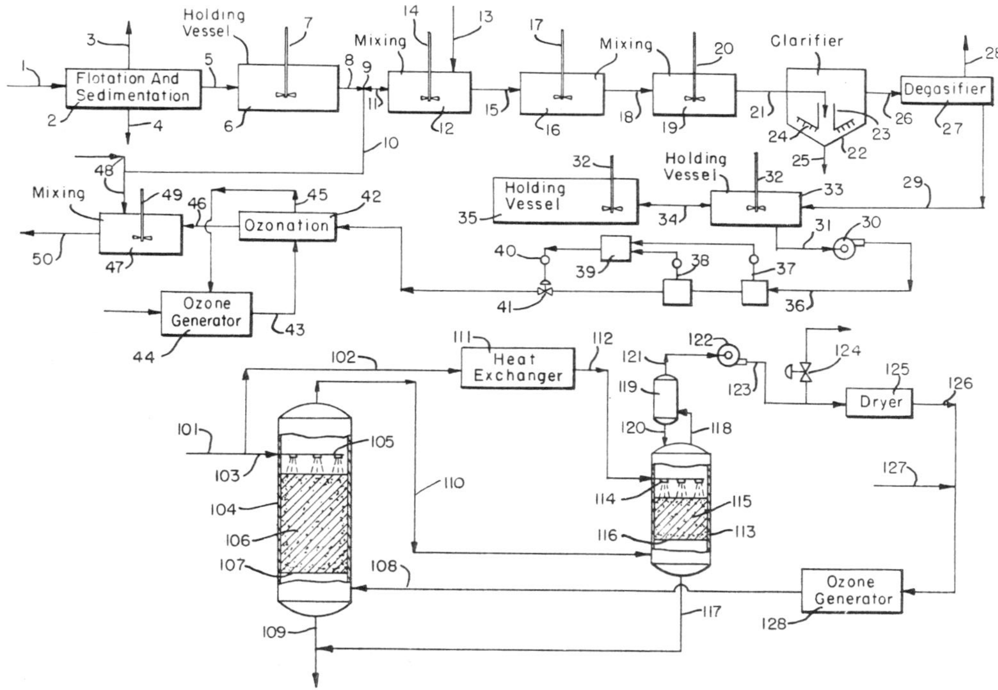

Source: H. Lapidot; U.S. Patent 3,855,124; December 17, 1974

via line 18 and conducted to vessel 19. A stream having a pH of at least about 12 is removed from vessel 19 by line 21 and is conducted to the clarifier 22 whereby the material flows through conduit 23. A slurry is removed from clarifier 22 through line 25. An aqueous waste stream is removed from clarifier 22 via line 26 whereafter it is conducted to degasifier 27. Gases are removed through line 28 and an aqueous waste stream is removed from the degasifier via line 29 whereby it is conducted to small holding tank 33.

The stream is pumped from tank 33 via line 31 by pump 30 and is analyzed for amount of material and for concentration of organics by flow measurer 37 and organics analyzer 38 respectively. These measurements are then multiplied in multiplier 39 whereby a flow controller 40 can regulate the opening or closing of valve 41 in accordance with the above measurements to provide a flow which introduces a predetermined amount of organics per unit time into the ozonation zone. Material will flow from small tank 33 to large tank 35 which is at the same level or will flow from the large to the small tank depending upon the relative flow rates of material into the small tank from line 29 and out of the tank through line 31.

The material entering the ozonation reactor through line 36 is contacted with ozone-containing gas entering through line 43. A gaseous stream is removed from ozonation reactor 42 via line 45 and is conducted to an ozone generator 44 whereby the ozone-containing gas is generated and removed therefrom and conducted back to the ozonation reactor. An ozone-treated aqueous product is removed from ozone reactor 42 and is conducted to a mixing vessel 47 via line 46. An acid such as hydrochloric acid is added to mixing vessel 47 through line 48. A product having a pH between the range of about 6 and about 9 is removed therefrom through line 50.

As shown in the lower view in Figure 48, line 101 splits up into line 102 and line 103, while line 103 connects to a packed tower 104. Packed tower 104 contains a liquid distributor 105, packing 106, packing support 107, a liquid outlet 109 and an overhead outlet line 110. Overhead outlet line 110 connects to packed tower 113. Line 102 connects to heat exchanger 111 which is provided with an outlet line 112 which connects to packed tower 113. If a degasifier is used and as a consequence the temperature of the water is about 5° to 10°C or below, then the heat exchanger 111 becomes unnecessary.

Alternatively, the degasifier 27 may be deleted and only heat exchanger 111 be used, which could result in a more economic operation since only a portion of the total water stream is conducted through this heat exchanger. Packed tower 113 is provided with a liquid distributor 114, packing 115, packing support 116, liquid outlet 117, and overhead outlet line 118. Liquid outlet line 117 connects to liquid outlet line 109 from the packed tower 104. Overhead line 118 connects to deentrainer 119 which contains liquid outlet line 120 and gas outlet line 121.

Liquid outlet line 120 connects to tower 113. Outlet line 121 connects to compressor pump 122 which is provided with outlet line 123. Line 123 is provided with valve 124 and connects to dryer 125. Dryer 125 is provided with an outlet 126 which in turn combines with line 127 and then line 126 connects to ozone generator 128. Ozone generator 128 is provided with an outlet line 108 which connects to packed tower 104.

The aqueous waste stream from the holding zone 33 is conducted through line 101 whereby a predominant portion of the stream enters the packed tower 104 via line 103 while the remainder of the wastewater stream is conducted to heat exchanger 111 via line 102. The wastewater is distributed to the packing of the first packed tower through the distributor 105 and flows downward thorugh the packing in the tower.

An ozone-containing gas is conducted to the first packed tower and flows upwardly through the packings in the tower. An ozone-treated product is removed from the first packed tower through line 109. A vapor stream is removed as the overhead from the first packed tower through line 110 and is introduced into the bottom portion of the second packed tower 113. Cooled liquid is removed from the heat exchanger 111 through line 112 and is conducted to the second packed tower and distributed into the tower by the distributor **114**. The liquid stream flows downward through the tower through the packing **115** while the vapor stream from line **110** flows upwardly through the packed tower. An ozone treated liquid product of reduced COD is removed from the packed tower through line **117** whereinafter it is combined with the product from packed column **104** at line **109**.

An overhead vapor stream is removed from the second packed tower through line **118** and is conducted to deentrainer **119**. Liquid is removed from deentrainer through line **120** and a vapor is removed through line **121**. The liquid is returned to tower **113**. The vapor stream is conducted to compressor pump **122** and exits pump **122** through line **123**. A portion of this vapor stream is bled from the system by opening valve **124**. The gas is then conducted to a dryer **125**. The dried gas is removed from the dryer through line **126**. Oxygen is added to the vapor stream by way of line **127**. The gas is sent to an ozone-generating device **128**. The ozone-containing gas is removed from the ozone-generating device through line **108** and is fed to the bottom portion of the first packed tower **104**.

AQUEOUS WASTE CHARACTERIZATION

The general process flow diagrams in the preceding pages indicate the major wastewater generation points for individual processes where information could be obtained. Flow rates and compositions of process wastewater streams at points of origin were not available according to EPA since the companies surveyed have rarely monitored these streams except where excessive losses of a particular component have been of concern, such as in the wastewaters from a distillation unit. Not only do wastewater streams emanate from direct process operations, from chemical reaction by-products and other contacts, but also a significant portion of wastewaters may come from the washdown of process vessels, especially where batch operations are preeminent as in the synthetic polymers industry, from area housekeeping, utilities blowdown and other sources such as laboratories and so on.

Raw wastewater flow ranges are shown in Table 10, and waste loads of BOD_5, COD, and suspended solids are shown in Table 11. These data are based on information provided by the companies contacted by EPA during the course of this study. Much of the data was provided as units per unit of production by the manufacturers and was not obtained from daily production rates and wastewater flows and concentrations.

Furthermore, it is known that much of the data on wastewater flows and raw waste loads has been derived from limited numbers of samples over short time periods. Because the synthetic polymers industry is based to a large extent on batch production methods and often on the commercial need to produce a large number of product types of a basic polymeric material, the wastewater flows and raw waste loads per unit of production were reported to vary from essentially no water use (water leaves with product or no water is used in manufacture) to nearly 300 m^3/kkg (36,000 gal/1,000 lb).

The major pollutant parameters for which data were obtained are BOD_5, COD, and suspended solids. Inspection of the ranges recorded in Table 11 shows that these pollutants vary by factors of up to 30 from low to high values for an individual polymer.

Other pollutants which may occur in the wastewater from various polymer manufacturing processes are listed in Table 12. These elements, compounds, and characteristics were developed by EPA from information obtained from industrial representatives, literature sources, Corps of Engineers Permit Applications for a number of plants in the plastics and synthetics industry, reviews with personnel in regional EPA offices, and internal industrial consultants.

Note also that in some instances, insufficient operating data on raw wastes and treatment were available to establish variabilities; as a result those chemical products that employ

TABLE 10: WASTEWATER LOADING FOR SYNTHETIC POLYMERS PRODUCTION

	Wastewater Loading			
	(gal/1,000 lb)		(m^3/kkg)	
Polymer	Observed	Range	Observed	Range
Acrylic resins		(1,700-5,600)		(14.2-46.7)
Amino (urea and melamine) resins				
Batch (liquid)	150		1.0	
Cellulose acetate resin	5,000	(2,000-50,000)	41.7	(16.69-417)
Cellulose ethers		(1,700-14,000)		(14.2-116.8)
Cellulose nitrate		(13,300-20,400)		(110.9-170.2)
Epoxy resins:				
Batch and cont (liq, solid and soln)		(2,200-4,200)	24.2	
Batch fusion (solid and soln)		(600-1,000)	7.1	
Ethylene-vinyl acetate copolymers		(275-300)		(2.3-2.5)
Fluorocarbon polymers		(2,200-18,300)		(18.4-152.7)
Nitrile barrier resins		(900-4,700)		(7.5-39.2)
Phenolic resins		(60-2,400)		(0.5-20.0)
Polyester resins (unsat) and alkyd resins		(38-1,440)		(0.3-12.0)
Lo Density polyethylene	2,130	(0-5,000)	17.8	(0-41.72)
Hi Density polyethylene-solvent	3,500	(0-3,700)	29.2	(0-30.87)
Polypropylene	1,000	(300-8,000)	8.3	(2.50-66.75)
Polyvinyl acetate	1,000	(0-3,000)	8.3	(0-25.03)
Polystyrene-suspension	1,100	(0-17,000)	9.2	(0-141.8)
ABS/SAN	2,060	(200-3,500)		(1.67-24.03)
Polyvinyl butyral		(7,800-14,200)		(65.1-118.5)
Polyvinyl chloride-emulsion	1,800	(300-5,000)	15.0	(2.5-41.72)
Polyvinyl ethers		(0-6,250)		(0-52.2)
Polyvinylidene chlorides	500 (Estimated)		4.2 (Estimated)	
Silicones		(1,000-33,500)		(8.3-279.5)

Source: EPA (7)(8)(9)

primary treatment only for treatment of plant wastes may show raw waste loads lower than guideline limits; it was announced that their variablities were the same as those of other products in the same major subcategory.

TABLE 12: OTHER ELEMENTS, COMPOUNDS AND PARAMETERS

pH	Chromium
Color	Copper
Turbidity	Lead
Alkalinity	Zinc
Temperature	Iron
Nitrogenous compounds (organic, ammonia and nitrates)	Cobalt
	Cadmium
Oils and greases	Manganese
Dissolved solids – principally inorganic chemicals	Aluminum
	Magnesium
Phosphates	Molybdenum
Phenolic compounds	Nickel
Sulfides	Vanadium
Cyanides	Antimony
Fluorides	Numerous organic chemicals
Mercury	

Source: EPA (7)(8)(9)

TABLE 11: SYNTHETIC POLYMERS PRODUCTION RAW WASTE LOADS

	Average Waste Loads, kg/kkg (lb/1,000 lb) production								
	BOD_5			COD			SS		
Polymer	Observed	Reported Range	Estimated	Observed	Reported Range	Estimated	Observed	Reported Range	Estimated
Acrylic resins		2-30			3-55			5-10	
Amino (urea and melamine) resins*	13			60					
Cellulose acetate	55	6-70			11-100			2-20	
Cellulose ethers		140-220			340-950			1-42	
Cellulose nitrate			55-110			75-275			35
Epoxy resins*:									
Batch and cont (liquid, solid and solution)	15-150	57-82		65-618	30-127			5-24	
Batch fusion (solids and solution)	0-25	57-82		0-100	30-127			5-24	
Ethylene-vinyl acetate copolymers			0.44-4.4			0.2-54	0-4.1		
Fluorocarbon polymers			0-6.6			4.4-44			2.2-6.6
Nitrile barrier resins			5-10			10-30			3-10
Phenolic resins*	20-72**	15-51***		52-188**	90-64***		1.2-21	0.5-7***	
Polyester resins (unsat) and alkyd resins		9-25			15-80			1-2	
Polypropylene		0-10			0-20				
LDPE	4.4	0.2-4.4		12	0.2-54		0.4	0-4.1	
HDPE	1.0	0-1			0-2.4			0-3.4	
ABS/SAN	20.7	2-20.7		33.5	5-33.5		8	0-30	
PVAcetate	1.4	0-2			0-3			0-2	
Polystyrene	1.0	0-2.2			0-6.0			0-8.4	
Polyvinyl butyral		30-200		40-400			NA		
Polyvinyl chloride	5.7	0.1-48		25	0.2-100		30	1-30	
Polyvinyl ethers	NA					10-40	NA		
Polyvinylidene chlorides			0			8			0.2
Silicones		5-110			15-200				50

*From survey by the Mfg Chemists Assn and Celanese Corp. **Assumes conc wastes from reactor does not appear in wastewaters. ***Presumed to include all raw waste load.

Source: EPA (7)(8)(9)

WATER POLLUTION CONTROL AND TREATMENT METHODS

As a prelude to the application of water pollution control and treatment methods, it is useful to determine major subcategories which characterize the industry. This has been done by EPA (7) as follows:

> Major Subcategory I — Low raw waste load (less than 10 units/1,000 units of product); attainable low BOD_5 concentration (less than 20 mg/l).
> Major Subcategory II — High raw waste load (greater than 10 units/1,000 units of product); attainable low BOD_5 concentration.
> Major Subcategory III — High or low raw waste load; attainable medium BOD_5 concentration (in the 30 to 75 mg/l range).
> Major Subcategory IV — High or low raw waste load; attainable high BOD_5 concentration (over 75 mg/l).

The attainable BOD_5 concentration in the effluent is influenced by both treatability and, for a specific wastewater treatment plant design, by variations in the influent concentrations.

In Major Subcategory I where reported BOD_5 raw waste loads are less than 10 units/1,000 units of product and where hydraulic flows ranged from 0.4 to 153 m^3/kkg (55 to 18,300 gal/1,000 lb), the influent concentrations ranged from 8 to 720 mg/l. While the influent concentrations varied over a 90-fold range, the effluent concentrations varied over a 3-fold range, i.e., 8 to 25 mg/l. This indicates that practicable wastewater treatment plants should be capable of retaining effluent concentrations in the vicinity of 15 mg/l when using properly designed and well operated biological systems.

The plants in Major Subcategory II are characterized by high raw waste loads, but the wastewaters can be treated to low attainable BOD_5 concentrations. Hydraulic flows varied from 14.2 to 116 m^3/kkg (1,700 to 14,000 gal/1,000 lb). Influent concentrations of from approximately 600 to 4,300 mg/l were reported. Although only one treatment plant was found in Category II and this was producing effluent concentrations of approximately 25 mg/l, it is known that the wastewaters from the other processes are readily treated by biological methods.

Major Subcategory III plants are characterized by high raw waste loads and observed flows from 0 to 170 m^3/kkg (0 to 20,400 gal/1,000 lb). Influent BOD_5 concentrations from 0 to 45,000 mg/l were found, and effluent concentrations varied from 15 to 80 mg/l indicating intermediate treatability of the wastewaters. One of the wastewater treatment facilities attained a BOD_5 removal of about 97% in a four-stage aeration basin indicating that medium BOD_5 concentrations are achievable.

Major Subcategory IV facilities have high raw waste loads with concentrations reported to be 2,200 mg/l at flows of from 7 to 40 m^3/kkg (900 to 4,800 gal/1,000 lb).

Estimates of BOD_5 concentrations from a one-stage biological system were in the vicinity of 225 mg/l. The supposition is made that practicable wastewater treatment technology, e.g., two-stage biological treatment, might reduce the effluent concentration of Category IV processes to levels comparable with the plants appearing in Major Subcategory III; however, attainable BOD_5 concentrations below these levels have not been documented.

Additional subcategorization within the above major subcategories was necessary to account for the wastewater generation which is specific to the individual products and their various processing methods. The separation of each individual product into separate subcategories simplifies the application of the effluent limitations guidelines and standards of performance by providing a clearly defined context for application of the numerical values. The advantages of this subcategorization appear to outweigh technical advantages that might be connected with product group characterization alone. The resulting major subcategories and component product subcategories are summarized in Table 13.

TABLE 13: INDUSTRY SUBCATEGORIZATION

Major Subcategory I	Major Subcategory II	Major Subcategory III	Major Subcategory IV
Ethylene-vinyl acetate copolymers	Acrylic resins	Alkyds and unsaturated polyester resins	Nitrile barrier resins
Fluorocarbons	Cellulose ethers	Cellulose nitrate	Phenolic resin
Polyvinylidene chloride	Cellophane	Polyvinyl butyral	Urea and Melamine resin (liquid)
Polyvinyl chloride	ABS/SAN	Silicones	Acrylics
Suspension	Rayon	Epoxy resin (batch, and cont, liquid, solid and solution)	
Emulsion		Epoxy resin (batch, fusion, solid and solution)	
Bulk		Cellulose acetate	
Polyvinyl acetate		Resin	
Polystyrene		Fiber	
Suspension		Resin and Fiber	
Bulk			
Polypropylene			
Polyethylene			
High density			
Solvent			
Polyform			
Low density			

Source: EPA (7)(8)(9)

Several other methods for subcategorizing the industry were considered. These included plant size, plant age, raw materials and products, and air pollution and solid waste generation. The utilization of municipal systems was considered as a method of characterization for the alkyd molding and unsaturated polyesters category, however, since the wastewaters are generally accepted into municipal systems and since pretreatment standards would be applicable, it was decided to establish a guidelines limitation as though the plants were treating wastewaters in private wastewater treating facilities.

The age of the plants in this industry is determined largely by obsolescence due to size or process changes and not physical age. Similar raw materials are often used to make dissimilar products. The impact of air pollution control systems using water scrubbing and the disposal of solid wastes is not sufficient to warrant segmentation. For these reasons none of the aforementioned factors was judged to have sufficient relationships with raw waste load generation or effluent compositions to warrant their use as a basis of categorization.

Technology for the control and treatment of wastewaters from the plastics and synthetics industry is not specific to the industry but can utilize any of the broad spectrum of technologies found in wastewater treatment. These technologies can be divided into the same three broad categories found in the rest of the synthetics and plastics industry. These are:

[1] Presently used wastewater treatment technology.
[2] Potentially usable wastewater treatment technology.
[3] Control of waterborne pollutants by in-plant or in-process practices.

The application of presently or potentially usable wastewater treatment technologies may be applied on selected bases to segregated streams or may be incorporated into a centralized wastewater treatment plant. Although categories [1] and [2], often denoted as end-of-pipe treatment, may be applied regardless of the manufacturing process, in-process control to prevent pollutants from entering water streams has a great potential for reducing the load of pollutants as well as the wastewater flows. The application of in-plant control technology falls into two broad categories: (1) process requirements and (2) plant practices. Process reguirements for water usage depend upon the types of reactions being carried out, the amounts of unreacted raw materials or undesired by-products that must be removed by water washing to attain product specifications, the removal of catalyst activators or other additives necessary to control the reaction or create the appropriate chemical characteristics, and the use of water for quenching, creating vacuum, or other operations that contact process streams.

The emission of pollutants into waste streams outside of the direct process operations may come from poor housekeeping practices or from the excessive usage of water for cleaning up spills, leaks, and accidental occurrences due to equipment failure or personnel error. Water used to control accidental occurrences or hazardous conditions, such as fires, etc., is employed very occasionally, and usually is not considered as contributing to the pollution loads in the wastewaters.

As indicated earlier, the survey found no wastewater treatment technologies unique to this segment of the plastics and synthetics industry. The wastewater treatment technology is generally similar to that of other industries. Obviously, application of basically similar technology, e.g., activated sludge biological treatment, often requires unique conditions for specific wastewater and results in considerable variation in performance characteristics such as efficiency of pollutant removal.

Biological treatment of the wastewaters from this segment of the industry appears to be the method chosen for effecting removal of soluble substances. Pretreatment before biological systems is often required whether the biological system is operated by the manufacturer or is a municipal sewage treatment plant. This treatment is predominantly neutralization for the control of pH prior to biological treatment. Primary treatment such as required in municipal treatment plants is not routinely necessary for these wastewaters; however, when significant amounts of oils or solvents do occur, the use of oil separators, skimmers, and settling basins or lagoons is used.

The effectiveness of a particular operational mode of the biological processes for removal of biologically degradable pollutants varies widely depending upon the characteristics of the wastewaters being treated. Consequently, it is impossible to generalize regarding the operating conditions applied in biological treatment other than to say that these are based on well understood principles. The design and operational characteristics of the biological wastewater treatment plants are paramount in determining the overall success in removing biologically degradable pollutants.

The applicability and limitation of biological treatment processes as well as physico-chemical processes to this segment of the plastics and synthetics industry will be found in Document 440/1-73/010 (9).

Alternative Treatment Technologies

The range of components used or needed to effect best practicable control technology currently available (BPCTCA), best available technology economically achievable (BATEA), and best available demonstrated technology for new source performance standards (BADT-NSPS) in this portion of the plastics and synthetics industry have been combined into nine alternative end-of-pipe treatment steps. These are as follows.

> Initial Treatment: For removal of suspended solids and heavy metals. Includes equalization, neutralization, chemical coagulation or precipitation, API separators, and primary clarification.
>
> Biological Treatment: Primarily for removal of BOD. Includes activated sludge (or aerated stabilization basins), sludge disposal, and final clarification.

Multistage Biological: For further removal of BOD loadings. Either another biological treatment system in series or a long-residence-time polishing lagoon.

Granular Media Filtration: For further removal of suspended solids (and heavy metals) from biological treatment effluents. Includes some chemical coagulation as well as granular media filtration.

Physical-Chemical Treatment: For further removal of COD, primarily that attributable to refractory organics, e.g., with activated carbon adsorption.

Liquid Waste Incineration: For complete treatment of small volume wastes.

Phenol Extraction: For removal of phenol compounds, e.g., from epoxy, acrylics, and phenolics wastes.

Zinc Removal and Recovery: For two-stage precipitation and recycle of zinc used in production of rayon.

Municipal Treatment: Conventional municipal treatment of industrial discharge into sewer collection systems. Primary settling and secondary biological stages assumed.

In addition, specific techniques may be used for the removal of specific pollutants when they are present.

Copper: The most widely accepted and economically feasible method for the removal of relatively low concentrations of soluble copper from wastewater streams is precipitation followed by sedimentation and filtration.

Under alkaline conditions, copper will tend to precipitate out of solution and form solid particles composed of the various oxides, hydroxides, and carbonates of copper. Alkaline conditions are generally accomplished by the addition of lime to the wastewater. Typically, a solids-recirculation clarifier is employed to promote the formation, growth, and sedimentation of the precipitated particles. Coagulants were often added to the wastewater in order to encourage the agglomeration of precipitated particles to such a size where they may readily settle.

The theoretical minimum solubility for copper in the pH range employed in the lime treatment process is on the order to 0.01 mg/l, but this level is seldom attained due to slow reaction rates, poor separation of colloidal precipitates, and the influence of other ions in solution. Most reported effluent copper concentrations from the lime precipitation process are on the order of 0.5 to 1.0 mg/l. If lime precipitation is followed by filtration, concentrations on the order of 0.25 mg/l are attainable. Lower levels can be achieved by subjecting the effluent from the lime precipitation step to carbon adsorption.

Ion exchange can be employed as an alternative to the lime precipitation process. While ion exchange has been reported to produce lower effluent concentrations (0.03 mg/l) than that achievable by lime precipitation, it usually entails much higher capital and operating costs. Ion exchange becomes more practical when the waste streams are relatively small and contain high concentrations of copper.

Lead: The most commonly employed process for the removal of soluble lead from wastewater is precipitation under alkaline conditions followed by sedimentation and filtration. Under alkaline conditions (usually created by the addition of lime) lead will precipitate out of solution and form solid particles of lead carbonate and lead hydroxide. As with the removal of copper, it is often necessary to also add coagulants to produce precipitates of sufficient size. Precipitation followed by sedimentation has been reported to produce effluent lead concentrations on the order of 0.5 mg/l. If the sedimentation step is followed by filtration, an effluent lead concentration of 0.03 mg/l may be achieved. Ion exchange while reported to produce lower lead concentrations, usually entails a much higher capital and operating cost than lime precipitation.

Mercury: The most promising and technically proven processes currently available for the removal of low concentrations of soluble mercury from large wastewater streams are ion exchange or sulfide precipitation.

In the ion exchange process, the wastewater, after sedimentation for the removal of any free mercury, is passed through a proprietary ion exchange resin. Mercury is removed as the mercuric chloride complex anion. A second stage ion exchange step serves as a polishing step and reduces the mercury concentration down to very low levels. Concentrations of less than 0.005 mg/l have been reported.

In the sulfide precipitation process mercurous and organic mercury compounds must first be oxidized to the mercuric ion. Lime and sulfide are then added along with coagulant aids in order to promote the formation of mercuric sulfide precipitates. The precipitates are removed from the wastewater stream by means of sedimentation and filtration, as in the copper and lead removal processes. This process has been reported to be capable of producing an effluent mercury concentration of 0.1 to 0.3 mg/l. In addition to ion exchange and sulfide precipitate several newly developed organic adsorbents and complexing agents have shown promising results in laboratory tests.

The entire technology for the removal of mercury has not been developed very far in terms of full scale actual plant operation.

Fluoride: Precipitation with lime is the standard technique for the reduction of high concentrations of soluble fluoride. The fluoride is precipitated as calcium fluoride. While the theoretical solubility limit for calcium fluoride is approximately 8 mg/l at pH 11, effluent concentrations of less than 20 mg/l are seldom achieved due to the slow reaction rates, difficulty in separating colloidal particles of calcium fluoride, and the interference of other ions.

The addition of alum and other coagulants encourages the formation of larger and more readily removable precipitates. Where effluent fluoride concentrations lower the 20 mg/l are required, various adsorptive techniques must be used.

In such processes the fluoride containing water is passed through contact beds of hydroxylapatite or activated alumina. Adsorption on activated alumina has been reported to be capable of producing effluent fluoride concentrations as low as 1.0 mg/l. The use of adsorptive techniques has largely been confined to the treatment of municipal drinking water containing undesirably high concentrations of fluoride.

Cyanide: The process most frequently employed for the treatment of wastewater containing cyanide is destruction by chlorination under alkaline conditions. In this process the cyanide may be partially oxidized to cyanate or totally oxidized to carbon dioxide and nitrogen, depending on the chlorine dosage.

Theoretically, if sufficient chlorine is added and sufficient contact time provided, complete oxidation of cyanide should be achievable. In reality, the presence of small quantities of soluble iron often causes the formation of extremely stable ferrocyanide complexes which prevent the complete oxidation of cyanide.

In recent years ozonation has shown to be effective in oxidizing cyanides. There are indications that ozonation is more effective than alkaline chlorination in attacking the more difficult to oxidize metal complexes of cyanide. Both the alkaline chlorination and ozonation processes can become prohibitively expensive if the cyanide containing wastewater also contains large quantities of oxidizable organic material which will unavoidably be oxidized along with the cyanide.

Oil and Grease: Oil and grease are usually present in wastewaters both in the suspended and emulsified form. Particles of suspended oil and grease are generally removed by means of gravity separators. Such separators can typically remove 90 to 95% of the suspended oil, but are totally ineffective in removing emulsified oil. To remove emulsified oil, the emulsion must first be broken by chemical means consisting of the addition of acids and/or coagulant salts such as alum. So treated, the oil can be removed by flotation or filtration. The concentration of oil and grease in the treated effluent depends largely on the degree of success in breaking the emulsion. Generally an oil and grease concentration of less than 30 mg/l should be achievable by emulsion breaking and gravity separation. If filtration is employed, a concentration of 10 mg/l should be achievable. Depending on conditions, varying amounts of emulsified oil may be removed along with other biodegradable material in standard biological treatment processes.

Table 14 shows the type of treatment and technology currently employed and the observed average treated effluent loadings for plants producing various plastics obtained from EPA plant inspections. The effluent loadings are given in kg/kkg (lb/1,000 lb). (E) is estimated.

TABLE 14

Product	Technology	- - - - -Effluent Loadings - - - - - -		
		BOD_5	COD	SS
Acrylic resins	Equalization, Trickling filters, Polishing lagoons	0.70	-	-
Acrylic resins	Neutralization, Equalization, Bio-oxidation (design)	3.1	30.8	-
Amino (Urea) resins	No separate treatment facilities encountered. Most plants discharge to municipal systems or are part of a major complex.	0.08(E)	0.40(E)	0.16(E)
Amino (Melamine) resins	Equalization, 2-stage bio-oxidation, Clarification, Polish (may be used for amino resins alone)	0.06(E)	0.30(E)	0.12(E)
Cellulose acetate	Equalization, Chem. treatment, Settling, Activated sludge clarification	1.7	14.0	2.8
Cellulose nitrate	Neutralization, Sedimentation, Spray oxidation	3.34	13.7	4.4
Epoxy resins	See Amino resins above. (Oil Sep., Neut., Chem. coag., Flotation, Bio-ox., Clarifier may be used for epoxy alone)	0.16(E)	0.80(E)	0.32(E)
Ethylene-vinyl acetate	Skimming, Bio-oxidation	0.07	0.25	0.15
Phenolics	See Amino resins above (Settling, Neut., Chem. coag., Equal., Clarif., Act. carbon may be used for phenolics alone)	0.55(E)	2.8(E)	1.1(E)
Polyesters (unsat) and alkyds	Settling, 4-stages of bio-oxidation	0.09	0.47	0.21
High density polyethylene	Screen, Chemical Treatment, Aeration pond	0.18	1.0	1.19
Low density polyethylene	API Separator, Equalization, Aerobic lagoon	0.13	0.87	0.26
Polypropylene	Screen, Equalization, Chem. treatment, Activated sludge, Polishing pond	0.33	0.66	0.57
Polyvinyl acetate	Equalization, Chem. treatment, Activated sludge, Clarification, Polishing pond	0.08	0.60	0.09
ABS/SAN	Equalization, Activated sludge, Aerobic sludge, Clarification	0.184	1.83	0.52
Polyvinyl butyral	Equalization, Activated sludge	2.6	13.6	6.5
Polyvinyl chloride	Equalization, Chemical treatment, Settling, Activated sludge, Clarification	0.14	1.0	0.80

(continued)

TABLE 14: (continued)

Product	Technology	Effluent Loadings BOD_5	COD	SS
Silicones				
Multiproduct	Neutralization, Bio-oxidation	8.6	46.0	25.0
Fluid product	Neutralization, Sedimentation, Skimming, Filtration	1.1	-	1.1

Source: EPA (7)(8)(9)

WATER POLLUTION CONTROL ECONOMICS

Various EPA publications (7)(8)(9) summarize the costs (necessarily generalized) of end-of-pipe treatment systems either currently in use or recommended for future use in synthetic polymers production facilities. Costs have been estimated for all product categories even though specific guidelines or standards were not recommended. Lacking specific effluent requirements, appropriate control technologies were assumed which were consistent with existing knowledge of waste composition.

In order to reflect the different treatment economics of existing versus new plants, large versus small plants, free-standing versus joint treatment facilities, or municipal versus industrial facilities, costs have been developed typically for more than one plant situation in each product subcategory. These product-specific analyses are presented in a series of tables in various EPA publications (7)(8)(9).

In addition two EPA publications (11)(12) have been devoted solely to an economic analysis of proposed effluent guidelines for the plastics and synthetics industry.

Table 15 (12) shows the costs associated with pollution abatement control to be achieved by 1977, best practical control technology currently available (BPT), and the cost associated with best available technology economically achievable (BAT), to be achieved by 1983, shows a range of 0.1% of the current selling price to 10.7%. Also shown are the data for best design technology currently available for new plants (BDT).

Costs of this magnitude are insignificant compared to the increases in costs which are now and will result from increased raw material prices. EPA believes (12) that in all but two cases, the cost of meeting these controls can be added to the sales price so that there will be no effect on the profitability, growth, employment, production, the community or balance of trade. In general, the producers have the capital and technology to achieve the standards set, according to EPA (12).

In two segments, saturated and unsaturated polyester resins, the costs of the initial investment to install treatment facilities to meet BPT, BAT and BDT, guidelines will have some effect and may cause some companies to shut down. This added expense may also prevent some potential producers from entering these two industry segments. The cost of adding the necessary water pollution control equipment is only one factor in these situations. Added to this are the costs associated with gathering the necessary technical data and employing experts to assess the information and determine what treatment technology is needed. These costs will be a deterrent to some small producers. The effect will be heightened by the necessity to meet other federal standards such as air pollution, solid waste disposal, safety and noise control regulations. None of these factors would affect major resin producers. They may, however, affect the small – i.e., less than 50 employee – company.

The producers which will face the greatest difficulty will be the small manufacturer of unsaturated polyester resins (approximately two to three direct dischargers and seven plants in municipal systems). These companies, faced with raw material shortages, costs

TABLE 15: COST IMPACT OF POLLUTION CONTROL ON SYNTHETIC POLYMERS PRODUCTION

	Capital Cost Per Plant ($1,000)			Annualized Cost (¢/lb)*			Annualized Cost as % of Sales Prices*		
	BPT	BAT	BDT	BPT	BAT	BDT	BPT	BAT	BDT
Methyl methacrylate	155-1,020	179-1,176	155-1,020	0.07-0.28	0.07-0.28	0.07	0.1-0.4	0.1-0.4	0.1
Ethylcellulose	360-530	750-1,386	360-530	0.55-1.0	1.65-2.84	0.60	1.1-2.0	3.3-5.7	1.2
Cellulose nitrate	1,113	2,260	1,113	0.4-0.85	0.5-1.8	0.45	0.8-1.7	1.0-3.6	0.9
Ethylene vinyl acetate	26-47	53-101	26-47	0.03-0.30	0.06-0.93	0.03	0.2-2.0	0.4-6.2	0.2
Fluorocarbon	44-145	44-145	44-145	0.32-1.95	0.32-1.95	0.32	0.1-0.6	0.1-0.6	0.1
Nitrile barrier	82	161	82	0.06-0.36	0.06-0.78	0.18	0.1-0.6	0.1-1.3	0.3
Unsaturated polyester and alkyd	198	198	198	0.08-0.38	0.08-0.76	0.16	0.4-1.9	0.4-3.8	0.8
Polyvinyl butyral	1,010	2,759	1,010	0.28-1.47	0.28-7.50	0.56	0.4-2.1	0.4-10.7	0.8
Polyvinyl ether	29	58	29	0.20-0.70	0.20-1.80	0.30	0.2-0.7	0.2-1.8	0.3
Polyvinylidene chloride	6-32	19-87	6-32	0.05-0.11	0.05-0.38	0.05	0.1-0.2	0.1-0.7	0.1
Silicones	1,696-2,480	4,176-5,965	1,696-2,480	0.60-1.20	1.70-3.50	0.70	0.6-1.2	1.7-3.5	0.7

*Including Capital Charges.

Source: EPA (15)

to meet other federal standards, and increasing water pollution control costs might elect to shut down - certainly they will not expand. This situation represents only a very minor portion of the industry segment, and any dislocations would have only a minor impact on the entire industry but would be serious to the companies involved.

Production capacity and industry growth would not be seriously affected since the plants represent less than 1% of the industry capacity. Possibly 30 employees might be affected, but this would not have any significant community effect. There would be no balance of trade effect since these producers do not export.

SYNTHETIC RUBBER MANUFACTURE

A summary of the environmental impact of polybutadiene manufacture as one example of synthetic rubber manufacture is presented in Table 16. This table was taken with permission from a Manufacturing Chemists Association publication (4).

TABLE 16: DATA FOR MANUFACTURE OF 1,000 POUNDS OF POLYBUTADIENE

Raw Materials	
Material process additions (solvent 25, catalyst 0.5, modifier 5.0, antioxidant 2.3)	32.8 lb
Energy	
Natural gas	2,330 scf
Water Volume	3,330 gal.
Process Solid Wastes	0.10 lb
Process Atmospheric Emissions	
Hydrocarbons	41.0 lb
Waterborne Wastes	
BOD	0.41 lb
COD	0.83 lb
Oil	0.07 lb
Suspended solids	1.25 lb
Transportation	
Rail	400 ton-miles
Truck	100 ton-miles

Source: Manufacturing Chemists Association (4)

AIR POLLUTION

Generally, air pollution control devices are not required by the industry. Odor problems do exist at some plants, but these are controlled by devices which are either dry or which do not impact on the wastewaters of the plant. Air pollution control is not a subject of consequence to the synthetic rubber industry.

WATER POLLUTION FROM SPECIFIC PROCESSES

For the purpose of establishing effluent limitations guidelines and standards of performance, the synthetic rubber industry has been categorized, on the basis of processing techniques, product type, and wastewater characterizations, into three separate subcategories: emulsion crumb; solution crumb; and latex.

All three subcategories generate wastewaters which contain the same general constituents. However, the concentration and loading of these constituents, termed raw waste load, vary between the subcategories. The significant wastewater constituents are COD, BOD, suspended solids, dissolved solids, and oil and grease. Latex production wastewaters, although lower in flow per unit of production than the other two categories, have the highest raw waste loads.

The wastewater parameters selected to be the subject of the effluent limitations are COD, BOD, suspended solids, oil and grease and pH. These parameters are present in the wastewater as a result of organic contamination. Heavy metals, cyanides and phenols were not found in significant quantities (less than 0.1 mg/l) in synthetic rubber process wastewaters.

Existing control and treatment technology, as practiced by the industry, emphasizes end-of-pipe treatment rather than in-plant reductions. This is because in-plant modifications which might lead to improved wastewater management could affect processing techniques or quality of the final product.

Current treatment technology for both emulsion crumb and latex plants involves primary clarification with chemical coagulation of latex solids, followed by biological treatment. As an alternative to chemical coagulation, air flotation clarification of primary and secondary solids is successfully practiced.

Biological treatment systems include activated sludge systems and aerated lagoon and stabilization pond systems. Best practicable control technology currently available for emulsion crumb and latex plants has been defined as that achieved by chemical coagulation and biological treatment.

Current treatment technology for solution crumb requires conventional primary clarification of rubber solid fines followed by biological treatment. Existing biological treatment systems employ aerated lagoon and stabilization pond systems or activated sludge plants. Best practicable control technology economically achievable for solution crumb production facilities has been defined as comparable to primary clarification and biological treatment.

Best available technology economically achievable technology for the three subcategories has been defined as equivalent to dual-media filtration followed by activated carbon treatment of the effluent from the biological treatment system to achieve acceptable COD removal.

Standards of performance for new sources are identical to best practicable control technology currently available. The proposed effluent limitations and standards of performance for plants within the three synthetic rubber subcategories are summarized on the following page (13).

Best Practicable Control Technology Currently Available and Standards of Performance for New Sources

	Emulsion Crumb Plants	Solution Crumb Plants	Latex Plants
	kg/kkg (lb/1,000 lb)		
COD	8.00	3.92	6.85
BOD	0.40	0.40	0.34
Suspended solids	0.65	0.65	0.55
Oil and grease	0.16	0.16	0.14
pH	6.0 to 9.0	6.0 to 9.0	6.0 to 9.0

Best Available Technology Economically Achievable

	Emulsion Crumb Plants	Solution Crumb Plants	Latex Plants
COD	2.08	2.08	1.78
BOD	0.08	0.08	0.07
Suspended solids	0.16	0.16	0.14
Oil and grease	0.08	0.08	0.07
pH	6.0 to 9.0	6.0 to 9.0	6.0 to 9.0

Wastewater characterization data was obtained from literature, EPA documents, and company records. Plant visits were made to selected plants to confirm existing data and fill the data gaps. Figures 49, 50 and 51 are generalized flow diagrams of emulsion crumb, solution crumb, and latex production facilities, respectively; they indicate the location of water supply and wastewater generation.

Data on total effluent flow and characteristics include utility wastewaters. It is virtually impossible to determine meaningfully total plant effluent flows and characteristics exclusive of utility wastes. It should be noted here that utility wastewaters are amenable to treatment by the existing treatment facilities in use and commonly practiced by the industry.

Emulsion Crumb Production

Of the several methods of polymerization employed to produce synthetic rubber, the two most commonly used processing techniques are polymerization in homogeneous solution and polymerization in emulsion. Solution polymerization may be considered to include bulk polymerization where excess monomer serves as a solvent. Emulsion polymerization may be considered as the bulk polymerization of droplets of monomers suspended in water. Emulsion polymerization is performed with sufficient emulsifier to maintain a stable emulsion. Solution polymerizations generally proceed by ionic mechanisms. Polymerization initiators which operate by ionic mechanisms are usually too reactive to be stable in water, emulsion polymerization systems are initiated by agents which produce free radicals.

Emulsion polymerization is the traditional process for the production of synthetic rubber. Since World War II (and for the foreseeable future), the bulk of synthetic rubber has been produced via emulsion polymerization. The use of emulsion polymerization systems is common because both high conversion rates and high molecular weights are possible. In addition, other advantages are: a high rate of transfer of the heat of polymerization through the aqueous phase; ready removal of unreacted monomers; and high fluidity even at high concentrations of polymer. The majority (more than 90%) of styrene-butadiene rubber (SBR), the principal synthetic rubber, is produced by emulsion polymerization. The emulsion polymerization process is used to produce either rubber latex or rubber crumb. Crumb is solid and is usually formed into 75 pound bales.

FIGURE 49: GENERAL WATER FLOW DIAGRAM FOR AN EMULSION POLYMERIZED CRUMB RUBBER PRODUCTION FACILITY

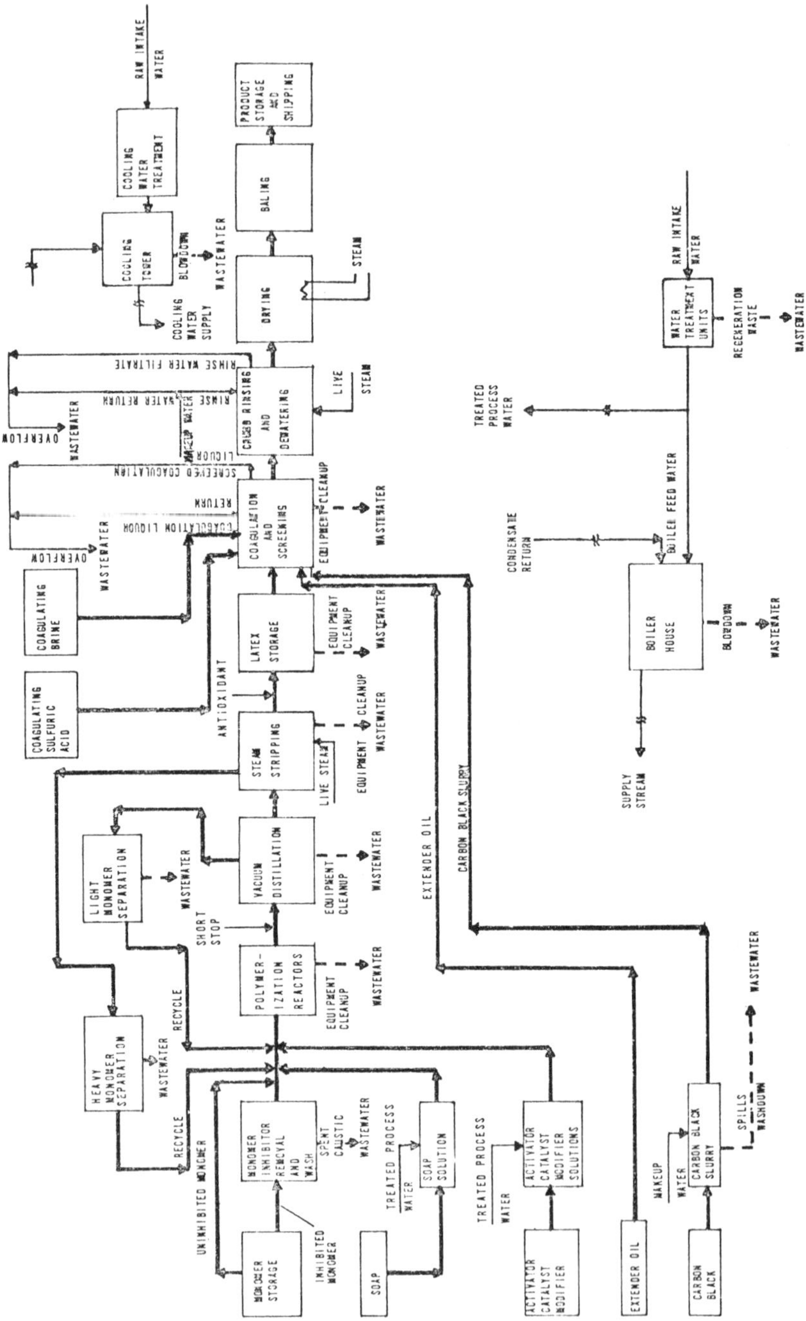

Source: EPA (13)

Figure 49 shows a generalized materials flow diagram for the continuous production of crumb SBR by the emulsion polymerization process. This schematic is essentially typical of all emulsion processes. In the typical production facility, operation is 24 hours per day, 365 days per year. Each plant consists of several production lines where different process recipes can be applied and various types of SBR can be produced, including nonextended, oil extended, and carbon black masterbatch varieties.

Styrene and butadiene (monomers) are either piped to the plant from adjacent suppliers, or shipped in by tank car or tank truck. The monomers are stored in a tank farm which is diked to retain major monomer spills and leakages and, in the case of fire, to control the spread of flaming liquid. The fresh monomers are piped to the plant from the tank farm and, if necessary, passed through a caustic soda scrubber before mixing with recycle monomers. Some monomers, such as butadiene, have inhibitors added to prevent premature polymerization during shipment and storage. These must be removed before the monomer can be polymerized. The inhibitor is removed in the caustic scrubber by the circulation of a caustic soda solution, approximately 20%. The caustic soda solution is discarded periodically or can be subjected to continual make-up and blowdown.

Soap solution, catalyst, activator, and modifier are added to the monomer mixture prior to entering the polymerization reactors. The soap solution is used to produce an emulsion of the monomers in an aqueous medium. The principal ingredients of this solution are generally a rosin acid soap and a fatty acid soap. The catalyst is a free radical initiator and can be a hydroperoxide or a peroxysulfate. The catalyst initiates and promotes the polymerization reaction. The activator assists in generating the free radicals more rapidly and at lower temperatures than by thermal decomposition of the catalyst alone. The modifier is an additive which adjusts the chain length and molecular weight distribution of the rubber product during polymerization. It is necessary that all the above solutions be made with high quality water. Usually city or well water is deionized for the preparation of the solutions.

The polymerization proceeds stepwise through a train of reactors. The reactor system is capable of producing either cold (40° to 45°F, 0 to 15 psig) or hot (122°F, 40 to 60 psig) rubber. The cold SBR polymers, produced at the lower temperature and stopped at 60% conversion, have improved properties when compared to hot SBR's. The hot process is the older of the two. For cold polymerization, the monomer-additive emulsion is cooled prior to entering the reactors, generally by using an ammonia refrigerant cooling medium. Depending on the polymerization temperature, the medium could be chilled brine or chilled water.

In addition, each reactor has its own set of cooling coils, usually containing ammonia refrigerant, and is agitated by a mixer. The residence time in each reactor is approximately one hour. Any reactor in the train can be by-passed. The reactor system contributes significantly to the high degree of flexibility of the overall plant in producing different grades of rubber. The overall polymerization reaction is ordinarily carried to no greater than 60% conversion of monomer to rubber since the rate of reaction falls off beyond this point and product quality begins to deteriorate. The product rubber is formed in the emulsion phase of the reaction mixture. The reaction mixture is a milky white emulsion called latex.

Shortstop solution is added to the latex leaving the reactors to stop the polymerization at the desired conversion. Two common shortstop ingredients are sodium dimethyl dithiocarbamate and hydroquinone. The stopped latex is held in blowdown tanks prior to the stripping operation. The blowdown tanks act as flow regulating holding tanks.

Recovery of the unreacted monomers and their purification is an essential step in economic synthetic rubber production. Butadiene, which has a lower boiling point than styrene, is first vacuum stripped from the latex. The stripping operation is generally carried out in a vacuum flash tank at about 80° to 90°F. The butadiene vapors are compressed and condensed before entering a receiver. A very small quantity of water collects in the receiver and is discharged periodically. The condensed butadiene is recycled to the feed area and mixed with fresh monomer prior to the polymerization step. Styrene recovery from the latex usually takes place in perforated plate stripping columns. These operate with steam injection at approximately 140°F. The steam-styrene vapor mixture is condensed and sent to a receiver where the styrene and water are decanted. The top styrene layer is recycled to the monomer feed stage; the bottom layer of the receiver, which is styrene-laden water, is discharged. Both the vacuum and steam strippers foul periodically with rubber solids. These must be removed by hand, followed with both steam or water jets. This cleaning operation puts the stripper out of commission and produces large quantities of wastewater.

An antioxidant to protect the rubber from attack by oxygen and ozone is added to the stripped latex in a blend tank. The latex is now stabilized and, as a result, different batches, recipes, or dilutions can be mixed. These mixing operations take place in the blend tanks. The latex is pumped from the blend tank to the coagulation step where dilute (pH 4 to 4.5) sulfuric acid and sodium chloride solution are added. The acid brine mixture is called the coagulation liquor and causes the rubber to precipitate from the latex. Theoretically, precipitation will occur with a coagulation liquor consisting of any combination of electrolyte and dilute acid. However, the quality and intended end use of the rubber limit the choice of coagulants. For example, some types of hot SBR which are used as insulation covering on electrical wire are coagulated with an acid-polyamine solution in order to produce a rubber with low electrical conductivity.

As mentioned earlier, rubber can be extended to improve its properties by using oils and carbon black. Carbon black and oil can be added to the latex during the coagulation step to produce a more intimate mixture than can be obtained by the subsequent addition of these materials to the crumb rubber as is the case with conventional rubber compounding. Wastewaters generated subsequent to the masterbatch operation (addition of carbon black) are usually black due to colloidal carbon black particles. The oil is added as an aqueous emulsion, and carbon black is blended into the latex as an aqueous slurry (approximately 5% by weight). There are various types of extending oils; some are staining and others nonstaining. Rubber extended with nonstaining oil will not mark surfaces and is required for some nontire uses. If a nonstained rubber is to be produced, not only must the extender oil be nonstaining, but also lighter-colored soaps, shortstops, and antioxidants must be used.

The coagulated crumb is separated from the coagulation liquor on a shaker screen. The coagulation liquor is recycled after make-up with fresh acid and brine and blowdown of part of the diluted liquor. The screened crumb is resuspended and washed with water in a reslurry tank. This operation serves to remove extraneous compounds from the rubber, particularly residual coagulation liquor. The crumb rubber slurry is then dewatered, generally using a vacuum filter, and the filtrate wash water is recycled to the reslurry tank for reuse with fresh water make-up and overflow. The overflow is necessary to blowdown accummulating rubber solids and contaminants. The coagulation liquor blowdown and crumb slurry water overflows are usually passed through separators. These facilities, called crumb pits, are generally outside the processing building and trap the floatable crumb rubber. The clarified underflow is discharged to the main treatment facility.

The rinsed and filtered rubber crumb is finally dried with hot air in a continuous belt or screen dryer. After drying, the rubber is weighed and pressed in bales and stored prior to shipment. Normally rubber bales weigh 75 lb and are wrapped in polyethylene film. The balers are operated hydraulically with oil or water as the hydraulic fluid. Due to the jarring baling action and the high hydraulic pressures, fluid leaks are frequent and, in the case of oil-driven balers, the leaked oil should be prevented from entering the plant drain system.

In addition to the processing operations described above, other operations are carried out regularly, though not necessarily continuously, which generate considerable quantities of wastewater. These include equipment cleanout and area washdown operations. Principal equipment cleanouts include the polymerization reactors, blowdown tanks, butadiene flash tanks, styrene stripping columns, and latex blend tanks. In most cases, high volumes of wastewater are produced that are laden with uncoagulated latex solids and are characterized by a milky white appearance. When the flash tanks and stripping columns are cleaned,

TABLE 17: SUMMARY OF POTENTIAL PROCESS-ASSOCIATED WASTEWATER SOURCES FROM CRUMB RUBBER PRODUCTION VIA EMULSION POLYMERIZATION PROCESSING

Processing Unit	Source	Nature of Wastewater Contaminants
Caustic Soda Scrubber	Spent caustic solution	High pH, alkalinity, and color. Extremely low average flow rate.
Monomer Recovery	Decant water layer	Dissolved and separable organics.
Coagulation	Coagulation liquor overflow	Acidity, dissolved organics, suspended and high dissolved solids, and color.
Crumb Dewatering	Crumb rinse water overflow	Dissolved organics, and suspended and dissolved solids.
Monomer Strippers	Stripper cleanout rinse water	Dissolved organics, and high suspended and dissolved solids. High quantities of uncoagulated latex.
Tanks and Reactors	Cleanout rinse water	Dissolved organics, and suspended and dissolved solids. High quantities of uncoagulated latex.
All Plant Areas	Area washdowns	Dissolved and separable organics, and suspended and dissolved solids.

Source: EPA (13)

the wastewaters contain rubber solids, due to premature coagulation of the latex, in addition to uncoagulated latex. Area washdowns are frequent, and the wash waters pick up primarily latex, rubber solids, and oil. The carbon black slurrying area is generally contaminated with carbon powder. Area washdowns and storm run off typically pick up the carbon, resulting in a fine carbon suspension.

It is opportune at this point to review the potential wastewater sources in a typical emulsion plant. Table 17 summarizes the principal wastewaters and the nature or appearance of their constituents.

Table 18 lists the total effluent flows for plants producing various emulsion crumb rubber products based on a unit of production. This data was obtained by plant visits by EPA (13). Although three plants were sampled, six cases of emulsion crumb production were studied. The wastewater contributions of other facilities included solution crumb production and nonrubber commodities.

It can be seen from Table 18 that, for similar products, separate plants appear to have different effluent flows. However, different products at the same plant seemingly produce identical wastewater flows. This is due to the following distinct facts:

[1] The water use practices in one plant for different emulsion crumb products are based on one technology, namely that of the company's process design and engineering; and

[2] The inability of the sampling team to discern small differences in effluent flows for different products at the same plant.

It can also be noted that there is no significant trend in wastewater generation rate between the various types of emulsion crumb rubber product (nonextended, hot, oil extended, and carbon black extended). The average effluent flow rate for emulsion crumb is 16,000 l/kkg (2,000 gal/1,000 lb) of production.

Table 18 also summarizes the raw waste loads for the six cases. It can be seen that the parameter with the highest concentration is COD. The BOD values are generally much lower. The high COD to BOD ratio is indicative of the high resistance of many of the constituents in the wastewaters to biological oxidation.

The raw suspended solids concentration in the emulsion crumb wastewaters were determined after separation of the rubber fines in the crumb pits. Since all emulsion crumb plants have separation pits, this raw waste load data is applicable to the industry. Much of the suspended solids contribution is due to uncoagulated latex solids. The concentration of oil does not appear to be related to the degree to which the crumb rubber is oil extended. The oil analysis cited is really carbon tetrachloride extractables and will also include insoluble monomers.

Another significant parameter in emulsion crumb wastewaters is total dissolved solids. This is due to the use of salt in the crumb coagulation process and associated rinse overflows. Surfactants are another characteristic produced by the emulsifying agents. The level of surfactants in the wastewater is considerably lower than the parameters reported in Table 19.

Table 19 presents the major constituent loadings of the principal wastewater streams in an emulsion crumb plant. The most significant parameter is total dissolved solids which is produced by the acid and brine coagulation liquors. The coagulation liquor and crumb rinse overflows, along with the utility wastes, provide the bulk of the total dissolved solids in the plant effluent. It can be seen that the quantity of surfactants produced are much lower than the other parameters. Surfactants are generated in appreciable quantities only by waste streams included in Table 18. The suspended solids are much higher than in the total effluent since the crumb pits remove much of the suspended solids in the crumb rinse overflow. Removals better than 95% are common. Oil entrained in the rubber is also removed along with rubber crumb solids.

TABLE 18: RAW WASTE LOADS FOR EMULSION CRUMB RUBBER PLANTS*

Plant	Product	FLOW L/kkg (gal/1,000 lb)	COD kg/kkg (lb/1,000 lb)	BOD kg/kkg (lb/1,000 lb)	SS** kg/kkg (lb/1,000 lb)	OIL kg/kkg (lb/1,000 lb)
J	SBR and NBR Part Oil and Carbon Black Extended	15,000 (1,860)	11.98	N.A.	3.73	2.09
K	SBR Part Oil Extended	18,500 (2,220)	22.23	2.13	2.30	0.13
K	SBR Oil Extended	18,500 (2,220)	19.76	2.13	11.31	3.54
L	SBR Oil and Carbon Black Extended	16,500 (1,980)	8.72	2.84	3.94	0.48
L	SBR "Hot", Non-Extended	15,500 (1,860)	29.24	2.84	N.A.	1.31
L	SBR Non-Extended	15,500 (1,860)	25.87	2.84	11.94	1.45
Median Value			19.63	2.56	6.64	1.5

*Includes utility wastewaters.

**Raw waste load determined downstream of crumb pits, where the suspended solids and oil levels are reduced.
N.A. Data not available.

Source: EPA (13)

TABLE 19: RAW WASTE LOADS OF THE PRINCIPAL INDIVIDUAL WASTEWATER STREAMS IN AN EMULSION CRUMB RUBBER PLANT

Wastewater Stream	COD - kg/kkg (lb/1000 lb.)	BOD - kg/kkg (lb/1000 lb.)	SS[1] - kg/kkg (lb/1000 lb.)	TDS - kg/kkg (lb/1000 lb.)	OIL[1] - kg/kkg (lb/1000 lb.)	SURFACTANTS (lb/1000 lb.)
Monomer Recovery	0.66	0.14	0.08	1.26	0.11	0.0001
Coagulation Liquor Overflow	1.30	N.A.	N.A.	46.25	0.10	N.A.
Crumb Rinse Overflow[2]	6.39	0.46	33.44	42.33	1.46	0.0077
Sub-Total	8.35	0.60	33.52	89.74	1.67	0.0078

[1] Raw waste load determined prior to crumb pit, where the suspended solids and oil levels are reduced.

[2] In one case, the crumb rinse overflow is combined with the coagulation liquor overflow and discharged as one combined stream.

N.A. Data not available.

Source: EPA (13)

When comparing the subtotal parameter values of Table 19 with the average total effluent loads of Table 18, it can be seen that the three streams listed in Table 19 are the major contributors to the total effluent.

The spent caustic scrub solution is an extremely low flow rate wastewater which has very high COD, alkalinity, pH, and color characteristics. It is not, however, a significant waste stream when combined in the total effluent. It is usually bled-in at low flow rates into the effluent.

Area washdown and equipment clean-out wastewaters are highly loaded with COD and suspended solids, and, by nature, are intermittent in flow. They cannot be characterized because they are generated on an irregular basis and have greatly variable concentration loadings.

Chromium and zinc are present in low concentrations (0.1 mg/l) in the final effluent. They are present due to cooling water treatment, and in some cases can be eliminated or reduced by substitution of chromium-free corrosion inhibitors. Heavy metals from catalysts and other reaction ingredients are not present in measurable concentrations in emulsion plant wastewater effluents.

Solution Crumb Production

As pointed out earlier, solution polymerization is a newer, less traditional process for the commercial production of crumb rubber in the U.S. Solution polymerization systems permit the use of stereospecific catalysts of the Ziegler-Natta or alkyllithium types which have made it possible to polymerize monomers, such as isoprene or butadiene, in a suitable organic solvent so as to obtain the cis structure (up to 98%) characteristic of the natural rubber molecule and with a high degree of regularity.

Rubbers with the cis structure are desired since they are usually rubbery, whereas the trans configuration is more rigid and similar to plastics. cis-Polybutadiene, for example, has higher abrasion resistance than the usual SBR type and is being used mainly to extend and partially replace both SBR and natural rubber in tires. Reports indicate that tread wear is improved by up to 35% in a 50–50 blend of polybutadiene and SBR.

A relative newcomer on the rubber scene is based on the cheap monomers, ethylene and propylene. Although not stereoregular, these polymers can be produced in solution plants with similar catalysts. The polymer chain, based on ethylene and propylene, does not contain sufficient unsaturation for conventional curing. The incorporation of a third monomer, usually a diene (thus EPDM—ethylene propylene diene monomer), adds unsaturation and facilitates conventional curing.

The production of synthetic rubbers by solution polymerization processes is a stepwise operation, and, in many aspects, is very similar to production by emulsion polymerization. There are distinct differences in the two technologies, however. For solution polymerization, the monomers must be extremely pure and the solvent (hexane, for example) should be completely anhydrous. In contrast to emulsion polymerization, where the monomer conversion is taken to approximately 60%, solution polymerization systems are polymerized to conversion levels which are typically in excess of 90%. The polymerization reaction is also more rapid, usually complete in one to two hours.

Figure 50 is a generalized materials flow diagram for the production of crumb SBR by a solution polymerization process. The processing steps shown are essentially typical of all solution polymerization processes. As in the case with emulsion plants, solution plants comprise several processing lines where different types of rubber for distinct end uses can be produced (including nonextended, oil-extended, and carbon black masterbatch varieties). Plant operation is typically 24 hours per day, 365 days per year.

FIGURE 50: GENERAL WATER FLOW DIAGRAM FOR A SOLUTION POLYMERIZED CRUMB RUBBER PRODUCTION FACILITY

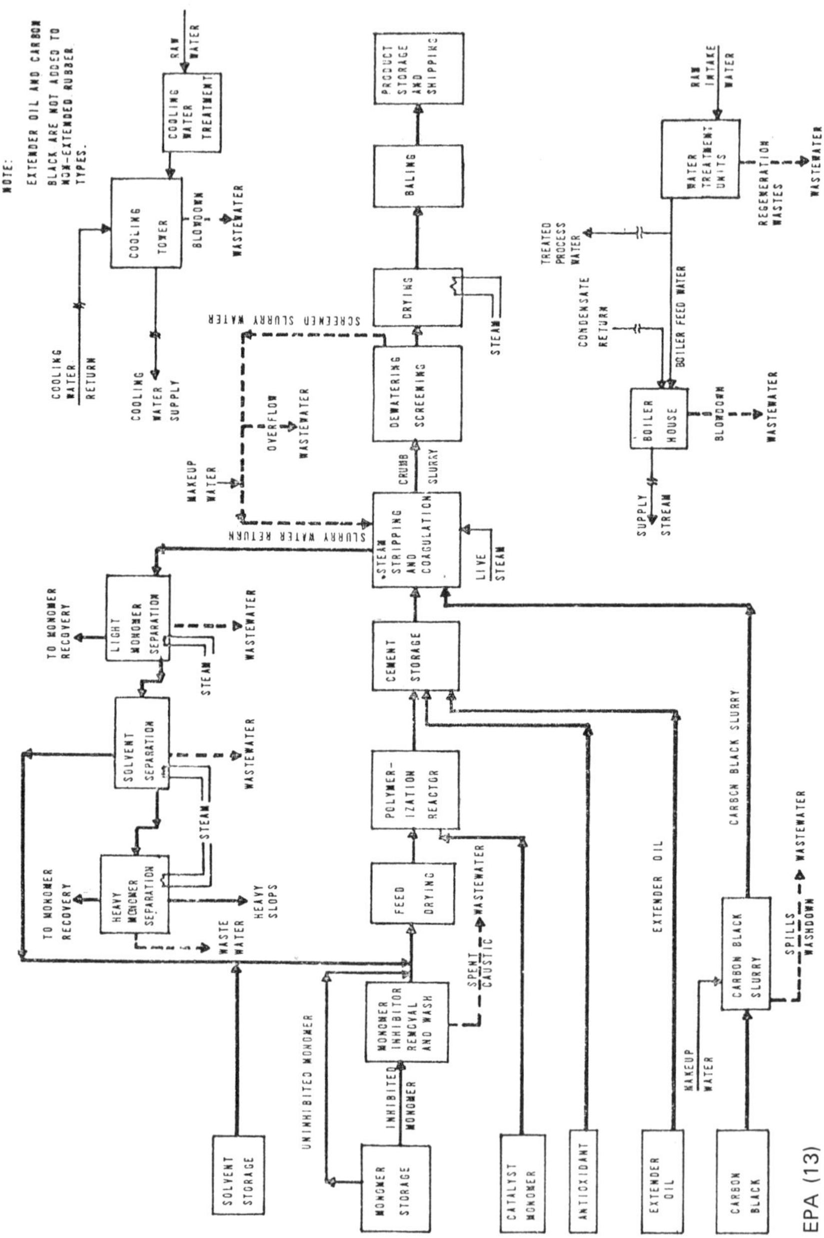

Source: EPA (13)

The fresh monomers are pumped to the plant from the tank farm. Inhibited monomers are passed through a caustic soda scrubber to remove the inhibitor. The monomers are then sent to fractionator drying towers where extraneous water is removed. Fresh and recycled solvent (for example, hexane) is also passed through a drying column to remove water and extraneous light and heavy components. The light- and heavy-components build up in the system as unwanted by-products or unrecovered monomer during the polymerization step and must be removed. The purified solvent and monomers are then blended. The mixture is generally termed the mixed feed. The mixed feed can be further dried to remove final traces of water using a desiccant column.

The dried mixed feed is now ready for the polymerization step and catalysts can be added to the solution (solvent plus monomers). The catalyst systems used vary. Typically they are titanium halide plus alkylaluminum combinations or butyllithium compounds. The catalysts can be added to the mixed feed just prior to the polymerization stage or to the lead polymerization reactor.

The blend of solution and catalysts is polymerized in a series of reactors. The reaction is highly exothermic and heat is removed continuously by either an ammonia refrigerant or by chilled brine or glycol solutions. The reactors are similar in both design and operation to those used in emulsion polymerization. The mixture leaves the reactor train as a rubber cement, i.e., polymeric rubber solids dissolved in solvent.

A shortstop solution is added to the cement after the desired conversion is reached. The stabilized cement is pumped to cement storage tanks prior to subsequent processing. At this point other ingredients, such as antioxidants, can be added. If the rubber is to be oil extended, oil can be added to the cement. The oil is usually blended with the cement at some point between the storage tanks and the steam stripping operation.

The rubber cement is pumped from the storage tank to the coagulator where the rubber is precipitated into crumb form with hot water under violent agitation. Wetting agents (surfactants) can be added to promote the control of crumb size and to prevent reagglomeration. In addition to coagulation, much of the solvent and unreacted monomer are stripped overhead. For carbon black masterbatch rubbers, the carbon black slurry is added to the coagulator in much the same manner as for emulsion crumb rubber.

The resultant crumb slurry passes to a series of strippers where steam stripping drives off the remaining monomers and solvent. The strippers are generally a flash tank or agitated kettle strippers. The steam, solvent, and monomer vapors are condensed and sent to a decant system. The bottom decant layer, saturated in monomers and solvent, is discharged. The organic layer is sent to a multistage fractionator (described earlier). Light fractions are removed in the first column and generally consist of unreacted light monomer (for example, butadiene). This is normally reclaimed at the monomer supply plant. The second column produces purified solvent, a heavy monomer-water fraction, and extraneous heavy components.

The heavy monomer (for example, styrene) is condensed, decanted, and recycled. The bottom water layer is discharged. The purified solvent is dried and reused. The heavy extraneous component stream is a waste which can either be decanted before disposal or can be incinerated as a slop oil.

The stripped crumb slurry is separated and further washed with water on vibrating screens. The slurry rinse water is recycled in part to the coagulation stage with water or steam makeup. The remaining portion of the slurry rinse water overflows and is discharged. This water contains floating crumb rubber fines and is generally passed through a crumb pit before discharge. The crumb fines are trapped in the pit. The screened rubber is passed through an extruder-dryer for further dewatering and drying. Dewatering and drying can also be carried out with a rotary filter and hot air oven dryer. The dried rubber is pressed into 75 lb bales and is usually wrapped in polyethylene for shipment. Balers, identical to those employed in emulsion processing, are used in solution polymerized rubber production. Oil leaks are a potential problem.

In addition to the processing operations described above, area washdowns occur. These are frequent and produce large volumes of wastewater which can be contaminated with dissolved organics, floating organics, oils, and suspended solids. Since the majority of the processing steps are operated on a strict water-free basis, there is little need for equipment cleanout operations with water. The processing units which are kept free of water are cleaned out with solvent when necessary.

This cleaning solvent is stored separately and is used solely for the cleanout operation. Process pumps, handling in particular the dried mixed feed prior to and during the polymerization stage, use a nonaqueous fluid (usually an oil) as a seal in lieu of water to prevent contamination of the process streams with water. Leaking fluid is a potential source of oil which can be picked up by area washdown waters. The carbon black slurrying area is a source of wastewaters laden with carbon fines. The main wastewater sources in a typical solution polymerization plant are summarized in Table 20.

TABLE 20: PROCESS-ASSOCIATED WASTEWATER SOURCES FROM CRUMB RUBBER PRODUCTION VIA SOLUTION POLYMERIZATION PROCESSING

Processing Unit	Source	Nature of Wastewater Contaminants
Caustic Soda Scrubber	Spent caustic solution	High pH, alkalinity, and color. Extremely low average flow rate
Solvent Purification	Fractionator bottoms	Dissolved and separable organics.
Monomer Recovery	Decant water layer	Dissolved and separable organics.
Crumb Dewatering	Crumb rinse water overflow	Dissolved organics, and suspended and dissolved solids.
All Plant Areas	Area washdowns	Dissolved and separable organics, and suspended and dissolved solids.

Source: EPA (13)

Table 21 presents the total effluent wastewater flows for facilities producing various solution crumb rubber products. The flow data is given in terms of liters per metric ton (l/kkg) of production. Five plants were visited and eight types of solution crumb product were sampled. Some plants are multiproduct facilities, and the contributions of the solution crumb facilities were accounted.

Table 21 shows that there is no discernible difference in the effluent flows between types of product. There appears to be more correlation between products at the same plant site. This is similar to the findings for emulsion crumb rubber production.

One plant (Plant M) has a considerably lower effluent flow than all the other facilities. The apparent reason for this difference is the use of a special rubber-finishing process which generates very little or no wastewater. The average effluent flow for solution plants is similar to emulsion plants, and typically approximates 16,600 l/kkg (2,000 gal/1,000 lb) of production.

Table 21 also presents the raw waste loads for the four main parameters. It can be seen that the constituent levels are approximately one-half of those present in emulsion crumb wastewaters. This supports literature and company data which indicate that the solution production processes are cleaner than their emulsion counterparts. The main factor behind this is the absence of coagulation liquor and uncoagulated latex. The COD to BOD ratio is high which indicates that a considerably portion of the raw wastewater components is not readily biologically oxidizable.

TABLE 21: RAW WASTE LOADS FOR SOLUTION CRUMB RUBBER PLANTS

Plant	Product	FLOW L/kkg(gal./ 1,000 lb)	COD kg/kkg(lb/ 1,000 lb)	BOD kg/kkg(lb/ 1,000 lb)	SS[1] kg/kkg(lb/ 1,000 lb)	OIL kg/kkg(lb/ 1,000 lb)
K	SBR Oil Extended	10,500 (1,260)	4.04	0.09	0.81	N.A.
K	SBR Carbon Black Extended	17,800 (2.137)	20.80	0.18	2.20	N.A.
L	PBR Oil Extended	28,500 (3,421)	18.40	1.55	5.72	2.43
L	SBR Non-Extended	14,700 (1,765)	13.28	0.82	1.79	1.43
M	PBR Non-Extended	3,400 (408)	0.17	0.06	0.05	0.07
N	IR Non-Extended	11,900 (1,429)	3.61	1.37	N.A.	0.01
N	PBR Part Oil Extended	11,900 (1,429)	3.01	1.37	5.37	2.32
O	PBR, IR EPDM Part Oil and Carbon Black Extended	29,000 (3,481)	5.33	3.57	3.71	0.23
	Median Value		9.03	1.13	2.81	1.08

[1] Raw waste load determined downstream of crumb pits, where the suspended solids and oil levels are reduced
N.A. Data not available.

Source: EPA (13)

The total dissolved solids content of solution crumb wastewater is considerably lower than for emulsion crumb plants. This again is mainly due to the absence of the coagulation liquor. Surfactant concentrations in the total plant effluent are low. Surfactants are used to deagglomerate the crumb rubber during coagulation and rinsing.

The solvent recovery systems do not produce any significant effect on the COD or BOD content of the effluent. One plant (Plant M) has considerably lower loadings than the others. This is probably due to the fact that rubber for nontire use is produced at this plant. This rubber is used to manufacture impact-resistant resins, and its quality and production controls are extremely critical. In addition, special finishing equipment appears to be used.

The crumb rinse overflow produced at a solution crumb plant is similar to that produced at an emulsion crumb plant, with the exception that uncoagulated latex is not present. The suspended solids, mostly crumb rubber fines, are similar to those in emulsion crumb rinse overflows; the crumb pits produce the same reductions.

The monomer and solvent recovery wastes are comparable to the monomer recovery wastes from emulsion plants. Although heavy slops are produced in some plants, these are usually disposed of by drumming or incineration. Since monomer purities must be high, recovered butadiene, for example, is returned to the monomer supply plant and has no impact on the solution crumb rubber wastewater.

Equipment clean-out wastewaters are less of an environmental problem in a solution plant because much of the processing equipment must be kept dry or water-free. Area washdowns are similar in volume, but do not contain latex. These washdowns do pick-up rubber solids and oil from pumps and machinery areas.

The spent caustic scrub solution, where used, is identical to that used in emulsion crumb production. In plants where emulsion and solution crumb rubber is produced, the same caustic scrub system is used for both facilities.

Catalysts and other reaction ingredients do not produce discernible quantities of heavy metals or toxic constituents. Chromium and zinc in cooling tower blowdown are present in some plant effluents, but in concentrations of 0.1 mg/l or less in the final effluent. These can be eliminated or reduced by using chromium-free corrosion inhibitors.

Latex Production

In addition to solid crumb rubber, emulsion polymerization is also used to produce latex rubber. Latex production follows the same processing steps as emulsion crumb production with the exception of latex coagulation and crumb rinsing, drying, and baling. Only about 5 to 10% of SBR is used as latex, but approximately 30% of the nitrile rubbers (NBR) enter the market as latex. Commercially available SBR latexes contain about 45 to 55% solids, although some can be as high as 68%. Most NBR latexes are in the 45 to 55% solids class. The polymerizations are taken essentially to completion (about 98 to 99% conversion) as opposed to emulsion crumb rubber production where conversion per polymerization pass is approximately 60%. As a result, the recovery of unused monomer is not economical. Process economics are directed towards maximum conversion on a once-through basis.

Figure 51 is a generalized materials flow diagram for the production of latex SBR by emulsion polymerization. The steps shown are typical of all latex production processes. Although latex plants are generally operated 24 hours per day and 365 days per year, the production runs for each recipe or type of latex are shorter than in emulsion or solution crumb rubber plants because latex consumption is on a smaller scale and latex consumers are usually outside companies with varying product needs.

FIGURE 51: GENERAL WATER FLOW DIAGRAM FOR AN EMULSION LATEX RUBBER PRODUCTION FACILITY

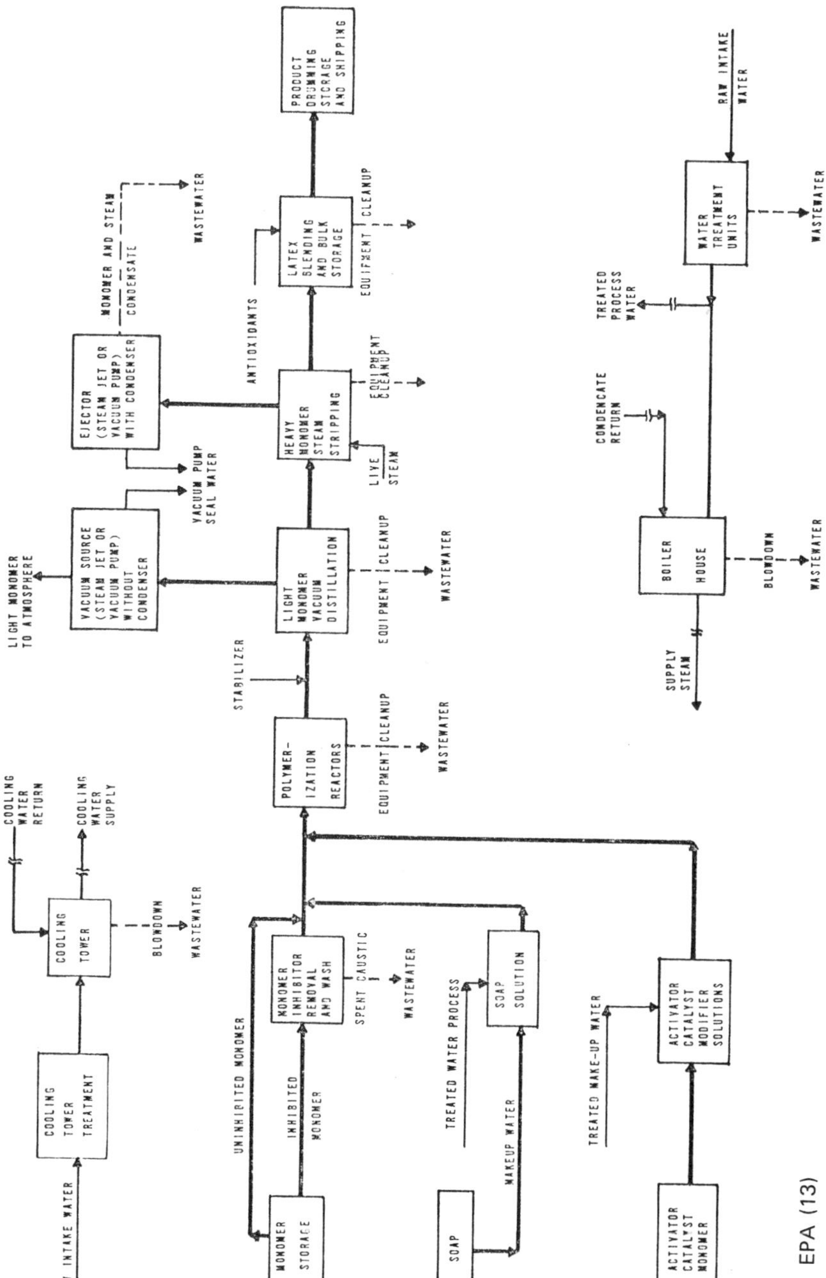

Source: EPA (13)

By contrast, the majority of crumb rubber is made for tire manufacture, is consumed by major tire companies, and is produced by their own synthetic rubber producing divisions. This has the effect of limiting the number of types of product and recipe, rationalizing production schedules, and, in the final analysis, leading to long production runs. Latexes are used to manufacture dipped goods, paper coatings, paints, carpet backing, and many other commodities.

The monomers are piped from the tank farm to the processing plant. Monomer inhibitors are scrubbed out by using caustic soda solution. Soap solution, catalysts, and modifiers are added to the monomer to produce a feed emulsion prior to feeding to the reactors. The water used in the preparation of the above solutions is generally deionized city or well water. The number of reactors in the reactor train is usually smaller than that used for emulsion crumb production. The temperature is generally kept at approximately 40° to 45°F and, therefore, most latexes are made by the cold process. When the polymerization is complete, the latex is sent to a blowdown tank for intermediate storage or holding. Stabilizers are usually added to the latex at this point to stop the polymerization and to stabilize the latex.

The latex passes from the blowdown tanks to a vacuum stripper where the unreacted butadiene is removed. The butadiene is vented to the atmosphere. The vacuum is pulled with either a vacuum pump or steam jet. The excess styrene is stripped from the latex in a steam stripper. The steam and styrene are condensed and sent to a receiver. The bottom water layer is decanted off and discharged. The styrene layer is not recycled but can be containerized and sent to disposal. The stripped latex is passed through a series of screen filters to remove unwanted large rubber solids. The latex is finally stored in blending tanks where various additives (for example, antioxidants) are mixed with the latex. The latex is shipped from the blending tanks by tank car or tank truck, or it is drummed ready for dispatch.

Since short production runs are common to the industry, the major wastewaters generated in a synthetic latex plant stem from equipment cleanout operations. When production is switched from one type of rubber to another, reactors, blowdown tanks, strippers, and filters require cleaning for the new product. In addition, tank cars and tank trucks owned or leased by the plant require cleaning after each trip. Area washdowns are frequent inside the processing buildings and at the vehicle loading-unloading areas. All the above wastewaters will contain oils, dissolved organics, and high concentrations of latex solids. Table 22 summarizes the origins and nature of the principal wastewater sources generated in a typical synthetic latex plant.

TABLE 22: PROCESS-ASSOCIATED WASTEWATER SOURCES FROM LATEX PRODUCTION VIA EMULSION POLYMERIZATION PROCESSING

Processing Unit	Source	Nature of Wastewater Contaminants
Caustic Soda Scrubber	Spent caustic solution	High pH, alkalinity, and color. Extremely low average flow rate.
Excess Monomer Stripping	Decant water layer	Dissolved and separable organics.
Tanks, Reactors, and Strippers	Cleanout rinse water	Dissolved organics, suspended and dissolved solids. High quantities of uncoagulated latex.
Tank Cars and Tank Trucks	Cleanout rinse water	Dissolved organics, suspended and dissolved solids. High quantities of uncoagulated latex.
All Plant Areas	Area washdowns	Dissolved and separable organics, and suspended and dissolved solids.

Source: EPA (13)

Table 23 lists the total effluent flows for latex rubber plants. Only two plants are presented, but the similarity between the data values is good. Latex plants are generally part of larger complexes, and flow data for latex operations is difficult to obtain. The flow from latex plants appears to be lower than from either emulsion crumb or solution crumb facilities. The major flow contributions at latex plants originate with equipment cleaning, area washdown operations, and waters from vacuum pump seal systems.

TABLE 23: RAW WASTE LOADS FOR LATEX RUBBER PLANTS

Plant	Product	FLOW L)kkg(gal) 1,000lb)	COD kg/kkg(lb/ 1,000 lb)	BOD kg/kkg(lb/ 1,000 lb)	SS kg/kkg(lb/ 1,000 lb)	OIL kg/kkg(lb/ 1,000 lb)
P	SBR and NBR	14,900 (1,790)	36.37	5.61	6.70	N.A.
Q	SBR	12,000 (1,500)	33.52	5.01	5.63	0.33
Average Value			34,95	5.31	6.17	0.33

N.A. Data not available.

Source: EPA (13)

The raw waste loads of latex plant wastewaters are considerably higher than either emulsion or solution crumb plants. Equipment cleanout and area washdowns are frequent due to smaller produce runs, and considerable quantities of uncoagulated latex are contained in these wastewaters. The high COD to BOD ratio is typical of all synthetic rubber subcategories and underlines the resistance to biological oxidation of the wastewater constituents. Oil concentration is lower than in emulsion or solution crumb facilities and is contributed by separable monomers, such as styrene, in the wastes. The suspended solids in the effluent are due mainly to uncoagulated latex. Total dissolved solid levels are lower than for emulsion plants because of coagulation liquor stream. Surfactants are present, but in much lower concentrations than the other parameters.

Tank, reactor, and filter cleaning produce considerably quantities of wastewater. These are characterized by high COD, BOD, and suspended solids. In addition, unloading and product loading areas and general plant areas are frequently washed down. The characteristics of these wastes are similar to those produced by usual equipment cleaning in this industry. Vacuum pump seal waters contain small quantities of organics which produce moderate levels of COD from the vacuum stripping operation. The stripping condensates contain condensed monomers. Most of these monomers are decanted from the water and reused. The water layer overflow from the decanter has high COD and BOD concentrations. Spent caustic scrub solution is an extremely low flow waste and has similar characteristics to spent solutions produced in emulsion crumb and solution crumb plants.

WATER POLLUTION CONTROL AND TREATMENT METHODS

Table 24 summarizes some of the wastewater control and treatment steps practiced at various exemplary plants visited in the course of an EPA survey of the industry (13).

In-Plant Control

Since the synthetic rubber industry is highly technological, involving many proprietary and confidential processing techniques, many potential in-plant wastewater control methods would call for radical changes in processing or product quality. Such techniques are obviously not feasible. However, some potential control methods deserve mention so that their applicability may be evaluated.

TABLE 24: WASTEWATER CONTROL AND TREATMENT TECHNOLOGIES AT EXEMPLARY SYNTHETIC RUBBER PLANTS

Plant	Product[1]	Process	Production Capacity (metric ton/year)	Control Measures	Primary Effluent Treatment	Secondary Effluent Treatment
J	Crumb SBR Crumb NBR	Emulsion	390,000 10,000	Crumb Pits, Carbon Slurry Pit	Chemical Coagulation and Clarification	Aeration, Settling and Stabilization Lagoons
K	Crumb SBR	Solution Emulsion	130,000 200,000	Crumb Pits, Carbon Slurry Pit	Air Flotation Primary Clarification	Aerated Lagoon and Air Flotation Secondary Clarification
L	Crumb SBR Crumb, hot SBR Crumb PBR Crumb SBR	Emulsion Emulsion Solution Solution	120,000 3,700 52,000 10,000	Crumb Pits	Chemical Coagulation and Settling Ponds	Aeration Lagoon and Stabilization Ponds
M	Crumb PBR	Solution	85,000	Crumb Pit, Dry Desolventizing Process, Spent Caustic Pit with Slow Bleed Discharge	Oil Separator	None
N	Crumb IR Crumb PBR	Solution Solution	65,000 110,000	Crumb Pits	Equalization Basin	Activated Sludge
O	Crumb PBR Crumb IR Crumb EPDM	Solution Solution Solution	56,000 50,000 25,000	Crumb Pits, Carbon Slurry Pit	Settling Lagoons	Stabilization Lagoon
P	Latex SBR Latex NBR	Emulsion Emulsion	18,000 3,000	Excess Monomer Decant System	Coagulation and Settling Pit	Activated Sludge
Q	Latex SBR	Emulsion	21,000	Excess Monomer Decant System	Chemical Coagulation and Clarification	Discharge to a Municipal Sewage Treatment Plant

[1] SBR: styrene-butadiene rubber, NBR: acrylonitrile-butadiene rubber, PBR: polybutadiene rubber, EPDM: ethylene-propylene diene rubber, IR: polyisoprene rubber

Source: EPA (13)

It was observed that some crumb rubber plants generate crumb rinse overflows which have a lower loading of rubber fines than other plants. Generally, however, such losses cannot be reduced with finer in-plant screens since they are a function of both the type and the coagulation properties of the rubber. One plant did use a proprietary method to finish the rubber cement in which a water slurry is not used. This system eliminates the crumb slurry overflow and the contained rubber fines. It is not necessarily applicable wholesale to crumb rubber production, but does merit investigation by industry.

Most emulsion crumb rubber processes use an acid and brine coagulation liquor. One plant, however, coagulates the latex with an acid polyamine liquor which reduces the quantity of total dissolved solids discharged in the coagulation liquor overflow. The use of this type of coagulation liquor is not always possible, but if employed could significantly reduce the total dissolved solids in the final effluent.

Several plants are converting vacuum systems from steam jet ejectors to vacuum pumps for efficiency and wastewater reasons. In order to maximize the wastewater benefits derived from the use of vacuum pumps, the seal water should be recycled. An overflow is generally required from the seal water recycle system; this overflow is normally slightly contaminated with oil but has a better quality than the steam jet condensate.

In some plants, the caustic soda solution used to remove inhibitors from some monomers (notably butadiene) is replaced batchwise. The spent caustic soda solution, usually 10 to 20% sodium hydroxide, should not be discharged batchwise. It should be containerized and bled into the total plant effluent, thereby diluting its high pH, alkalinity, COD, and color contributions.

The usual method is to slurry the carbon black for addition to the rubber with water. One plant visited employs a steam grinding-slurrying technique which reduces carbon black spillage and consequently washdown and runoff wastewaters laden with black fines; this technique avoids the need for carbon black settling pits and the associated pit cleaning costs.

Latex spills and leakages occur from time to time in all emulsion crumb and latex plants. In most cases, the spill is washed to the nearest plant drain using a water hose. In many cases, this produces unnecessary washdown water and dilutes the latex so that subsequent treatment by coagulation is much more difficult. An alternative technique is to coagulate the latex in situ with alum, for example, and remove the coagulated rubber solids with scrapers. The volume of subsequent washdown water required is less and the latex solids in the washdown water are greatly reduced.

As a result of the high hydraulic pressures involved and the continual jarring action of the balers, oil leaks are frequent. Back-welding the hydraulic lines, although more expensive as an initial equipment cost, does significantly reduce the occurrence of baler oil leaks and can produce appreciable savings in baler oil usage. In addition, plant floor drains should be sealed and, if necessary, retention curbing installed to keep leaked oil from leaving the baler area. Balers using water as the hydraulic fluid are also available and are being used in some plants; oil leakage with this type of machine is obviously eliminated.

End-of-Pipe Treatment

Emulsion Crumb Plants—Primary: It is normal practice for crumb rubber producers to recycle part of the crumb rinse water. The remainder of the crumb rinse water is discharged in order to blow down accumulating fine rubber solids, dissolved solids and organics. The rinse water discharge or overflow is clarified before final treatment in a crumb separation pit. The trapped rubber solids are removed periodically by scoop. A very common shortcoming of these separators is that they are operated as single units and are not cleaned frequently enough. This results in short-circuiting followed by poor separation. In addition, when the pits are cleaned, the separated rubber solids are disturbed and rubber solids that had previously separated recombine with the pit effluent until the condition of the pit stabilizes. Dual pits would solve this problem: one pit would stay in operation while the other was cleaned and allowed to stabilize.

Since wastewaters from emulsion crumb plants contain considerable quantities of latex, it is necessary to coagulate the latex in order to achieve a good quality effluent. Chemical coagulation by itself is seldom sufficient, because the density of the coagulated rubber is normally close to that of water. Therefore, it is customary to add a sinker (clay or limestone) to the coagulation mixture to sink the coagulated rubber and effect the separation. For a small wastewater flow, chemical coagulation using (for example) alum, polyelectrolyte, and clay in a rapid mix tank can be followed by flocculation in a flocculator tank. Clarification can then be accomplished in a rectangular clarifier equipped with solids removal equipment. Larger wastewater flows can be treated in a reactor-clarifier with rim overflow and central sludge draw-off.

The collected sludge can be thickened and dewatered before disposal in a landfill. Dewatering studies on this type of sludge concluded that a plate-and-frame pressure filter performed well. The installed filter was automatically controlled for feed shut-off, filter opening, core blowout, filter closing, precoating, and feed restoration (refer to Plant J).

One plant (Plant K) uses chemical coagulation followed by air flotation for primary clarification. Instead of sinking, the rubber solids are floated to the clarifier surface with air bubbles and removed by surface solids removal equipment. This treatment facility is relatively new and has had start-up troubles, although they have been satisfactorily resolved. Air flotation in this application produces primary effluent of good quality. The collected surface solids are pumped to a sludge impoundment lagoon where they dry out. The use of this lagoon is limited and a long-term solids dewatering and disposal technique will have to be found.

Where adequate land is available, rubber solids separation has been achieved using primary settling ponds. Chemical coagulation of the solids prior to discharge to the ponds is usually necessary. The settled and floating solids (since both types are produced) are removed from the ponds periodically by vacuum truck or scoop.

Emulsion Crumb Plants—Secondary: Biological oxidation of the primary effluent is achieved in aerated lagoons or in activated sludge plants. Generally, a nutrient must be added. Both technologies obtain satisfactory oxidation of the dissolved contaminants; problems can arise in the clarification of the secondary effluent. Good secondary clarification of effluent from an aerated lagoon has been obtained with the aforementioned air flotation plant, which is a dual system with both primary and secondary air flotation clarifiers. If sufficient land is available, the effluent from the aerated lagoons can be clarified and stabilized in stabilization ponds. This type of facility is temperature dependent, of course, and performs better in warmer climates.

Clarifiers are commonly used for secondary clarification in activated sludge plants. They are generally adequate, but cases exist where high solids carryover is a problem. Secondary clarification can be assisted with coagulation chemicals in much the same manner as for primary clarification, but the additional chemical cost is high.

One plant in the industry, in an area where water is in short supply, is evaluating evaporation to remove wastewater contaminants. Satisfactory operation has not been achieved to date because of corrosion and fouling of the evaporator tubes. It is proposed that satisfactory operation can be achieved with the use of more effective corrosion resistant materials of construction as well as pretreatment to reduce the quantities of certain organics and sulfides.

Emulsion Crumb Plants—Advanced: After secondary treatment, emulsion crumb wastewaters still contain high levels of COD. A high COD level appears to be a common characteristic of secondary effluents from emulsion crumb, solution crumb, and latex plants, and indicates that certain constituents of the wastewaters generated in synthetic rubber plants are refractory to biological oxidation.

With the exception of the evaporation treatment described above, only one other study has been made of tertiary or advanced treatment of emulsion crumb rubber wastewaters. This was carried out on a pilot plant scale using activated carbon treatment. Approximately 70% of the COD remaining after secondary treatment was removed by the carbon.

Solution Crumb Plants—Primary: Primary clarification of the solution crumb plant wastewater is carried out in crumb pits. These pits are similar in design to those for emulsion crumb production facilities. To avoid resuspending the separated rubber solids, dual crumb pits should be used. Other forms of primary treatment are not required for solution crumb wastewaters, since uncoagulated latex is not present and the fine rubber solids separate readily.

Solution Crumb Plants—Secondary: Secondary treatment technology uses both activated sludge and aerated lagoon systems. Good BOD removals are achieved, but poor secondary clarification is a problem in most cases. The reasons for this are not certain. A high level of COD remains after biological treatment, indicating that much of the wastewater constituents are biologically refractory.

Solution Crumb Plants—Advanced: Advanced or tertiary treatment technologies have not been used on secondary effluents from solution crumb plants. It is probable that activated carbon treatment would give COD removals similar to those for emulsion crumb wastewater, since the raw wastewater constituents (for example, traces of monomer) are similar for both types of wastewater.

Latex Plants—Primary: Since latex plant wastewaters contain uncoagulated latex solids, primary clarification is assisted by chemical coagulation. In much the same manner as for emulsion crumb wastewaters, clarification can be effected in reactor-clarifiers or systems with separate rapid mix, flocculation, and clarification tanks. Latex wastewaters can also be clarified by air flotation.

Latex Plants—Secondary: Activated sludge plants are used for the secondary treatment of latex wastewater. High residual COD levels are a problem. These levels are higher than for either emulsion crumb or solution curmb plants, because the initial COD loading of the raw wastewater from latex plants is much higher. It is feasible that aerated lagoons and stabilization ponds will produce satisfactory oxidation and stabilization of latex wastewaters.

Latex Plants— Advanced: Advanced or tertiary treatment technologies have not been used on latex wastewaters. COD removals similar to those achieved by emulsion plants can be achieved for latex wastewater by using activated carbon columns.

Figure 52 shows a hypothetical end-of-pipe secondary wastewater treatment facility for emulsion crumb, solution crumb and latex-producing plants.

Polymer Corporation Process

A process developed by *R.C. MacKenzie; U.S. Patent 3,729,412; April 24, 1973; assigned to Polymer Corporation Limited, Canada* involves the dewatering of a sludge derived from the treatment of an industrial wastewater stream, wherein the sludge is mixed with a latex comprising rubbery polymer and, on addition of a coagulating agent, a coagulum is formed which is separable from the aqueous phase. The latex in question may be but is not necessarily a waste from a latex-producing plant. Thus, here is an example of a case where a latex waste may be disposed of by using the latex itself as the treating agent.

Petro-Tex Chemical Corporation Process

A process developed by *P.M. Merchant, Jr.; U.S. Patent 3,775,312; November 27, 1973; assigned to Petro-Tex Chemical Corporation* involves wastewater treatment from polychloroprene (neoprene) manufacture.

FIGURE 52: HYPOTHETICAL END-OF-PIPE SECONDARY WASTEWATER TREATMENT FACILITY FOR SYNTHETIC RUBBER PLANTS

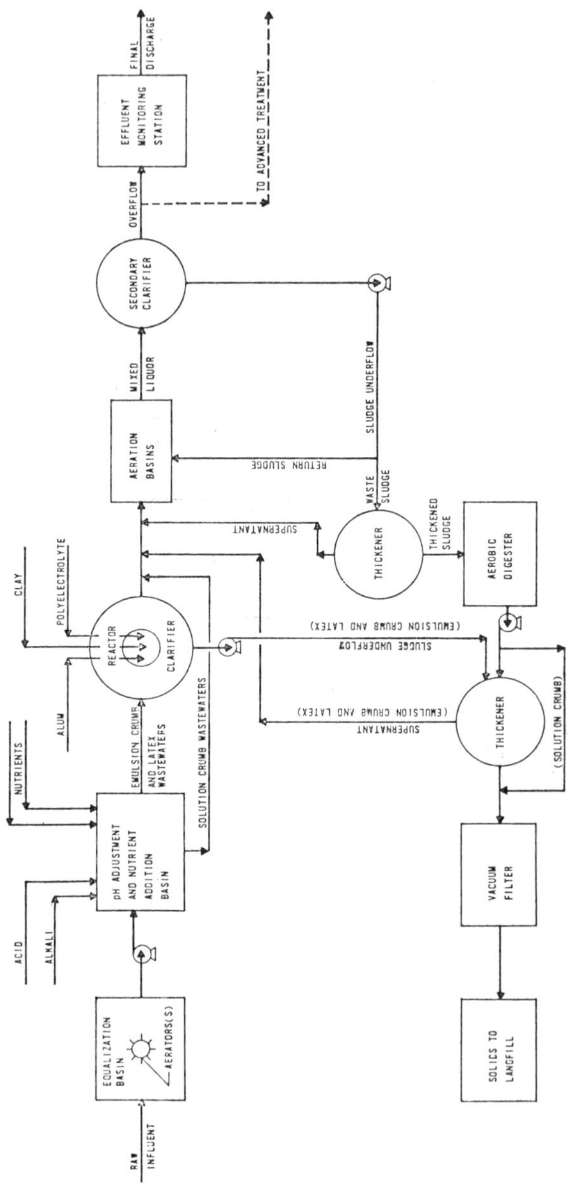

Source: EPA (13)

Neoprene, or polychloroprene, is conventionally polymerized in emulsion polymerization utilizing a combination of emulsifying agents. Conventionally, emulsifying agents are rosin acid salts and various secondary emulsifiers. Some of the emulsifying agents may remain in the polymer but certain water-soluble emulsifiers are substantially removed before final isolation and processing of the polymer. These emulsifiers that are removed are sometimes referred to as secondary emulsifiers. For example, the polymer may be recovered by coagulation of the latexes and thereafter the water-soluble emulsifiers removed before final milling.

The water-soluble emulsifiers are removed for example, by washing the polymer with warm water on a wash belt. This washing may be assisted by extraction by use of solvents. Normally, the washing requires the use of large volumes of water and if the emulsifying agents are biodegradable this water may be processed by biooxidative degradation; however if the wash water contains nonbiodegradable emulsifying agents this presents a major problem because of the restrictions on the releasing of significant amounts of organic compounds into streams and rivers. It is an object of this process to remove certain nonbiodegradable emulsifiers from neoprene process water or other water containing these emulsifiers. Separation of nondegradable emulsifiers which are salts of the condensation product of naphthalenesulfonic acid and formaldehyde from process water is achieved in this process by contacting the process water with an amine having a molecular weight of at least 60. The process water and/or the amine is reduced to a pH of 7.0 or less prior to contacting.

WATER POLLUTION CONTROL ECONOMICS

Tables 25 through 27 inclusive summarize wastewater treatment costs for solution crumb, emulsion crumb and latex plants respectively as taken from a detailed EPA study (13). Additional information of the economics of various effluent treatment alternatives is available in a report prepared by Arthur D. Little, Inc. (14). This report has also been issued in the form of an economic analysis of proposed effluent guidelines for synthetic rubber manufacture by EPA (15).

TABLE 25: SOLUTION CRUMB RUBBER SUBCATEGORY

Estimated Wastewater Treatment Costs for Different Degrees of Treatment based on a Typical Production Capacity of 30,000 metric tons/year

	Treatment or Control Technologies[1]	
	A	B
Investment[2]	$810,000	$1,182,000
Annual Costs		
Capital Costs	$ 81,000	$ 182,000
Depreciation	162,000	236,000
Operating and Maintenance Costs (excluding energy and power costs)	68,000	145,000
Energy and Power Costs	4,000	6,000
Total Annual Costs[2]	$315,000	$ 569,000

Parameters (kg/kkg production)	Raw Wastewaters	Effluent Quality A	Effluent Quality B
COD	6.00	3.94	2.08
BOD	0.80	0.40	0.08
Suspended Solids	2.20	0.65	0.16
Oil and Grease	0.25	0.16	0.08

[1]Technology A is Primary Clarification of Crumb Rinse Wastewaters followed by Activated Sludge Secondary Treatment.
Technology B is Technology A followed by Dual-Media Filtration plus Activated Carbon Treatment to produce 65% overall COD reduction.

[2]August 1971 Dollars.

Source: EPA (13)

TABLE 26: EMULSION CRUMB RUBBER SUBCATEGORY

Estimated Wastewater Treatment Costs for Different Degrees of Treatment based on a Typical Production Capacity of 128,000 metric tons/year

	Treatment or Control Technologies[1]			
	A	B	C	D
Investment[2]	$1,180,000	$2,002,000	$2,907,000	$2,993,000
Annual Costs				
Capital Costs	$ 118,000	$ 200,000	$ 291,000	$ 299,000
Depreciation	236,000	400,000	581,000	599,000
Operating and Maintenance Costs (excluding energy and power costs)	147,000	230,000	349,000	396,000
Energy and Power Costs	10,000	20,000	25,000	29,000
Total Annual Costs[2]	$ 511,000	$ 850,000	$1,246,000	$1,323,000

Parameters (kg/kkg production)	Raw Wastewater	Effluent Quality			
		A	B	C	D
COD	20.00	12.80	8.00	5.04	2.08
BOD	2.13	2.00	0.40	0.08	0.08
Suspended Solids	7.20	1.28	0.65	0.16	0.16
Oil and Grease	1.40	0.32	0.16	0.08	0.08
Flow ($L/10^3$ kg production)	16,600				

[1] Technology A is Chemical Coagulation followed by Clarification.
Technology B is Technology A followed by Activated Sludge Secondary Treatment
Technology C is Technology B followed by Dual-Media Filtration plus Activated Carbon Treatment to produce 75% overall COD reduction.
Technology D is Technology B followed by Dual-Media Filtration plus Activated Carbon Treatment to produce 90% overall COD reduction.

[2] August 1971 Dollars.

Source: EPA (13)

TABLE 27: LATEX RUBBER SUBCATEGORY

Estimated Wastewater Treatment Costs for Different Degrees of Treatment based on a Typical Production Capacity of 10,000 metric tons/yr.

	Treatment or Control Technologies[1]			
	A	B	C	D
Investment[2]	$361,000	$637,000	$784,000	$784,000
Annual Costs				
Capital Costs	$ 36,000	$ 64,000	$ 78,000	$ 78,000
Depreciation	72,000	127,000	157,000	157,000
Operating and Maintenance Costs excluding energy and costs)	33,000	57,000	100,000	117,000
Energy and Power Costs	1,000	3,000	3,000	3,000
Total Annual Costs[2]	$142,000	$251,000	$338,000	$355,000

Parameters (kg/kkg production)	Raw Wastewater	Effluent Quality A	B	C	D
COD	33.52	14.75	6.95	4.31	1.78
BOD	5.51	4.77	0.34	0.07	0.07
Suspended Solids	5.63	0.93	0.55	0.14	0.14
Oil and Grease	0.70	0.34	0.14	0.07	0.07

[1] Technology A is Chemical Coagulation followed by Clarification.
Technology B is Technology A followed by Activated Sludge Secondary Treatment.
Technology C is Technology B followed by Dual Media Filtration plus Activated Carbon Treatment to produce 87% overall COD reduction.
Technology D is Technology B followed by Dual Media Filtration plus Activated Carbon Treatment to produce 95% overall COD reduction.

[2] August 1971 Dollars.

Source: EPA (13)

PLASTICS PRODUCTS INDUSTRY

The plastics products industry as defined in this volume can be divided into three types of activities: compounding, the process of mixing the polymer with additives or colorants; fabricating, the process of transforming the resulting compound or uncompounded polymer to a finished or semifinished plastic or plastic containing end item such as film and bottle, or tubes and sheets; and converting, the process of modifying a fabricated plastic product by decorating it, cutting, or sealing it.

Another activity that is included under processing is reprocessing, which is the operation required to recycle plastics. Reprocessing is the process of converting scrap plastic to a useful material. Normally, this requires remelting the scrap to reshape it. Scrap is often generated during the manufacture of plastic items, industrial scrap. It may be in the form of contaminated compound, film trimmings, strands, or large chunks of plastic from the molding machinery. If the scrap plastic (SP) is not reprocessed nor reprocessable, it becomes nuisance plastic (NP).

The product obtained by reprocessing scrap plastic is called secondary resin. This material competes with virgin prime resin and, particularly virgin off-grade resin. Prime resin is a resin that only has been subjected to the minimum number of heat histories required in its production, and it meets the required specifications. Off-grade resin is similar to prime, except that as prepared it did not meet the original customer's specifications. Generally, off-grade has a wide specification range and is used to fabricate plastic products that do not require raw material with tightly controlled specifications. Resin produced as the manufacturer changes his manufacturing line from one grade to another, the so-called transition material, and that produced in experimental runs, also become off-grade resin.

To recycle or reprocess plastics, they must be homogeneously one plastic type and free of foreign matter. In this report, these are called monoplastic products. If the plastic item contains more than one plastic compound, as in a laminate, it is a multiplastic, and if the end item contains both plastic and nonplastic materials, it is considered a composite. Figure 53 presents in graphic form an interesting summary of the environmental impact for container manufacture as one example of the plastics products industry.

Some idea of the overall environmental impact of the manufacture of specific plastic products may be gained from Table 28 adapted (with permission) from a Manufacturing Chemists Association publication (4). The study was actually made by Midwest Research Institute, Kansas City, Missouri, under contract to the Manufacturing Chemists Association.

FIGURE 53: SUMMARY OF ENVIRONMENTAL IMPACTS FOR CONTAINER MANUFACTURE

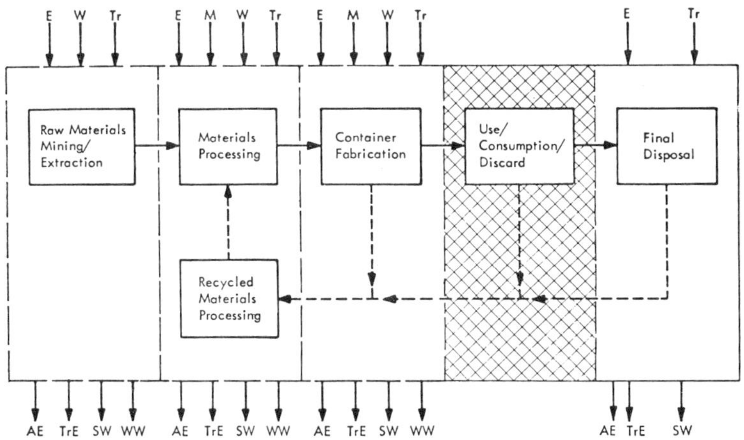

SUMMARY OF INPUT/OUTPUT CATEGORIES

INPUT	OUTPUT
E = Energy (in all forms)	AE = Atmospheric Emissions
M = Virgin Materials (consumed and unconsumed)	TrE = Transportation Effluents (for each fuel type)
W = Water	SW = Solid Wastes
Tr = Transportation to Next Operation (including all modes, all fuels in each mode)	WW = Waterborne Wastes

TABLE 28: DATA ON ENVIRONMENTAL IMPACT OF FABRICATION OF VARIOUS PLASTIC PRODUCTS

Per 1,000 lb of:	Energy, Electric (kwh)	Water Volume (gal)	Process Solid Wastes (LG)
PVC half-gallon bottles	495	1,772	1.0
1 gal HDPE milk cartons	358	455	1.0
1 gal polyethylene produce bags	239	1,780	1.0
LDPE lids	338	1,780	1.0
ABS dairy tubs	495	1,772	1.0
HIPS 9-fluid ounce soft drink vending cups	174.5	243	1.0

Source: Manufacturing Chemists Association (4)

AIR POLLUTION

Air pollution problems presented by the plastic products industry consist primarily of removing processing chemicals such as solvents, softeners and plasticizers from plant exhaust air streams.

A process developed by *W.R. Evans, Jr.; U.S. Patent 3,232,029; February 1, 1966; assigned to Celanese Corporation of America* involves the recovery of organic solvents from gaseous media in such operations as the manufacture of cellulose ester sheets and films where solvents are evaporated in product dryers.

A suitable device for the conduct of such a process is shown in Figure 54. An incoming gaseous medium carrying a vaporized organic solvent which was evaporated in a drying device is passed through heat transfer chamber **20**, in which the gaseous medium comes into contact with coils **21**, through which a coolant flows countercurrent to the flow of the gas through the heat transfer chamber. Heat is transferred from the gas to the coolant resulting in a cooling of the gas and some condensation of the vaporized solvent.

The floor of the chamber is inclined so that condensed solvent flows into receptacle **22**. The gas is then passed into condensation chamber **23** where it is subjected to a spray of coolant from sprayers **24**, the coolant having been previously refrigerated by refrigeration means **25**. The coolant preferably comprises the organic solvent composition in the liquid state. The organic solvent condenses and is incorporated into the coolant accumulating in the receptacle. From the receptacle, the coolant is driven by driving means such as a pump **26** through coils **21** where it undergoes the previously described heat transfer with the incoming gas from the drier. Due to the heat transfer in the chamber and in condensation chamber **23**, the coolant exiting chamber **20** at **27** is at a higher temperature than the gas exiting condensation chamber **23** at **28**.

FIGURE 54: APPARATUS FOR SOLVENT RECOVERY FROM CELLULOSE ESTER SHEET AND FILM DRYING OPERATIONS

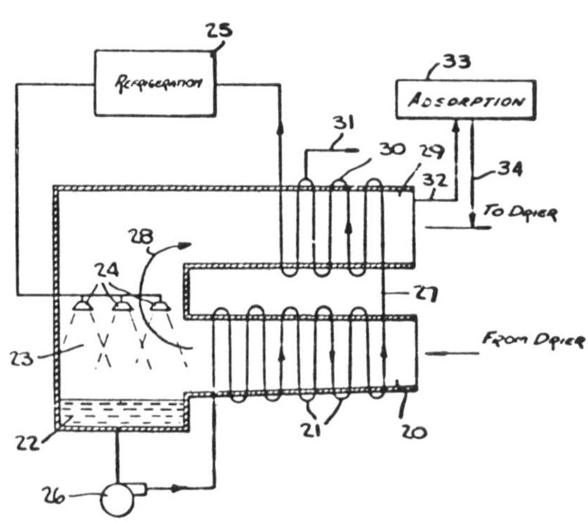

Source: W.R. Evans, Jr.; U.S. Patent 3,232,029; February 1, 1966

This exiting gas is then passed through heat transfer chamber **29** where the exiting coolant is passed through coils **30** countercurrent to the flow of gas through chamber **29**. Heat is now retransferred from the coolant to the gas, thereby cooling the coolant which is then recycled to refrigeration means **25** prior to being recirculated. The heated remaining gas leaving the chamber **29** may then be recycled to the drying device. Solvent may be removed from the coolant circulatory system at any point by any suitable means such as tap **31.**

The remaining gas (the gas leaving chamber **29**) after the condensation and removal therefrom of the solvent may still contain residual amounts of solvent. If desired, residual solvent may be removed from this gas by passing a portion of the gas leaving chamber **29** through line **32** into chamber **33** containing an absorbent such as activated carbon to adsorb the solvent vapors after which such portion of the gas may be returned to the dryer through line **34**. The solvents may then be recovered from the adsorbent, e.g., as set forth in U.S. Patent 2,856,331.

A process developed by *R.B. Perry; U.S. Patent 3,538,613; November 10, 1970; assigned to E.I. du Pont de Nemours and Company* is a process for the recovery of cellulose softener by scrubbing exhaust air with aqueous softener solution and returning softener solution to the treating bath, and an apparatus therefor.

Figure 55 shows a suitable form of apparatus for the conduct of this process. In the figure, a regenerated cellulose web **1**, passing over rollers through treating tank **2**, is heated by the warmed softener **3**, causing substantial quantities of the water previously held by the film to be discharged into the tank, diluting the softener solution. The excess, diluted fluid is taken off by pump **4** and transported to meter **5** which regulates the quantity of diluted softener solution passing into the reclamation cycle. Upon entering the reclamation cycle, the softener is pumped by pump **6** through recirculating lines **7** to an absorption unit **8**.

The absorption unit is in the exhaust ductwork of dryer **9**, the exhaust air **10** passing concurrently through the absorption unit, propelled by fans **11** and **12**. The diluted softener solution enters the absorption unit and is dispersed therein by spray banks **13** and **14**, impinging on absorption media **15** and **16**. The softener solution is then collected in a sump in the bottom of the absorption unit with the aid of entrainment separator **17** whereafter it is either recycled in the reclamation cycle or passed back to tank **2** through pipe **18** where it is again used to treat the regenerated cellulose film passing through the tank. The following is a specific example of the operation of this process.

Example: Regenerated cellulose web is treated in an apparatus of the type illustrated by consecutively passing it through the softener bath and the drying chamber. Approximately 14,000 cubic feet per minute of cellulose web dryer exhaust air **10** is exhausted via dryer discharge fan **11** directly to an absorption unit **8** where propylene glycol is removed and water evaporated into the exhaust air. The absorption unit comprises the preferred herringbone pattern absorption plate. This humidified effluent air is then discharged from the absorption unit to the atmosphere by exhaust fan.

The exhaust air, as it leaves the web dryer, has a relative humidity of about 25% and contains 1.2×10^{-4} lb of propylene glycol per cubic foot of air. Thus, 1.68 pounds of propylene glycol per minute is passed to the absorption unit via the dryer effluent air stream. The maximum effluent air velocity through the absorption unit is 220 fpm while the maximum air velocity through the absorption medium slots is 4,000 feet per minute.

Concurrently, 360 gpm of aqueous softener solution is recirculated via the recirculating spray supply line **7** from the sump of the absorption unit through the spray head banks **13** and **14** onto absorption plates **15** and **16** in the amount of 5.5 gpm per square foot of absorption media. Circulating pump **4** pumps 15 gpm of softener solution containing 11% propylene glycol into the inlet side of absorption unit circulating pump **6**. Propylene glycol is removed from the dryer effluent air stream at the gas-liquid interface of the absorption media **15** and **16** so that the humidified effluent air, as it leaves the absorption unit, contains 0.2×10^{-4} lb of propylene glycol per cubic foot of air.

FIGURE 55: APPARATUS FOR SOFTENER RECOVERY FROM REGENERATED CELLULOSE FILM MANUFACTURE

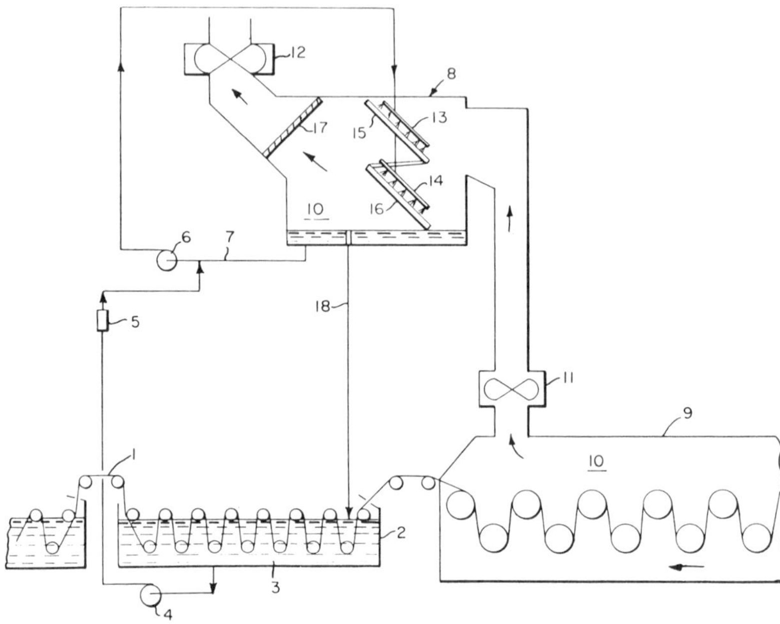

Source: R.B. Perry; U.S. Patent 3,538,613; November 10, 1970

Thus, 1.4 lb of propylene glycol per minute is passed from the effluent air stream to the softener solution scrubbing liquid. Water is evaporated from the scrubbing liquid in the absorption unit into the dryer effluent air at the rate of 15 lb/min. This enriches the solution in the sump of the absorption unit such that softener solution containing 13% propylene glycol is returned to the softener tank. The degree of softener enrichment of the solution from the softener tank is such that approximately 80% of the propylene glycol content of the web dryer effluent air stream is removed.

A process developed by *W.P. Davis, J.J. Golner and S.S. Feinstein; U.S. Patent 3,852,392; December 3, 1974; assigned to Borden, Inc.* is one in which hot polyvinyl chloride film is blown in a confined area so as to eliminate escape of plasticizer mist. The plasticizer mist generated is passed through a suitable scrubber, filter or other such known device to remove the plasticizer. The heated air component of the mist, free of plasticizer, is then returned to the process building to recover the heat content, or discharged to the atmosphere as desired.

Figure 56 shows a suitable form of apparatus for the conduct of this process. A conventional extruder **10** is commonly used in the manufacture of polyvinyl chloride film. Polyvinyl chloride in pellet or powder form is fed into the feed hopper **12** located at one end of the extruder. The polyvinyl chloride is then fed through the extruder and heated therein until it reaches its formative plastic state and in such condition fed into a 90° elbow **14** approximately secured to the head end **16** of the extruder. A die, generally designated at **18**, is provided with an annular orifice **20** from which the molten mass of polyvinyl chloride merges as a hot viscous thermoplastic tubing **22**. Misting of the plasticizer takes place here.

FIGURE 56: APPARATUS FOR PLASTICIZER MIST REMOVAL FROM PVC FILM BLOWING DEVICE

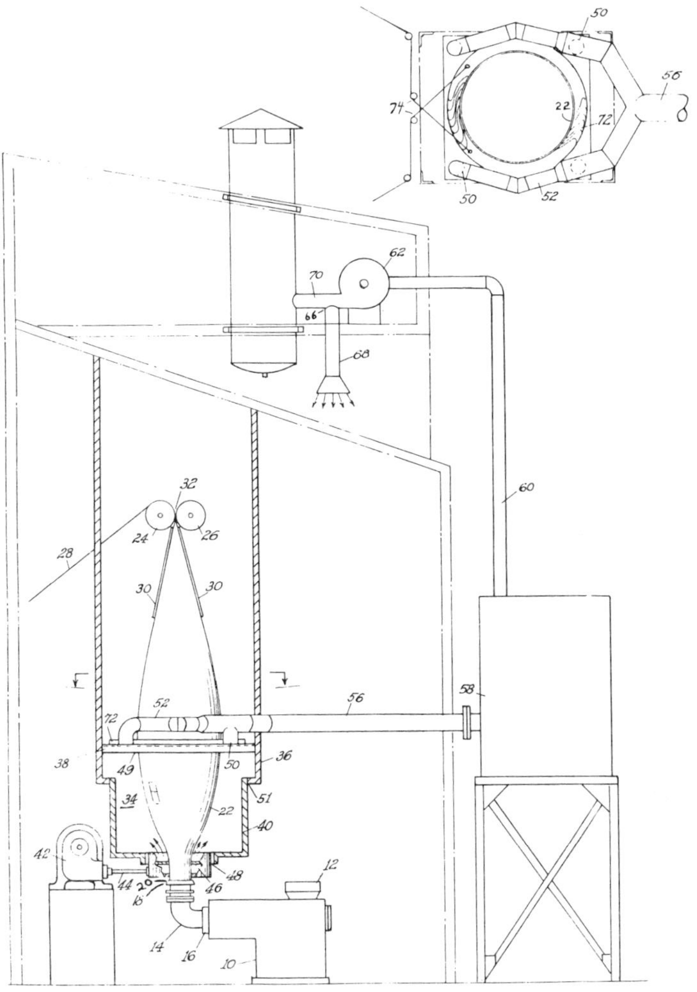

Source: W.P. Davis, J.J. Golner and S.S. Feinstein; U.S. Patent 3,852,392; Dec. 3, 1974

The die **18** is connected to an air supply whereby air is introduced interiorly of the tubing to inflate the same. The air supply is provided with a valve (not shown) so that when the desired quantity of the air has been introduced within the tubing, further supply thereof can be prevented. In the event the quantity of air within the tubing decreases, as for example, by leakage or otherwise, the requisite amount of air can be added by proper manipulation of the valve. The extruded and the inflated tubing mechanism and their application to polyvinyl chloride are widely known in the art and are, therefore, not described in detail here. The inflated tubing **22**, which generates plasticizer mist upon emerging from orifice **20**, is withdrawn upwardly from the die orifice **20** in a substantially vertical direction by a pair of counterrotatable nip rolls **24** and **26** disposed in spaced relationship above the die **18**. These nip rolls are rotatably mounted with the peripheral speed of the rolls **24** and **26** being preferably adjusted to stretch the tubing while it is in the formative plastic state. These rolls also serve to collapse the tubing passing there between into a flattened, ribbon-like material **28** which is wound up on a wind-up reel not shown.

Assisting the rolls **24** and **26** in flattening the tubing **22** are converging guide means intermediate to die **18** and these rolls. The guide means comprises collapser plates **30** which contact the tubing after it has been inflated to its predetermined final diameter and which are arranged in an inverted V-shape to form a converging path of the nip point **32** of the rolls. It will be noted that the inflating medium which is entrapped between the nip point of the rolls and the die constitutes an isolated air bubble which remains substantially stationary while the tubing advances from the die through the collapser plates to the nip.

An area **34** has been set up consisting of the housing **36**, plenum **38**, lower enclosure **40** and iris **72**. This is the area in which all of the generated plasticizer mist is confined. This confined area prevents the escape of the mist to the surrounding process area outside of the confined area **34**. The housing, plenum and lower enclosure also serves to eliminate the entrance of larger amounts of air into the tubing forming area **34** thus decreasing the amount of mist laden air to be cleansed. The confined area is maintained at a pressure less than atmospheric by an exhaust system to be discussed later. The subatmospheric pressure prevents plasticizer mist from passing to the atmosphere through small openings in the confined area.

The hot viscous thermoplastic tubing **22**, as it leaves the die of the extruder **10**, is cooled by a stream of refrigerated air. The air is refrigerated by cooling means **42** which can be of any conventional type for chilling air. The air is usually chilled to about 60°F in cooling means **42** conveyed through conduit **44** to an annular air ring **46**. This annular air ring encircles the die at its discharge end and projects into the lower enclosure through an opening **48**.

This opening allows the annular air ring and the die to be easily removed and inserted into the lower enclosure. The opening can either be such as to provide a tight fit with the air ring or preferably an opening providing a gap of from ¼" to 3". This width prevents an undue amount of air from entering the confined area, allows easy removal of the air ring and the die and prevents a fluctuating pressure in the confined area. Air discharge from the air ring becomes laden with plasticizer mist by contact with the hot tubing and such mist is continuously exhausted from chamber **34** through four exhaust ports **50** located in the plenum.

The hot (120° to 160°F) plasticizer mist laden air which enters the ports passes through exhaust ducts **52** thence to main duct **56**. The main duct carries the mist laden air to the mist removal system **58**. The mist removal system may be any of those conventionally used, such as, an incinerator, electrostatic precipitator, scrubber, centrifuge, impingement device or other suitable mechanical filter.

The systems for removing the plasticizer mist from air are widely known in the art, and, therefore, will not be described in detail here. The cleansed air leaves the mist removal system through duct **60** to blower **62** thence through duct **70** to control point **66**. At control point **66**, the cleansed air may be diverted to the process work area through duct **68** or duct **70** to the outside of the process building.

An iris diaphragm **72** is mounted on top of the plenum to control the quantity of air drawn into chamber **34** from around the periphery of the tubing. A quantity of air is required at the iris diaphragm to prevent escape of the plasticizer mist carried along with the cooled tubing in its passage upward to collapser plates **30**. The iris diaphragm is an adjustable diaphragm of thin plates that can be turned by an adjusting mechanism generally shown at **74** so as to change the diameter of the central opening. The iris diaphragm mechanically operates like the conventional iris diaphragm in a camera does to regulate the light opening behind the lens of the camera. The central opening is preferably adjusted to provide an air space of from ¼" to 3" between the iris diaphragm and the inflated tubing **22**. An air space of less than ¼" between the iris diaphragm **72** and the tubing often creates a problem of pulsating air pressure in the chamber in which the tubing is formed. The pulsations result in undesirable variations in the size and thickness of the tubing. An air space of greater than 3" is undesirable in that it allows a larger amount of air to be drawn into the tubing forming area **34**. The air space dimensions given refer to the commercial film forming unit now generally used.

The lower enclosure **40** can be removed in part from the extruder **10** with relative ease. When the movable portion **40** is in its normal operating position, it encircles the air ring **46**, and provides a tight fit with upper tower enclosure **36**. In a preferred form, an annular air opening **51** provides a space of approximately ¼" to ½" in width between the top of the lower enclosure and the bottom of the tower enclosure through which a controlled amount of clean air passes to form an air seal against plasticizer mist leakage into the process work area.

Another type of problem which can occur in the plastics product industry is reactant (or monomer) emission. Thus, as pointed out by *E.B. Baker; U.S. Patent 3,833,712; Sept. 3, 1974; assigned to Borden, Inc.*, many industrial process which employ phenol or low condensed phenolic resins are confronted with toxicity and odor problems arising from the evolution of phenol and phenol derived volatile substances into air.

Low molecular weight phenolic resins in particular can evolve considerable amounts of such phenolic substances during their transfer, drying or curing, either by themselves or in products incorporating them. Particular industries which are troubled by this problem include those concerned with the manufacture of mineral or glass wool products, separators for lead storage batteries, various laminated products using phenolic adhesives, and the like.

Attempts have been made to reduce phenolic air pollution by removing the undesirable substances as solutions in water or other solvents, but such dilution is not a real answer. It merely converts an air pollution problem into a water pollution or liquid pollution problem. Nor is it satisfactory to resort to the common disposal method involving combustion of the effluent gases from a production unit; this is uneconomical and in addition squanders scarce resources such as natural gas.

According to this process, however, it has been found that atmospheric pollution by volatile phenolic contaminants in industrial effluent gas streams is substantially reduced by absorbing the contaminants in an aqueous solution of a water-soluble alkaline phenol-formaldehyde resin, condensing the mixture of absorbing resin and absorbed contaminants to form an augmented resin, optionally repeating the absorbing and recondensing steps finally to form a useful by-product resin suitable as a component of an adhesive.

WATER POLLUTION

Water pollution problems presented by the plastic products industry are relatively small in magnitude, compared to the solid waste problem, for example.

SOLID WASTE GENERATION FROM SPECIFIC OPERATIONS

In the discussion of specific operations which follow, two abbreviations will be used which are defined as follows:

NP–Nuisance plastic	That portion of total plastics production that has no value and is usually found in the disposal area.
SP–Scrap plastic	That plastic which has value and is generated by the resin producer, compounder, fabricator or converter in his or her operation. Usually this material requires reprocessing; i.e., it must be subjected to an additional heat history.

Compounding

In the narrow sense, compounding is mixing; it signifies the dry mixing or blending of powders, but more broadly, it signifies mixing in a melt to produce pelletized or diced resin. Dry blending is mostly used to prepare PVC powders, but other compounds are prepared similarly. Because of the heat sensitivity of PVC, many fabricators prefer to use it in powder form instead of pellets to minimize degradation during fabrication. Dry blend compounds are more economical, particularly if compounding is done by the fabricator.

Dry blending requires that the ingredients are: free-flowing; high in bulk density, i.e., not too fluffy; free of volatile constituents; free from a tendency to separate into their original state; and free from a tendency to cake or agglomerate. An important factor is particle size. Typical PVC resins for dry blending have a particle size ranging from 3 to 6 mils. A narrow particle size distribution is preferred.

Dry blending, e.g., in dry coloring, is frequently accomplished in mixing equipment such as a ribbon blender or a vertical auger mixer. For more difficult blending, as in the mixing of impact modifiers with resin, a high shear mixer is required. The most widely used units resemble a high speed household mixer with a vertical beater shaft driven from underneath. The intensive stirring usually heats the material being mixed.

Pellet compounding is usually accomplished by extruding multistrands or tapes of plastic that, after cooling, are cut to produce pellets. Another method involves melt blending in a Banbury mixer and then feeding the melt to a two roll mill to produce a plastic sheet. The sheet is then dice cut into pellets. In both systems, mixing is usually accomplished during the melt processing, although for extremely difficult mixing problems, dry blending may precede melt blending.

The Banbury mixer is a very effective mixing device. It has a mixing chamber equipped with special rotor blades that are somewhat elliptical in cross section, and the clearance between the blade and chamber is small. Polymer and additives are charged to the mixing chamber and melted by the vigorous mixing action of the blades; if required, the chamber can be heated. In many cases, the pellet phase is omitted, and the sheet stock from the mixer is used to fabricate an end item directly. If extrusion is used in the compounding process, then normally a number of special screw designs are employed to combine adequate mixing with melting.

Often a particular compounding unit is selected for a product area. The more viscous the molten mixture, the more intense the compounding process must be. This is noticeably true when particulate and especially fibrous fillers are added to polymers. Adequate mixing is very important, because improper compounding can yield inferior products.

As regards scrap plastic from compounding, in the Banbury compounding operation drippings from the Banbury mixer and the mill, as well as stock from aborted runs are potential sources of NP. In dry blending, agglomerates of the powder can occur. These are removed by a screen and are usually discarded as NP.

Extrusion compounding leads to scrap plastic that is in the form of large chunks or strands. This scrap is generated by the purging operation which is required to clear the extruder, as one changes from one product line to another (transition material). Faulty operation of the extruder (overheating, etc) or the inadvertent contamination of the feed, all lead to the generation of scrap or nuisance plastic.

Whenever possible, the scrap is reused in the compounding equipment, or it is sold to a reprocessor. NP generation varies with the type of operation. A company involved in custom compounding producing a wide range of products, particularly colored ones, can generate 1 to 2% NP, whereas a resin producer who does compounding in line with his polymerization equipment will generate as little as 0.2% NP.

Injection Molding

Of the many fabricating processes for plastic compounds, the best known and most widely used is probably injection molding. It is used primarily for thermoplastic compounds, and polyolefins and polystyrene in particular. The current trend, however, is toward injection molding of thermosetting compounds and thermoplastics containing fillers. Injection molding is basically a process of heating a plastic material into a melt in a heating chamber and forcing this melt through a nozzle under high pressure into a closed mold which abuts firmly against the nozzle. The melt feeds into the mold cavity through small diameter runners and gates, and in the mold the melt is cooled and solidified into the shape of the cavity.

Originally injection machines consisted of a simple heating cylinder with a plunger for moving the molten mass. However, in recent years, the ram or plunger of the injection machine has been replaced by a screw, which is more efficient in plasticating the resinous materials. The reciprocating motion of the screw duplicates the function of the plunger (Figure 57). A few designs include combinations of these two principles; the most popular are the screw plunger and the plunger-plunger types of machines. However, over 75% of all machines currently sold are basic single stage reciprocating screw machines, and the percentage is increasing.

While manufacturers still sell single stage plunger machines, these are used infrequently, because they do not provide good mixing. Improper mixing leads to problems of temperature control, poor pigment dispersion, and degradation of the resin. However, plunger machines (or some variation thereof) could perhaps be attractive for molding products with recycled plastics to yield mottled or marbleized color patterns. Here poor mixing is an asset.

FIGURE 57: RECIPROCATING SCREW INJECTION MOLDING MACHINE

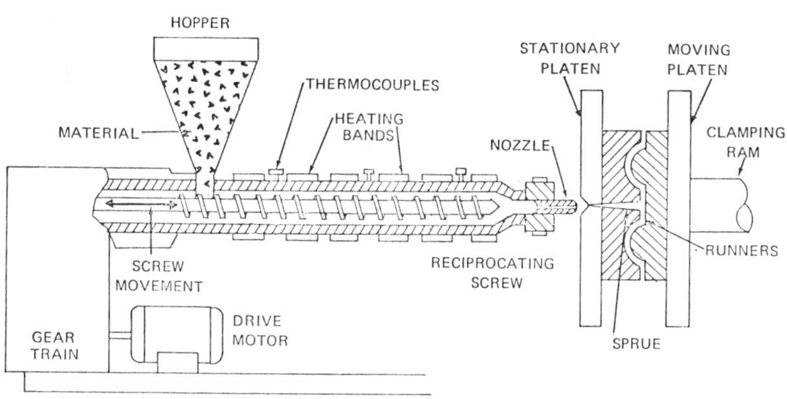

In a reciprocating screw injection molding machine plastic material is melted by heat and friction between the rotating screw and the barrel.

Source: J. Milgrom (16)

The injection molding process is very versatile, and the technology is well developed. It produces three dimensional parts to close tolerances. But most significantly, the process is capable of extremely high outputs per hour and is usually most economical when production volume is high. Molding machines will accept virtually all injection molds, within the proper range of sizes, of course.

Cooling is a critical part of the injection molding cycle. It normally accounts for 75% or more of the total cycle time. Simple products, such as thin wall, throw away drinking cups are molded in cycle times of about 1 to 1½ sec with ½ to ¾ sec for cooling. Most parts required by industry are thicker and thus require considerably longer cycle time, from 10 to 20 sec.

Injection molding of thermosetting materials is more difficult than molding of thermoplastics. Under heat, thermosetting plastics, as mentioned earlier, first soften and then harden to an infusible state. Thus, it is essential that no softened thermosetting material be allowed to remain in the heating chamber long enough to set. Reciprocating screw type machines overcome this problem by liquefying the thermosetting plastic material just as it goes through the injection nozzle into the mold, but not before. Thermoset injection molding in the United States is used primarily for small parts, while molding machines for thermoplastics are constantly getting larger. The largest tonnage thermoplastic press delivered to date is a 5,000 ton unit.

The major disadvantage of injection molding is that it has a relatively high mold cost and, therefore, requires large production volume to warrant this type of expenditure. Another disadvantage is that undercuts are difficult or impossible to mold without having complex moving parts in the mold. Because all molten material must pass through the wall areas formed by the mold, thin walled structures are also difficult to mold. However, other processes, like blow molding and thermoforming, discussed below, can produce products with much thinner wall construction.

Injection molding is used for a wide range of end products utilizing a variety of resins, colors, shapes and special effects. Because it makes products to close tolerances in all three dimensions, it can produce highly engineered products. On the other hand, it also can produce low cost consumer disposable products. Wall thickness of parts is generally under 0.100", although parts with thicknesses up to several inches are injection molded. Typical injection molded thermoplastic products include: boxes, tubs, caps and closures, disposable cups, dishes and eating utensils, appliance housings, telephone sets, houseware items, refrigerator parts, automotive steering wheels, knobs, lenses, electrical components, sporting goods, toys and novelties.

Generally, the condition of raw materials used in the injection molding process has to be carefully controlled. Uniformity of granule size of the resin is important because nonuniformity leads to feed problems, pressure loss (in plunger type machines only) and temperature control problems. Air entrapment causes streaking and bubbles in the molded article. Furthermore, the ability of the injection molding process to tolerate long fibers or large filler particles is limited. Moisture can be a serious problem. Under the high temperature and pressure of the molding process, water vaporizes and causes bubbles.

Compounds having low melt viscosity (free-flowing) are required in order to permit flow in the passages of the molds. These are produced by using low molecular weight resins plus lubricants and processing aides. Most injection molding compounds range in melt index from 5 to 50 (as contrasted to a range of 0.1 to 2.10 for blow molding).

Melt index consistency is critical for injection molding. For example, a lot having a melt index of 4 should range only from 4.0 to 4.5. The process has the capability of using certain mixtures of resins and some mixing is practiced. The polyolefins LDPE, HDPE, ethylene viny acetate, and polypropylene are currently mixed to some extent. However, injection molding also has a tendency to produce single direction orientation which can produce laminar structures in mixed compounds.

In scrap plastic from injection molding, sprues and runners from the molded articles and scrap from rejects are usually reground and reprocessed in amounts ranging from 1 to 15%, but in many cases as high as 5% is used. Some products have been injection molded using 100% regrind; thus, a low percentage of nuisance plastics is generated. In most operations the average is less than 1% NP. These are in the form of dirty grindings and floor sweepings, large chunks (50 to 200 lb) from purging, and contaminated moldings.

If a variety of small runs are made with different colors, regrind is more difficult to use and NP generation can increase from 2 to 4%. Reject material from this operation varies from company to company. It can vary from 1 to 3%. Depending upon the particular operation, rejects can be recycled either in the fabricator's facilities or sent to a reprocessor. Only if the molded product is badly contaminated does it end up in the disposal area.

The tolerance of the injection molding process for scrap plastic as currently practiced is low. Large specks of unmelted or decomposed material can clog the injection nozzle, runners, gates, or narrow areas in the mold cavity. No filters are normally used in the process. However, screen packs (such as used in extrusion) could be designed into the process. Processing of contaminated scrap could be carried out with one or more automatic screen changers that could be built into an injection molding machine or into the systems for refining the scrap plastics.

On the other hand, the high pressures, compacting action and the melt state of the resin during the process do offer the possibility of utilizing resin contaminated with particulates and foreign resin. Thus, with modifications, the injection molding process potentially offers a good method of converting scrap plastics into usable products.

Film and Sheet Extrusion

Most film is made by the blown film process, which consists of extruding a tube of molten resin through a die with a thin annular (circular) opening (Figure 58). The plastic tube is expanded upon emerging from the die by means of internal air pressure, which is introduced through the center of the tube die. The bubble thus formed is cooled by an external air blast or by an internal cooling cylinder. The resulting film is then taken up onto rolls on a spindle type winding machine.

If desired, auxiliary converting operations can be carried out prior to the winding station, for example, splitting, folding, printing, perforating, or forming into bags on continuous webs. Bags may also be individually packaged or spindled instead of being wound up in web form. Many of these operations are usually performed as separate conversion operations by outside companies, but sophisticated machines are rapidly incorporating these functions directly into the blown film extrusion line. Film thicknesses between 0.7 and 2 mils are most frequently used, and speeds of 100 to 300 fpm are common for this type production.

A small amount (10 to 20%) of film is produced by the slit die extrusion method. This product is frequently referred to as cast film to differentiate it from blown film. The slit die extrusion process involves extruding molten resin from an extruder through a wide slit die with adjustable lips into a cooling system, such as a water bath, or onto chilled metal rolls and then onto a windup stand (Figure 59). Rapid chilling gives better film gloss and clarity.

The physical properties of cast film are also different from blown film; thickness variations are less and cast film has a low degree of orientation as compared to blown film. Sheet plastics are also manufactured by slit die extrusion. The term sheet is applied to thicknesses greater than 0.010". For some applications varying degrees of orientation are desired. This is accomplished by stretching the plastic film under accurately controlled conditions to create molecular orientation and then locking this orientation into the film by fast cooling. This locked in energy can be released when the film is heated; the film then pulls back toward its original unstretched condition.

FIGURE 58: BLOWN FILM PROCESSES

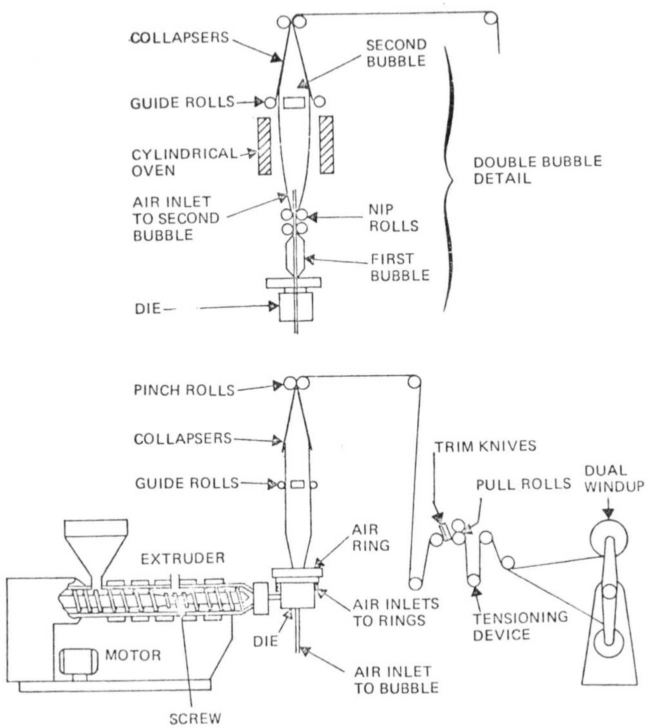

Blown plastic film is produced by extruding a circular profile (or tube) of molten plastic and applying air inside the tube to cause expansion of the molten plastic. After cooling, the plastic is wound into rolls as tubing or by slitting and trimmings, as single thickness film.

Source: J. Milgrom (16)

This is the common shrink film of commerce. Another type of oriented film is the heat set film, which is obtained by holding the stretched film at a high temperature just below its melting point (heat set temperature). Such films have improved properties but are not normally shrink films, although they will shrink if reheated above the heat set temperature. Polyester films (Mylar) are good examples of this type.

Cast, blown, or extruded film is oriented by passing it over heated differential speed rolls where it is stretched in the machine direction. The advanced roll, traveling faster than the feed roll, causes the film to thin down without substantially reducing its width. If the process is stopped at this point, the product is monoaxially oriented. To obtain biaxially oriented film, subsequent stretching in the transverse or cross machine direction is accomplished on equipment known as a tenter (stretching) frame, which consists of two endless chains equipped with gripper clamps and mounted on a Y frame and enclosed in an oven. Heat setting is usually accomplished at the exit end of the tenter frame.

FIGURE 59: FILM AND SHEET CHILL CASTING

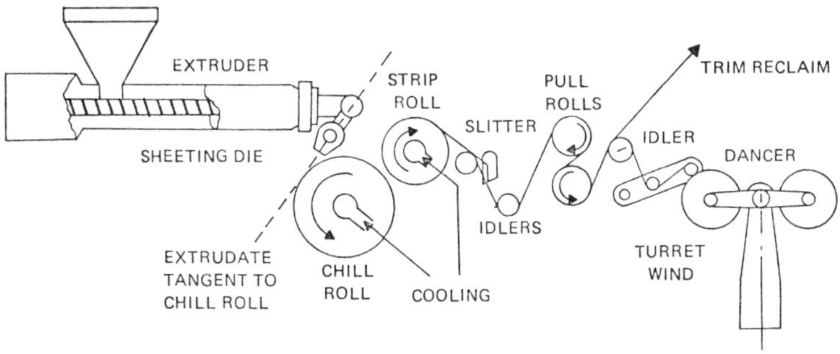

Plastic film and sheet are produced by extruding molten plastic through a slit die and chilling it on rollers prior to final wind up.

Source: J. Milgrom (16)

Monoaxial orientation of blown film is obtained on conventional equipment by using heated differential speed rolls prior to wind up. Most blown biaxial film is produced by the double bubble process (Figure 58). In this process, the film is blown into a small tube (or bubble), cooled quickly, and passed through pulling rolls known as nips. After passing through the nips, the tube is reheated to a temperature below its melting point and further expanded by internal pressure, while at the same time the longitudinal speed of the film is increased by another pair of pulling rolls.

Blown polyethylene film is used extensively in packaging and for construction and agriculture applications. This film is used to make garment covers, household wrapping, and storage bags, etc. In construction, it is used as a wind breaker and to waterproof underground concrete structures. In agriculture, it is used for mulching, hot house glazing, storage covers, and in numerous waterproofing applications.

Blown PVC film is used extensively as a meat and produce wrap, mostly in the unoriented form, but some monoaxially oriented PVC is also used. Clear nonshrink biaxially oriented polystyrene and polypropylene films are similarly used for many packaging applications. A typical high quality biaxially oriented film is the familiar Saran Wrap, which is a clear polyvinylidene chloride (PVDC) film.

Shrink wrap plastic packaging, which is produced by the conventional blown film methods described above, is a rapidly growing segment of overwrap packaging. Polyethylene, PVC, polypropylene and polyvinylidene chloride (Saran) are the most common films used for shrink wrapping. Other end uses include industrial palletizing, meat and produce wrap, and wrap for toys, games, records, newspapers and magazines, frozen poultry, six pack beer cans, and soft drinks, and numerous multipacks such as cosmetic sets and premium items.

The major plastics used in sheet extrusion are high impact polystyrene and ABS. Some polyethylene and PVC are also extruded into sheet. Polyethylene is mostly thermoformed into food containers and tote boxes, and clear PVC sheet is used for blister packaging. Other important uses for extruded sheet include signs, displays, refrigerator doors and panels, duct work, appliances, housing, snowmobile bodies, and some automotive parts. All film processes require high quality compounds with a low melt index (stiff flow).

Higher melt indexes are preferred for sheet extrusions. Any foreign matter in the compound, including high molecular weight resin particles or gel, can cause a bubble to break in the blown film process and a void or depression to appear in the slit die extrusion process. Furthermore, the particle size of all additives mentioned are required for proper processing, and special additives are required to meet FDA food grade requirements for film used in contact with food products. Printing requirements on packaging film present additional raw material problems. For example, certain antioxidant and other additives have been known to inhibit the rate of ink drying.

As regards scrap plastic from film and sheet extrusion, film fabrication generates a large amount of scrap plastic in the form of start up, tails, and reject film together with extruder purgings. In the production of polyethylene film this amounts, on the average, to 15% of production. Most is recycled by the fabricators who use as high as 25 to 30% regrind in standard film and up to 100% in applications such as construction and agricultural film. NP generation from polyethylene operation amounts to no more than 1 to 2%, mostly in the form of floor sweepings and badly contaminated rejects.

The fabrication of PVC film generates about 25% scrap. The larger PVC fabricators utilize most of their own scrap as regrind, and smaller fabricators usually sell the scrap to reprocessors. NP generation in PVC film fabrication averages 1.5% of production. In the manufacture of polypropylene film, which is mostly oriented, as much as 40% of the production becomes scrap. The poor quality scrap is often baled as film and sold to reprocessors, and the high quality scrap (about 25% of production) is repelletized for in-house recycling. The critical requirements of the orienting process limit the incorporation of recycled material to about 15% of the total. Thus, some of the pelletized scrap is also sold to reprocessors.

Handling reprocessed film scrap is difficult. Normally it is chopped into a confetti or fluff. This material can then be fed directly into the extruder using a special auger feeder, or it can be repelletized before mixing it with virgin resin. Generation of scrap plastic in the fabrication of sheet can run as high as 40% of production. But most of this scrap is easily recycled, and NP amounts to about 1% of total.

Film applications generally can not tolerate the melt viscosity variation and foreign contamination of most scrap from sources outside of the manufacturer's own shop. On the other hand, heavy agricultural and construction film can and does tolerate considerable amounts of scrap plastic. For NP to be utilized in these areas, however, the feed material has to be well mixed, uncontaminated with other resins, and subjected to heavy screening during reprocessing to eliminate particulate contamination (see section on Reprocessing later in this volume).

Sheet materials, being thicker than film, can tolerate more NP. One possibility for using NP would be to coextrude a thick core material of NP with thin coatings of virgin material on each side. However, such sheet would be opaque and could probably not qualify for FDA approval.

Extrusion Coating and Coating of Wire and Cable

The mechanical arrangement of slit die extrusion equipment readily permits coating to be carried out in line with extrusion. This can be accomplished by passing the substrate material over the top roll of a cooling stack and into the nip of the rollers along with molten resin from an extrusion die (Figure 60). A wide range of materials such as textiles, paper, nonwoven webs, plywood, and sheet metal are coated with plastic in this manner.

The operation is referred to as extrusion coating or laminating. Thick coatings are most often referred to as laminates, and thin ones as coatings. In the coating of wire and cable, the thermoplastic is extruded around a continuous length of wire or cable which passes through a cross head extrusion die. In a cross head die the material to be coated enters the die perpendicular to the initial direction of flow of the plastic. The coated wire is then wound on spools or drums after cooling.

FIGURE 60: EXTRUSION COATING

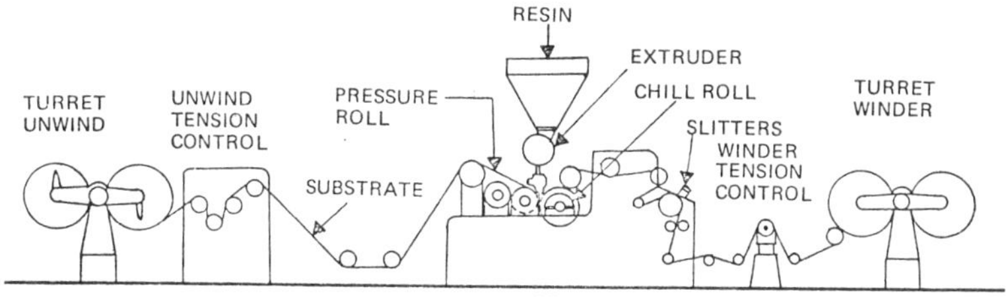

A thin film of plastic is extruded from a slit die directly onto a substrate such as paper, cloth or metal foil followed by conventional wind up on rolls

Source: J. Milgrom (16)

The most widely used plastic extrusion coated product is the familiar polyethylene coated paperboard used to manufacture the popular Pure Pak paper milk carton. Other related products are washable plastic wallpaper that has a clear coating of PVC extruded over the printed paper and some automotive and furniture fabrics which are produced by a one side extrusion of PVC onto a fabric substrate. Waterproof tarpaulin and other fabrics are produced by the two side coating of fabrics. Many packaging films are also produced by extrusion coating of polyethylene or polyvinylidene chloride (Saran) onto such substrates as PVC, cellophane, acetate, polyester, aluminum foil, and paper.

Wire and cable uses range from thin coatings of polyethylene on military field wire to multi-extrusions of several plastics in the manufacture of large diameter underground and high voltage cables. Most extrusion coatings require raw material specifications similar to those used for film and sheet extrusion. Because all coatings are thin, melt viscosity specifications are very critical. A small reject area in an extrusion coated product can cause the rejection of large rolls of the product, because commercial specifications usually limit the number of splices or cutouts permitted per roll.

Low density polyethylene is the major raw material used for extruded coatings on web substrates. Specific LDPE compounds with narrow specifications of melt index, density, and molecular weight distribution are required to accentuate specific properties such as clarity, heat sealability, and the need for FDA approval. Other resins utilized for web coating include PVC, polypropylene, copolymers of polyethylene, and nylon. PVDC is usually emulsion coated, not extrusion coated.

Wire and cable plastic coatings require high electrical performance characteristics. Compound formulas are very exact, and little regrind is used even within individual shops having equipment suitable to handle it. Polyethylene and PVC are used most widely for wire and cable coatings; PVC dominates the market.

In scrap plastic generation from extrusion coating, wire and cable coating regrind often ends up as NP. Extensive efforts are currently being made by this industry to recycle metal (mostly copper) scrap. In this process, the reject coated wire is finely chopped, and the copper is separated from the plastic in an air stream, taking advantage of the difference in specific gravity between the metal and the plastic. The plastic portion, amounting to 5 to 6% of production, is disposed of as NP. In extrusion coating on paperboard, NP levels average about 6%.

Due to the thin films used in extrusion coatings, it is doubtful that any NP could be utilized as a coating feed stock. Many wire and cable applications are unable at this time to utilize scrap because of the technical difficulties, however, certain applications, such as TV lead wire, etc., can use scrap if it is free of metal contamination. Other applications are limited because the government has imposed restrictive specifications excluding secondary material.

Pipe, Rod, Tubing and Special Profile Extrusion

These articles are formed as continuous extrusions through an annular die approximating the desired profile. Cooling is usually effected by a water bath or trough. Sometimes an elaborate operation is required to shape the rough plastic from the die into the desired dimensions of the finished product.

Pipe applications range from low specification drain, waste, and vent (DWV) pipe having no internal pressure requirements to those used for potable water in pressure systems. The resins used in pipe applications are PVC, styrene-rubber, polyethylene, and ABS, in the order of their poundage consumption. Rod plastic, after being machined and cut, is used as parts of complex products ranging from simple housewares to complex appliances. Rods are available from plastic supply houses in sizes ranging from ⅛" up to 8" in diameter.

Plastic tubing is available in sizes ranging from small diameter dip tubes for aerosols and plastic bottles to garden hose. There is no distinct line of demarcation between pipe and tubing. A circular extrusion having one or more of the following properties: small outside diameter, thin wall, flexible material, and no inside pressure is usually called tubing. Large diameter, rigid wall, circular extrusions for electrical conduit are also often referred to as tubing, yet they may be very similar in construction to the DWV pipe described above. Typical special profiles include house siding, automotive and building trim, lighting fixtures, window channels and ducts for electrical wiring.

The scrap generated during the extrusion phase of the manufacture of pipe, rod, tubing, and special shapes is similar to that from other extrusion processes. The amount of scrap that is reprocessed varies with the product, and color can be a problem. Most of the end uses for these products, however, require fabrication away from the site of extrusion. Because the volume is too small at any one location to warrant return to the extruder, the scrap from fabrication becomes NP. This can amount to 2 to 3% for most pipe end uses, to as high as 40 to 50% for parts machined from rod stock (including fabrication and converting).

For noncritical applications, pipe, rod and tubing offer good potential use areas for NP. The thick walls and typical dark colors marketed are complimentary to the processing of NP. Where a good exterior appearance is desired, coextrusion of an inside layer of NP combined with a virgin exterior is very practical. Currently, much PVC garden hose is coextruded using scrap plastic as an inner core. Special profiles used in many applications are dark colored and have heavy sections; this process area also can use NP. Again, the use of NP as an inner core is appropriate with profiles such as siding, etc.

Coextrusion

Profiles of all types are produced by coextrusion. As the term implies, the process consists of mixing two or more melt streams from separate extruders through one extrusion die. The melt streams can be different resins, different colors of the same resin, or different grades of the same resin.

The technology of this process is very sophisticated, particularly in the design of the die and the use of different resins. In the coextrusion process the various melt flow conditions (temperature, pressure, speed) from the independent extruders must be carefully matched. Coextrusion is used to make film, sheet, foam, tubing, pipes and coatings. Coextrusion of films is finding increased use in packaging applications.

Their main attraction is the fact that they are manufactured in one step to produce a single web having the advantageous properties of two or more plastics, without the cost of secondary operation of coating or laminating. For example, polyethylene/polypropylene/polyethylene coextruded films are widely used in the packaging of snack foods, because the inner layer of polypropylene offers a low water vapor transmission rate and the outer layers of polyethylene have superior heat sealability.

Another important coextruded film is polyethylene/polyvinylidene chloride/polyethylene. This is used in large quantity alone or as part of a foil laminate for both gas and vacuum packaging of cheese and soluble coffee. The need for very low oxygen transmission in these applications requires polyvinylidene chloride. Coextrusion is used in applications other than film. For example, white high density polyethylene and black polypropylene are currently coextruded into a heavy sheet and then thermoformed into containers for motor oil.

Most coextrusion processes are unable to reuse their own scrap, because the scrap is a mixture. Scrap, which can range as high as 20% or more in individual processes, ends up as NP (9 to 10% is an average value). Much of this is concentrated in convertors' plants. An important potential use of NP in this process would be to use it as a core material in a coextrusion with virgin plastic (or high grade scrap), where the latter material is used as the surface material. For example, three ply coextrusion processes could completely encapsulate NP in sheet, pipe, rod, tubing, and special profiles, except for the cut off areas.

Blow Molding

There are two major types of blow molding in commercial use in the United States: extrusion blow molding and injection blow molding. Extrusion blow molding currently predominates. In extrusion blow molding (Figure 61) a hot plastic tube is extruded, usually vertically, between two halves of a female mold. The closing of the two mold halves completely pinches off the tube (or parison, as it is called) at the bottom and partially pinches off the top around the neck area of the bottle.

Air is then blown through the inside nozzle of the extrusion die or a moving plunger through the neck of the bottle expanding the tube into the shape of the mold. Air also can be introduced into the tube via a needle which pierces the plastic tube in an area that is subsequently cut off in the finishing operation. The molds consist of water cooled cavities, the periphery of which is either completely or partially sharpened to a knife edge. Mold halves are mounted on platens which are part of a clamping mechanism. When the mold clamps around the hot tube (called a parison), the knife edges of the mold seal the tube wherever they touch.

After an appropriate cooling time, the mold mechanism opens and the product is removed. The portion of the original tube that remained outside of the knife edged pinch off is squeezed together and is still attached to the body of the blown part, however, the plastic at the knife edge pinch off is thin enough to permit the excess material, called flash to be pulled off by hand or removed in a die stamping machine. Some blow molding machines have been developed to separate the flash within the mold before or during the opening process.

A cold tube, blow molding process has been made available from Europe. In this process, polyvinyl chloride tubes of a predetermined diameter and length are produced by a standard extruder. The tubing is shipped to the blow molding location in cut form, and the tubes are automatically fed into a blow molding machine.

They are dropped onto spindles mounted on a belt, which carries the tubes through a heating oven. When properly heated, the hot tubing and spindles enter a molding station where the tube is pinched off and blown into a finished bottle. The process as practiced has many limitations, but new systems involving the cold tube principle are receiving serious attention.

FIGURE 61: EXTRUSION BLOW MOLDING

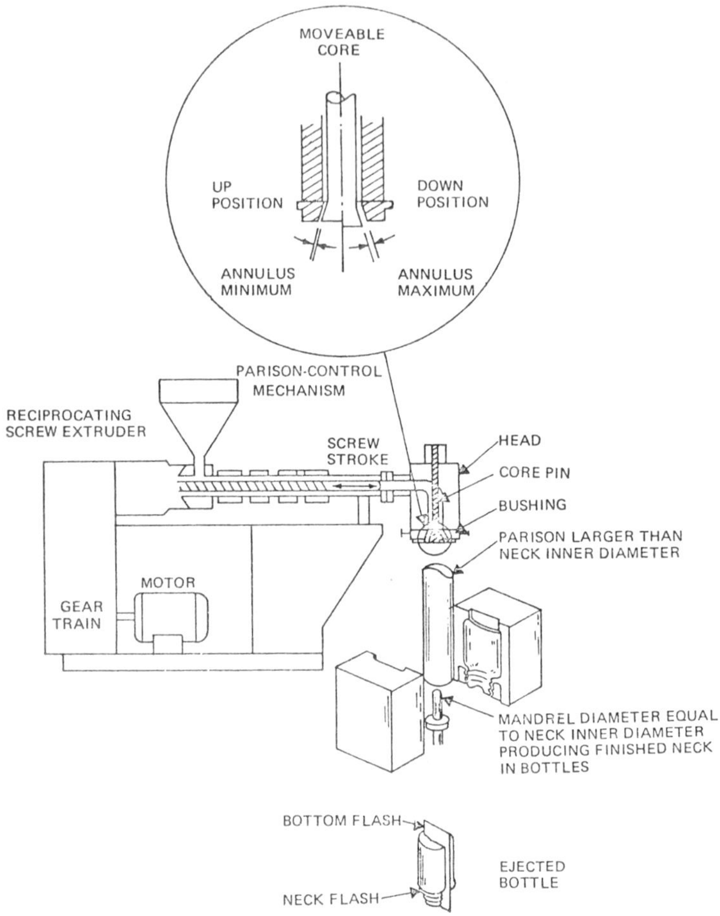

In extrusion blow molding a hot plastic tube is extruded between two halves of a female mold. The closing of the two mold halves pinches the parison at the bottom and partially pinches off the top. Air is blown through a nozzle or plunger inserted into the neck of the bottle expanding the plastic tube into the shape of the mold.

Source: J. Milgrom (16)

One important feature of this variation of the blow molding process is that it can produce biaxially oriented plastic bottles. The improved physical properties offered by orientation reduce the material requirements of the bottle by 20 to 35%, which offers the possibility of reducing the weight of plastic solid wastes. In the injection blow molding process the parison is molded in a conventional injection mold (Figure 62).

FIGURE 62: INJECTION BLOW MOLDING

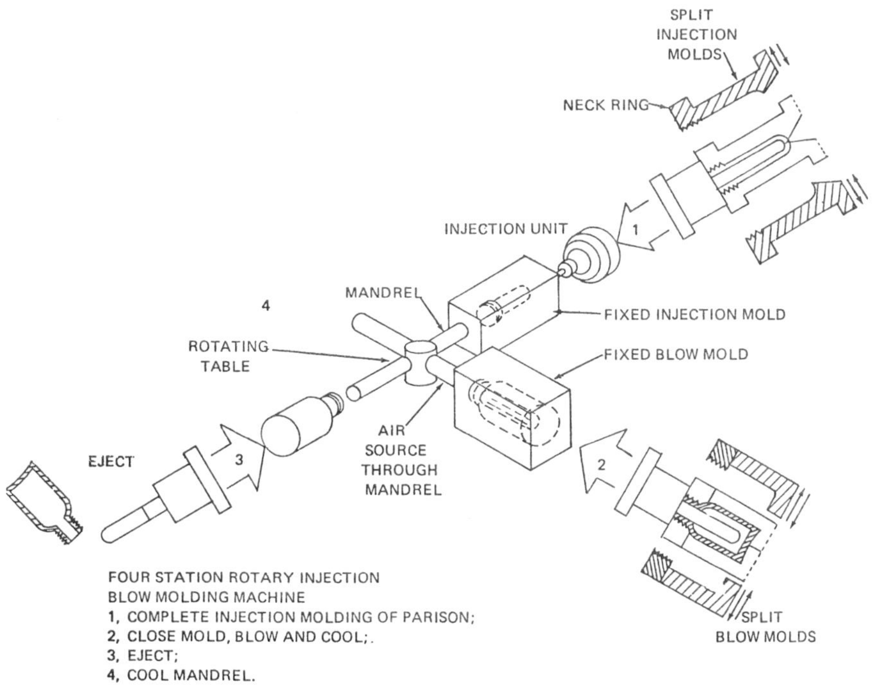

In this process the parison is molded in a conventional injection mold at station **1**. The test tube shaped parison is retained on the plunger of the injection mold and transferred into a blow mold at station **2**, where the parison is blown to the final bottle shape and cooled. Final cooling and ejection takes place at station **3**. Station **4** is used to cool the parison mandrels before the beginning of the next cycle.

Source: J. Milgrom (16)

The test tube shaped parison is retained on the plunger or core rod of the injection mold and transferred into a blow mold which has the bottle shape. The injection molded parison, still hot from the original molding, is blown into the final shape of the die by allowing air to enter the parison through the core rod. The Cheeseborough Ponds Vaseline jar is made this way.

The injection blow molding process, in contrast to extrusion blow molding, is basically a no scrap process. The entire mold charge is turned into a finished part during each cycle. Furthermore, orientation is also possible in the injection blow molding process. Both extrusion and injection blow molding processes permit the manufacture of thin (even paper thin) and medium thick wall constructions, but for very heavy walls (above ⅛") blow molding is somewhat impractical. One of the limitations of blow molding is that this process provides limited control over inside dimensions. Lack of precise dimensional control is not detrimental in bottle construction, but it does introduce technical limitations in the fabrication of some industrial products.

Composite bottles can be produced by all blow molding processes. These techniques are still in the development stages but offer improvements in product retention in packaging applications, for example, an inner wall of PVDC can be used for an oxygen barrier and an outlet wall of HDPE for strength.

The principal use of the blow molding process is for producing hollow shaped objects with one or more reentrant angles, e.g., the shoulder of a bottle shaped object. Some hollow objects (such as ordinary bowls and drinking cups) do not lend themselves readily to the blow molding process and are produced by injection molding. A bottle shape, on the other hand, could not be produced by injection molding because the steel plunger could not be removed around the reentrant angle of the shoulder.

In comparing injection molding with blow molding, another very significant technical difference exists. Blow molding is accomplished at pressures of from 25 to 100 psi between the plastic and the mold surface. Injection molding requires pressures of from 10,000 to 20,000 psi to push the molten plastic through long, narrow passages. The reduced pressure of blow molding results in lower internal stresses in blow molded items.

Approximately 80% of all blow molding is used to manufacture packaging items. Most of these are bottles or bottle like containers for the household. Some are industrial containers such as drums, carboys, and tanks. The household containers include bottles for cosmetics, toiletries, pharmaceuticals, chemicals, and foods such as milk.

The remaining 20% of blow molding is used to manufacture industrial products such as gas tanks, windshield wiper reservoirs, air ducts, visors for automotive use, floor waxer reservoirs, vacuum cleaner hose, hair dryer housings, lighting fixture globes, toilet floats, tool boxes, furniture seats and backs, Christmas tree ornaments, and numerous toys.

Practically any thermoplastic material can be blow molded: LDPE, HDPE, and rigid PVC are used in quantity in the blow molding of containers, toys, household, and industrial products. Lesser amounts of polystyrene are used. Blow molding today is expanding at a rate faster than that of the plastics industry as a whole.

Of all plastic materials, HDPE has been the workhorse of the blow molding industry. Extrusion blow molding requires a resin with stiff flow and a low melt index (0.2 to 2.0) and high molecular weight. The resulting parison must have a sufficiently high melt strength so that as it is dropped from the extrusion head it holds its shape prior to and during blowing. Blow molded items have the advantage of utilizing the low permeability, and high oxidation and ultraviolet resistance, etc., of higher molecular weight resins, as previously mentioned.

Melt viscosity control is very critical. Batch-to-batch and bag-to-bag variations can completely shut down a production line. Color control also is very critical. Lot-to-lot color consistency is required because in consumer packaging various lots are mixed on store shelves.

Blow molding permits a limited use of mixtures of resins. Some mixing of the polyolefins can be tolerated, but none of the other resins can be used as a mix. Incompatible resins (even different melt index material of the same resin) and degraded material and foreign particulate matter will cause streaks of varying thickness in the parison tube during extrusion of the tube.

Control of PVC blow molding is even more critical than polyethylene; specially designed screws and increased horsepower are required. PVC blow molding compounds contain more additives than HDPE compounds. This substantially increases the risk of migration of additives with subsequent subtle changes in flavor or odor (irrespective of the question of toxicity). Normally, PVC must be processed in the blow molding operation at temperatures in the range of 390° to 410°F, which requires the use of high temperature stabilizers. The extrusion blow molding process generates a large amount of scrap plastics which is proportional to the amount of pinch-off on a given part. The amount of SP depends upon the amount of excess width or length of parison that a particular molder might use. These two factors can be minimized by good engineering design. Products with a large percentage of pinch-off (e.g., outdoor seats and backs, watering cans, handled bleach bottles) generate 25 to 40% SP. Much of this SP is recyclable by regrinding. Regrind, however, can introduce contamination into the plastic part, and the molding properties of regrind change with every pass through the extruder.

The percent of regrind in an end item must be carefully controlled, otherwise the product specification changes and, at worst, product control can be lost. In the latter case, it is necessary to purge the blow molding machine and start with new batches of material to obtain proper plastic melt balance to bring the process under control. Additional scrap nuisance plastics are created when control is lost. Generally, nuisance plastic generation is low, about 1 to 1.5% with the polyolefins, but as much as 5% with PVC. This NP consists of badly contaminated regrind, floor sweepings and lumps from purging.

Because of the need for melt extensibility during the blowing phase, blow molding does not offer an attractive area for the use of highly mixed resins or NP. Any contamination, particulate matter, moisture, volatile hydrocarbons, or a variation in resin molecular weight, can cause fracture of the parison during the blow cycle. A fractured parison produces a reject and repeated rejects create a loss of production control. Thus, only very closely controlled scrap plastic would be suitable for this process.

However, if a sufficient volume of clean recycled materials (such as milk bottles) could be separated properly and cleaned of particulate material at a recycle refinery by multiple extrusion methods utilizing intensive filters (such as automatic screen changers), such material could be recycled in the blow molding process. Maintenance of color under these conditions could present a technical problem, but this might be solved by the development of products with low color requirements.

The high temperature of extrusion should destroy bacteria and volatile hydrocarbons, and the screen pack should eliminate large size particulate matter. Some small sized particulate matter should be sufficiently encapsulated within the polyethylene to be nonextractable, therefore, primary recycling can become a reality in certain instances.

Rotational Molding

Rotational molding or rotomolding consists of tumbling or sloshing powdered or liquid formulas of plastic compounds in hollow molds through a heating and cooling cycle. Rotation is usually through 360° on two axes (Figure 63). Some processes utilize less than 360° on one axis by tilting or rocking the mold to obtain an even distribution of the compound. There is a common misconception that rotational molding utilizes centrifugal force to distribute the material throughout the mold but, in fact, the speeds of rotation must be maintained below the centrifugal point, otherwise the powdered resin goes to one spot and fuses in a lump.

The process usually consists of three steps: a measured amount of compound is placed in a mold having a temperature below the softening point of the polymer; the mold containing the compound is inserted in an oven where the material softens and builds up on the walls while the mold is rotated; and the mold is subjected to sudden cooling.

The thickness of the wall is determined by the amount of compound charged into the mold, and the inside of the part consists of a smooth surface roughly paralleling the contours of the mold. The walls are generally of uniform thickness, but thickness can be varied somewhat by either insulating sections of the mold to form this wall or heating sections to increase thickness. The process normally produces details only on the mold side of the finished product. Textured, matte, or highly glossy finishes can be obtained. Three major types of heating systems are used for heating rotational molds: air convection, salt spray, and circulating oil.

FIGURE 63: ROTATIONAL MOLDING

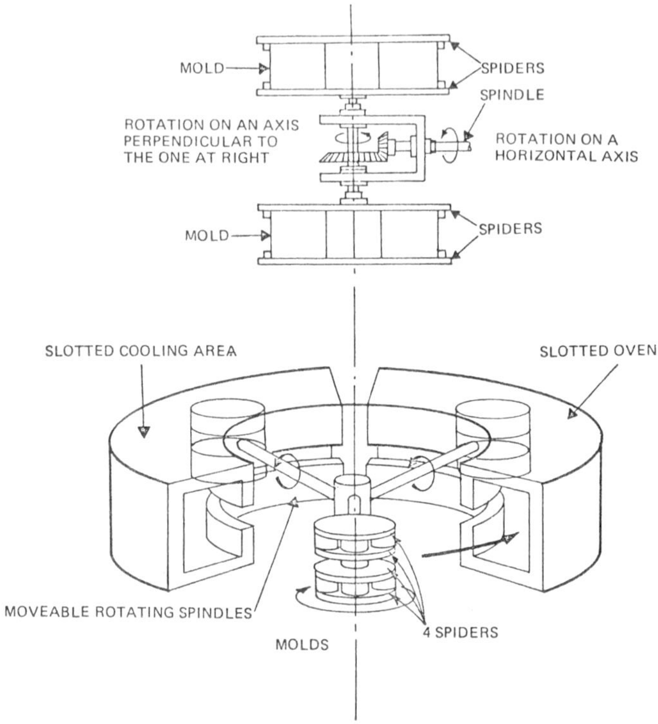

Rotational molding consists of three steps: a measured amount of compound is placed in a mold having a temperature below the softening point of the polymer; the mold containing the compound is inserted in an oven where the material softens and builds up on the walls while the mold is rotated; and the mold is subjected to sudden cooling.

Source: J. Milgrom (16)

Heated air circulated in ovens by air blowers is the most widely used heat source. With this method one has to avoid overheating the mold, otherwise the plastic product is degraded. In salt spray heating, molten heat transfer salts are pumped through spray nozzles, and the melt is sprayed over the mold throughout its rotational cycle. The rotational mold must be liquid tight at operating temperatures to prevent salt entering the mold and spoiling the finished part.

Circulating oil is the newest process. This process eliminates the use of complex and expensive ovens and cooling stations and is used mostly for single cavity molds of large parts. It has the disadvantage that intricate parts with projections, ledges, protrusions, etc., are difficult to design for proper oil circulation. The major advantages of rotomolding are: low mold cost and the formulation of strain-free parts with controlled wall thickness. Rotational molding is particularly suited for the production of very large parts, size being limited only by the capacity of the equipment. Rotational molding is used for all sizes of products but is outstanding for the production of medium to very large size thermoplastic items. The process affords a high degree of flexibility; products varying from relatively simple cylindrical and rectangular shapes to highly irregular shapes incorporating compound curvatures can be produced.

Rotomolding is particularly suitable for the manufacture of hollow bodies or items having reentrant angles or undercuts and wall thicknesses greater than 0.030". The process has an advantage over blow molding in that a uniform thickness can be produced on a very irregular shape without the detrimental thinning down of the plastic melt that is characteristic of blow molding. Blow molding can require as much as 25 to 35% more resin for a specific end item than rotational molding.

Rotomolding is economically attractive for: short runs, frequent color changes, thick walls, very large parts, and intricate designs. Typical rotomolded products (usually HDPE and LDPE) include tote bins, tanks for chemicals with a capacity as high as 1,600 gal, bins for storing food, machine guards, point-of-purchase displays, light globes, water and waste tanks, flower pots, hobby horses, apparel display forms, trash cans, and small gasoline tanks.

Semihollow parts, such as garbage cans, can be molded by covering the top of the mold cavity with an insulating material which reduces or eliminates the deposition of molding material in this area. Plastisol rotomolded products include doll parts, squeeze toys, playballs, artificial fruit, syringe bulbs, automobile sun visors and arm rests, lawn ornaments, duck decoys, bicycle seats, and auto battery water dispensers. These compounds can be used to produce simulated leather surfaces and flesh in toy dolls.

The first materials to be rotationally molded were vinyl plastisols. At molding temperatures the PVC and plasticizer combine to form a molten thermoplastic. It is important that no gases evolve when the change from liquid to solid occurs, because they would mar the appearance of the molded object by forming bubbles or other defects.

During the 1950s, low density polyethylene powders were converted into hollow items by rotomolding, and during the 1960s powders of high density polyethylene, polystyrene, nylon and other thermoplastic materials, including foam formulations were developed for rotational molding. Many plastic compounds are still specially ground to produce rotational molding powders; others are dry blends in their original powder form. PVC dry blending powders are used in rotomolding as well as PVC plastisols.

Compounds for rotational molding require special properties, for example, high flow characteristics (high melt index) are needed so that the material flows easily over the surfaces of the mold as the mold rotates. The preferred melt index for rotational molding HDPE is in the range of 5.0 to 10.0. Normal size pellets are too large for rotational molding. Powders with particle sizes ranging from 25 to 60 mesh are typically used, and particle distribution also can be very important.

Resin formulas for rotational molding require a narrow molecular weight distribution. The presence of both high and low molecular weight fractions causes difficulty. Low molecular weight materials melt at lower temperatures and, therefore, soften early in the cycle. High molecular weight fractions resist melting until the end of the cycle, thereby causing a grainy effect. Attempts to adjust the temperature of the oven to compensate for either the low or the high molecular weight fractions usually aggravates the problem.

The powdered compounds that are required for this process tend to pick up more moisture than pellets. Moisture absorbed or adsorbed by the powder can cause trouble in the rotational molding and overnight drying in a dehumidifying oven is required. In rotational molding, discrete amounts of compound are charged into the molds for each cycle; generally no regrind is used. Some products produce scrap plastic, for example, some is obtained from the removal of open sections and small amounts of trim flash at a mold parting line or insert.

Removed openings and product rejects are usually of good quality and can be reground for possible use in other processes. The quality of this regrind is not suitable for reuse in rotational molding due to its particle size and the presence of minor degradation and contamination. The nuisance plastic generation rate for rotational plastics is 1 to 3% of production. This consists mostly of floor sweepings, badly burned or contaminated products (such as with salt). In small rotational molding shops, the accumulation of scrap plastic is too small for economic salvage, therefore, it ends up as nuisance plastic to the extent of an additional 1%.

The highly complex nature of the rotational process, particularly its requirements in special resin properties, makes it a poor candidate for scrap plastic. One area where NP might be used, if a foamed plastic mass from NP could be developed, is as filler for empty rotationally molded products such as rocking horses, etc., to provide a product with better rigidity, improved strength, and additional weight.

Compression and Transfer Molding

For many years a clean cut distinction existed in molding terminology—injection molding processed thermoplastics, and compression and transfer molding processed thermoset material. A small amount of thermoplastics has always been compression molded, but it is seldom publicized. A major breakthrough in the techniques of injection molding of thermosets in the 1960s (discussed under Injection Molding) make the distinction very much more difficult at this time. The confusion is compounded by some of the new cold forming techniques such as forging, coining, drawing and stamping which also involve compression pressure to mold thermoplastics.

In conventional compression molding (Figure 64), the molding compound, which consists of a liquid thermoset resin mixed with such fillers as wood flour, cellulose, asbestos, or glass fiber to strengthen or give other added qualities to the finished product, is put directly into the open mold cavity either in the form of bulk material or as a preheated biscuit. The mold, which is normally heated, is then closed, pressure applied, and the compound caused to melt and flow throughout the mold. The heated mold causes the thermosetting material to undergo a chemical change which permanently hardens it into the shape of the mold.

FIGURE 64: COMPRESSION MOLDING

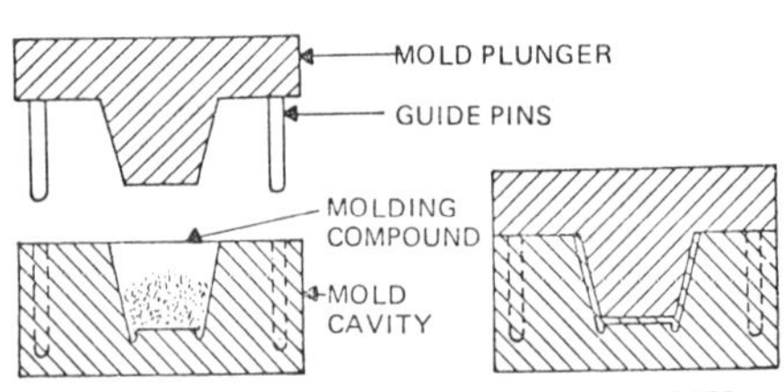

In compression molding, the resin powder is put directly into an open heated mold. Heat and pressure from the closing of the mold melt the material and cause it to fill the shape of the mold.

Source: J. Milgrom (16)

The three compression molding factors: pressure, temperature and time the mold is closed, vary with the design of the finished article and the material being molded. It is common to degass the mold (opening and reclosing the mold about ⅛" after it initially closes) to allow volatiles developed during curing to escape. Improper degassing produces rejects, because bubbles are formed that result in weak sections.

In transfer (or plunger) molding (Figure 65), the thermoset molding compound is charged initially to a separate chamber or transfer pot. When the material is fully melted, a hydraulic ram pushes it through runners and sprues to fill closed mold cavities. Transfer molding is used in preference to compression molding for intricate parts, parts with very thin walls or small deep holes, and parts with inserts that might be damaged during the conventional compression process. The disadvantages of transfer molding over compression molding include: additional equipment costs; higher scrap rates, all of which are NP; and slightly slower cycles.

Some special forms of compression molding are pulp molding and premix molding. In pulp molding, a porous form, approximately the shape of a finished article, is lowered into a tank containing a mixture of pulp, resins, and water. The mixture is then drawn through the porous form by vacuum, and when a sufficient thickness of pulp has been drawn, the pulp preform is removed and molded into final shape by conventional compression molding. In premix molding, long fiber material such as chopped fiberglass is mixed with liquid thermosetting resins.

Molded thermoset parts, as they are removed from the mold, have a degree of flash or excess material which escapes from the mold cavity during molding. Usually this cannot be recycled. This flash is removed by hand or by tumbling the molded parts. NP generation in compression and transfer molding is 2 to 5% of production.

FIGURE 65: TRANSFER MOLDING

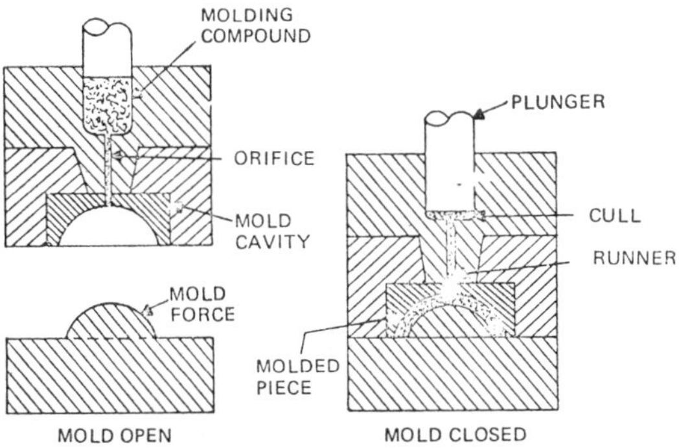

In transfer molding, a thermosetting material is softened in a separate chamber and pushed by plunger action from the hamper into a heated closed mold, while still in the molten stage. The product solidifies by chemical reaction within the mold, after which it is ejected while still hot.

Source: J. Milgrom (16)

In its present form, thermoset compression and transfer molding processes could use NP only as a filler material. However, it is conceivable that products and formulas could be developed utilizing NP thermoplastics combined with fillers and/or liquid thermosets in compression or transfer molding similar to pulp and premix molding.

Dip and Slush Molding

Slush molding is a process used to produce thin walled flexible products. Usually plastisol PVC is the raw material. In slush molding, a hollow mold of the proper contour receives an excess charge of the liquid plastisol. Heat applied to the mold changes the liquid suspension of resin and plasticizer to a solid solution on the heated mold surfaces. Because heat transfer through the suspension is poor, only the material at the mold surfaces is converted to solid. Excess material is dumped out and used again. Frequently, several coatings are required to produce the finished product. In some cases the last coat is a foamed plastisol to simulate, for example, a foam rubber or felt lining in a finished overshoe.

In the dipping process plastisols are coated onto the surface of products (particularly metal) by dipping the article into the plastisol solution and then drying it in air or in an oven. The product to be coated is usually pretreated to enhance adhesion of the plastisol. Products such as handlebar grips are produced by dipping onto mandrels treated to permit release of the plastic.

A similar plastic coating technique is that of thermofusion or the molding of plastic powders. When a steady stream of air is introduced under controlled conditions into the bottom of a column of powdered compound, the resulting mixture acts very much like a liquid. Hot items can then be dipped into this mixture, known as a fluidized bed, to obtain coatings directly from the powdered compound. Oven fusing is usually required to finish the process.

Slush moldings of PVC plastisols are used extensively to produce plastic overshoes, rubbers, bathing caps and similar products. Dip molding is used to produce skin-like products such as gloves. Both liquid plastisols and solid plastics (particularly polyethylene and nylon) are used to coat metal furniture and household items such as dish drainers, scissor-handles, etc. The specifications for plastisols for both slush molding and dip coating are similar to those used in rotational molding. Powders used for fluid bed coating are also similar to those used in rotational molding and have similar specifications.

The dip and slush molding process utilize discrete amounts of material in the operating cycle. There is no predetermined amount of cutoff requiring regrind considerations. On the other hand, in both processes with both liquids and solid compounds, the material is handled over and over before being consumed as a final product. In slush molding the liquid goes in and out of a mold. In dipping, the mandrel or product form goes in and out of a central storage area of raw material. The potential for contamination resulting in rejected parts is great in these processes. But in reject form the contaminated batches still represent a high grade of scrap plastic, and much of it is sold for use in other fabricating processes. These processes generate approximately 3% NP.

Casting

An essential difference between casting and conventional molding is that little or no pressure is used in casting. In casting, the molten plastic or the fluid monomer is poured into either open or closed molds, cured at varying temperatures, depending on the compound used, and removed.

Casting may be employed using either thermoplastic or thermosetting materials to make special shapes, rigid sheets, film and sheeting, rods and tubes. Thus, both finished and semifinished products can be made by this process. One of the best known casting techniques is monomer casting (often known as potting). In this process, liquid monomer plus a catalyst is poured into a mold and around the material to be encapsulated (potted).

The resin is then allowed to polymerize either at room temperature or in special ovens. Continuous casting of film and sheeting is also done on an endless metal belt. Monomer plus a catalyst or polymer in solution is spread to the desired thickness on the belt as it moves through an oven. When the reaction is complete, the film or sheet is stripped off the belt and cut.

An important commercial process is the casting of thermosetting polyurethane foam. In simplest form, two liquid components are mixed and stirred in a container and a measured quantity of the mixture is then poured into an open mold where foaming takes place. The reaction is complex, and control over the properties of the final product is subject to a great many pitfalls related to variations in temperature of the components, humidity of the ambient air, mixture ratio of the components, mold design parameters, and accuracy of mold filling.

The best known end products of casting are the sheets of transparent and highly colored methyl methacrylate plastic, well-known as Lucite and Plexiglas. The transparent materials are widely used as unbreakable glazing and decorative constructions, and the colored sheets are used for thermoforming outdoor signs, particularly those with inside illumination. And today, urethane castings are rapidly becoming popular as substitutes for wood carving in furniture and other conventional wood consuming products. Polyester casting resins (which do not foam) are also popular for wood substitute products and simulated marble. Potted (or encapsulated) electrical parts are widely used in the electronics and related fields. Also, many familiar hobby products such as rare coins, insects, flowers, etc., are encased in plastics for display purposes.

Methyl methacrylate monomer is the most widely used for thermoplastic applications; small amounts of styrene and nylon monomers are cast for special end uses. When thermoset materials are needed, the urethanes, epoxies, and polyesters are heavily used, not only as castings for wood substitutes but also for potting. All of these raw materials are liquid initially, and they undergo polymerization during casting.

Scrap from major thermoplastic casting operations is reground and used for special injection or compression molding products. Where special end uses have not been found, the scrap becomes NP. As expected, scrap from thermoset casting processes can not be reused. It all ends up as NP. NP amounts to 3 to 5% of the total volume of material cast. Because all casting processes involve a basic nonreversible chemical reaction, no NP can substitute for the basic monomer. Potentially, the only use area for this NP would be as a filler or perhaps a composite product yet to be developed.

Calendering

Calendering can be considered a very special type of extrusion. It is used mostly to produce PVC sheeting and film and similar web products. In calendering, the plastic compound is passed between a series of three or four large, heated, revolving rollers which squeeze the material into a sheet or film (Figure 66). The thickness of the finished material is controlled by the space between the final rolls. In this process the surface of the film or sheeting may be smooth, matted, or embossed, depending on the surface of the final rollers.

The calendering process usually utilizes a separate high shear mixer such as a Banbury mixer to plasticate the material prior to the calendering step. The material is then dropped from the mixer or otherwise conveyed into the feed rolls of the calendering stack. As the feed rolls turn, they, in effect, extrude the plastic into the nip (where the rolls meet) of the calender rolls.

Calendering also can be used to coat thermoplastics onto textiles or other supporting materials. In applying a plastic coating to a fabric or other material by the calendering process, the coating compound is passed through two top horizontal rollers on a calender, while the uncoated material is passed through two bottom rollers.

FIGURE 66: CALENDERING

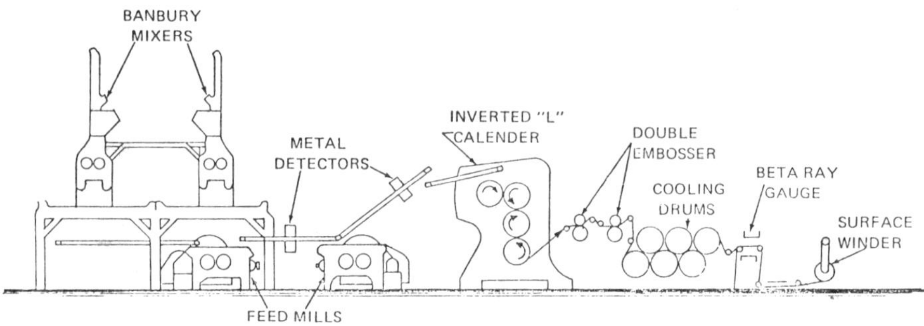

In this process, plastic resin is first mixed with additives in high shear Banbury mixers. The material drops from the Banbury in a semiplastic state and is fed between the rolls of the calender where it is pressed into sheet form

Source: J. Milgrom (16)

The product emerges as a smooth film which is anchored to the substrate when substrate and film pass between the same rolls. In addition to providing a product of high quality, the calendering process is very economical for very high volume production. Because of this very high volume requirement, few end uses can use the calendering process. The main product made on calender equipment is vinyl sheeting and coated vinyls, especially coated fabrics that are used for upholstery. The vinyl film and sheet are used in the manufacture of inflatables, footwear, purses, raincoats, tablecloths, etc.

PVC is the principal resin used in calendering operations, and it is used for products ranging from thin packaging materials 0.002" thick to floor tiles ⅛" or more in thickness. The process can handle a wide range of variations in resins, but batch-to-batch variations fed to the calender rolls must be consistent, particularly for thinner gauge sheet and film.

For example, variations in resin flow characteristics can produce an uneven thickness. The intensive premixing which occurs in the Banbury mixer prior to feeding to the calender rolls can accommodate some variability in resin viscosity, providing the final mix meets the product specifications. Any contamination of the compound with particles, whether foreign or in the form of polymer gels, can result in objectionable pimples, if the product is a thin sheet.

Scrap plastic is generated in the calendering process in the form of drippings from the mixers and the calender, and trim, tails, and front end strip from the calender rolls. Typically, no attempt is made to recover the drippings or badly contaminated or damaged sheet or film. This becomes NP. Levels of NP in this operation are usually less than 1%, if the product is an unsupported sheet.

Scrap from the calender is normally reused, and as much as 30% can be recycled; however, if the product is a coated substrate, such as a coated fabric, then the trim and scrap from poor starts can amount to 6 to 7%. This scrap is not easily recycled because it is a composite; it becomes NP. The manufacture of floor tile utilizes a considerable amount of scrap plastic today. However, as mentioned previously, this scrap has to be free of contamination with particulate matter. In addition to destroying the smooth surface of the tile, particulate matter could conceivably damage the expensive chrome plated calender rolls.

However, a calendering process using less expensive rolls could perhaps use some contaminated feedstock to produce a heavy sheet of plastic having a rougher surface than commercial floor tile. This product might be used for wall covering and similar applications.

Cellular Plastics

During World War II, attempts to produce lightweight bulletproof flotation material for boat hulls, buoys and floats lead to the development of expanded polystyrene (EPS). After the war the process was commercialized for producing three dimensional consumer products. The process of producing three dimensional products of expanded polystyrene foam is also commonly referred to as chest molding. In this process, beads of polystyrene containing a blowing agent are heated slightly by the molder in a preexpander so that they are partially expanded before they are introduced into a closed mold.

This preexpander is basically a steam chamber with mechanical agitation. Preexpansion of the raw material is adjusted to a density close to the density of the finished part. When the foaming action takes place the expansion volume goes into the holes and voids within and around the resin pellets.

The injection molding process (with modifications) also can be used to produce foam structures of polystyrene and other plastics. This variation is generally referred to as structural foams. Cellular structures have been fabricated by other techniques, including extrusion, casting, and rotational molding. Injection molding produces cellular structures with specific gravities in the range of 0.3 to 0.7. In general terms, structural foam molding can be defined as a process in which gas is mixed with molten resin.

Because the structural foam process is most frequently used for molding thick sections, it is characterized by long cycle times. Some process variations offer the additional advantage of using relatively inexpensive tooling compared to conventional high pressure injection mold tooling. These low pressure systems, however, do not produce as good a surface skin as can be obtained by high pressure systems. Most thermoplastic resins have been foamed by these techniques.

Thus far, impact polystyrene is the most commonly used resin in the manufacture of structural foam. The capital investment in equipment for this process is comparable to that of conventional injection molding. The structural foam processes offer the usual advantages of injection molding (low labor cost, continuous reliable production), but with the added potential for making thick, bulky, lightweight parts with less material. The processing costs are usually considerably lower than those for conventional injection molding.

Three types of blowing agents used to provide the gas for expansion: solid agents ranging from expensive azobisformamide to low cost bicarbonate of soda, which decompose at injection temperatures to give gaseous decomposition products; normally liquid agents (like pentane) which vaporize at injection molding temperatures; and gaseous agents such as nitrogen. The solid chemical blowing agents are used most frequently because of their convenience.

Expanded polystyrene foam, while fairly new, is rapidly becoming a volume item for meat, fruit, and produce trays, egg cartons, and as industrial packaging material. Vending machine cups are also a large volume use. In the form of foamboard, polystyrene foam is used as side wall and roof insulation in buildings.

Thin sheets of polystyrene foam are laminated in tar paper for use as inexpensive insulation material, and a loose spaghetti-like form is shipped partially expanded to be fully expanded in the field as a low cost, extremely low density cushion packing dunnage material. Special extruded shapes are also produced for use as dunnage. Most dunnage ends up as NP, when the package is unpacked. In still another packaging application, preexpanded styrene foam beads are mixed with an adhesive and blown around a fragile product after it has been encased in a polystyrene film bag and set into a corrugated paperboard box.

Thus, the beads can be made into an effective cocoon-like cushioning layer for packaged products. In furniture applications, expanded polystyrene can be made into chair shells that are inexpensive and compete with wood and metal shells. ABS expansion molded shells are too expensive for consumer or commercial seating, but are used for high performance seating such as airplace pilot ejector seats.

Because of its buoyancy, polystyrene foam is used extensively by boat manufacturers to replace airtight compartments, and by manufacturers of life rafts, aquaplanes, swimming rafts, navigation buoys and floating docks. It is also used in refrigeration and air conditioning appliances, as drip trays, compartment separators and section insulation, toys, novelties, displays, housewares, portable coolers, counter displays, carvings, signs, Christmas ornaments, and floral displays. Resin requirements for molded foam applications are very similar to those for high quality injection molded products.

EPS molding utilizes a discrete amount of material for each shot, a practice that keeps scrap plastic generation to a minimum. On the other hand, the expandable pellets are very light and difficult to transport. Spillage is very visible and difficult to dispose of. The light weight of the finished EPS product makes regrinding and reuse difficult. If one molds thick articles, one can regrind the scrap and incorporate between 10 and 15%. On the other hand, if one molds a thin walled product, no regrind can be incorporated. Thus, in molding foamed polystyrene between 5 and 10% ends up as NP.

Most extruded sheet foam processes use front end injection of blowing agents with general purpose polystyrene as the raw material. In this process, foam can be reprocessed in the extruder. If the extruder is also the thermoformer, as the majority now are (see following section on Thermoforming), the scrap from thermoforming can be (and is) reprocessed. If the sheet is made from expanded beads, one can sometimes grind some of the scrap and use a crammer to put it back into the extruder, however, this is not the usual practice. Here, NP generation can increase to 10 to 15% of production.

The EPS and sheet extrusion processes are sensitive to the quality of raw material, particularly the amount of blowing agent incorporated. The use of scrap plastic in these processes does not appear attractive. The structural foam process, however, could potentially use NP for low cost, low specification wood substitute products, for example, single trip shipping pallets.

Thermoforming

The thermoforming process is a post-forming process for extruded sheet; both solid and foam sheet are used. The basic process (Figure 67) is as follows: preheat the plastic sheet; position it over a male or into a female mold; and by vacuum, pressure, and/or mechanical action cause the sheet to take the form of the mold. Excess material is then trimmed off to obtain a finished product.

When the force is obtained by vacuum, the process is known as vacuum forming, and when higher pressures are required, they are obtained by sealing a chamber to the opposite side (usually the top side) of the sheet, and building pressure within the chamber with compressed air. This system is known as pressure forming. Vacuum and pressure forming are variations of the same basic technique, even though the difference in the formed parts may be significant. Thermoforming cycle times are short, and this is a major attraction of the process. Other advantages are the ability to obtain extremely thin walls, and the ability to use a wide range of production levels. The major limitation in the cycle time is the thermal conductivity of the plastic itself during the cooling cycle.

As an example of the high speed production capability of this process, cottage cheese tubs (16 oz) are produced in 20 to 30 cavity molds in 2 to 3 sec cycles. The entire production is automatic, from roll stock to the finished carton ready for final shipping. Thermoforming also is used to produce blister packages. A blister is formed in the shape of the item to be packaged, but slightly oversize to provide room for the product.

FIGURE 67: THERMOFORMING

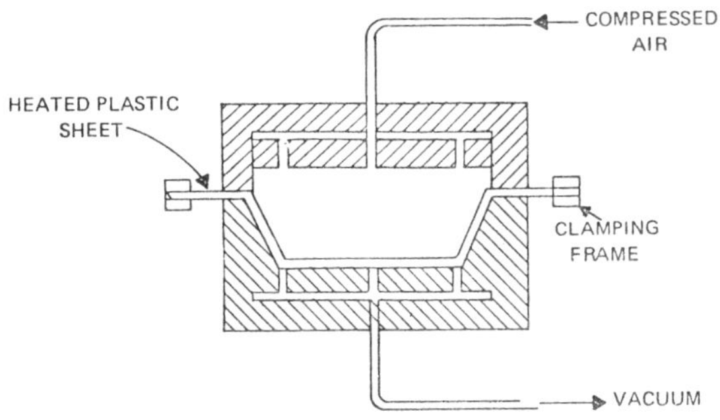

The basic thermoforming process consists of the following steps: preheat plastic sheet; position the sheet over a mold; and bring the sheet into contact with the mold by the use of vacuum between the mold and the sheet and/or pressure above the sheet.

Source: J. Milgrom (16)

Plastic materials such as acetates, butyrates, vinyls and oriented polystyrenes are usually used to produce the blisters. After they are formed, the blisters are trimmed and placed into fixtures, and the parts to be packaged are placed inside the blisters. Precoated cards are then sealed to the blister using heat and air pressure.

Thermoforming is also the processing technique used in skin packaging operation, which have found a wide variety of uses in the shipping and handling of fragile items. The skin pack process utilizes strong flexible films such as vinyl or polyethylene and/or film laminates in combination with corrugated cardboard or chipboard paper products.

In this process the articles to be packaged are placed onto the paper product, which is then placed in a machine where the plastic film is heated and drawn down by vacuum around and over the items to be packaged. At the same time, the plastic adheres to the paper. Thus, in skin packaging, the item to be packaged becomes the mold, and it is enveloped in the plastic film. Expensive forming tools are not required.

Thermoforming competes directly with injection molding in several areas. But each product, in fact, each product variation, has its own requirements. Thus, impact polystyrene cottage cheese containers, which are opaque to hide the appearance of the cheese curd, are thermoformed, whereas transparent containers for ice cream or candies require crystal polystyrene, and these are usually injection molded. Transparent, biaxially oriented polystyrene sheet (the only sheet form in which clear polystyrene is available) is difficult to thermoform and therefore costly. Conversely, injection molding of opaque, thin wall, impact polystyrene containers could not compete economically with thermoforming.

In areas where blow molding and thermoforming are both capable of producing a similar and competitive product, the advantages of thermoforming are: increased throughput per dollar of investment and processing stability. Although the thermal problems of producing uniform plastic sheet plus thermoforming are considerable, they are less troublesome than those of blow molding. Principal areas of application of thermoplastics are in the mass production of rigid, thin wall containers, and in blister and skin packaging.

Cigarette packages, thermoformed cottage cheese containers, foamed styrene trays, and single service drinking cups are typical examples. Some of the more recent new product areas are thermoformed oil cans, foamed polystyrene egg cartons and the spin welched-thermoformed noncarbonated beverage bottles. High production rates, complete automation, minimal downtime and reject losses, plus extremely high scrap recovery rates combine to make thermoforming the best available process for the manufacture of thin wall, simple shaped containers in high volume, when it is in line with the extrusion process.

One of the most difficult problems to overcome in thermoforming is control of trim. Both the amount of regrind and the amount of nuisance plastics are serious limitations of the process. Recycling is only possible if the thermoformer also fabricates the sheet. Otherwise the trim becomes NP, unless it is sold back to the sheet fabricator or to the reprocessor. Regrind averages 40% in many sheet fabricating operations. By careful layout and control of machine conditions, some cottage cheese container manufacturers have reduced the amount of regrind to about 30%, and new processes are being developed that will reduce the volume of regrind further.

The range of scrap recovery rates in thermoforming is very wide, depending upon the details of the process, the housekeeping of the processor, and his ability to use or sell the scrap. In addition to scrap from trim, scrap is also generated in some processes from grinding rejects. Nuisance plastic generation rates as high as 8 to 10% are cited by some as necessary to the process, but the leaders in the field are believed to be keeping loss rates down to the 1 to 2% level. Use of beta-ray gauges and other processing controls has been important in establishing positive control over scrap recovery by controlling sheet thickness and, therefore, thermoforming accuracy.

A process developed by *P.J. Boeke; U.S. Patent 3,551,242; December 29, 1970; assigned to Phillips Petroleum Company* is one in which recovery of waste plastic materials in industrial manufacturing operations is enhanced by a process which comprises applying a coating material such as polyvinyl chloride to a substrate such as polyethylene in a predetermined geometrical pattern. Subsequent manufacturing operations affect only the coated portion of the substrate. The noncoated substrate outside the processing zone is readily recovered in condition for reuse.

In one specific application of the process, polyethylene in bulk form is first extruded in the form of flat sheets. The flat sheets are then laminated with polyvinyl chloride by a transfer roll in a predetermined geometrical pattern that corresponds precisely to the geometry and physical dimensions of the thermoforming molds. In this case, the coating was applied to rectangular areas, spaced along the width of the sheet in rows of three, similar rows being successively imprinted on the sheet by the transfer roll.

The thermoforming process is then conducted using thermoforming apparatus wherein the sheet containing the thermoformed products is subsequently trimmed of the unlaminated polyethylene. In this case, single metal end polyethylene containers for motor oil are produced, and the coated areas prevent reaction of ultraviolet light with additives in the oil, and resultant formation of a vacuum in the head space of the container. The unlaminated polyethylene is now uncontaminated with the lamination materials and, as a result, a substantial economic saving is effected by its availability for reuse as crude material.

The saving is important in the case of thermoforming the described motor oil containers, because the thermoforming apparatus requires that large spaces be left between the coated areas on the roll which form the bodies of the containers. Hence, the quantity of wasted material is very substantial in the absence of practicing this process. Conventional processes laminated the entire bulk material, hence, the unthermoformed portions trimmed from the thermoformed portions were contaminated with the lamination materials and were therefore unavailable for reuse.

Laminating

In the laminating process, multiple layers of resin and filler are combined to produce a final product. The most common example is high pressure decorative laminated sheet such as that known as Formica. This type of sheet is produced by pressing resin impregnated paper and other sheet materials (including an outside printed pattern sheet) in a hot multi-platen press. Thermosetting plastics are most generally used; they bind the reinforcing materials such as cloth, paper, wood or fiberglass. In producing flat high pressure laminates (Figure 68), impregnated sheets are stacked between two highly polished steel plates and subjected to heat and high pressure in a hydraulic press. This cures the plastic and presses the plies of material into a single piece of the desired thickness.

High pressure tubing is produced in a similar process. Resin treated reinforcing sheets are wrapped, under tension and/or pressure, around a heated rod, and the assembly is then cured in an oven. In producing formed shapes, the reinforcing material is cut into pieces that conform to the contour of the product, fitted into the mold, and cured. Whatever the final shape, the first step in high pressure laminating is to impregnate the reinforcing materials with plastics. Textile fabrics (particularly glass cloth) impregnated with thermosetting resins are contour laminated. Spherical structures of high strength are obtained by laminating continuous filament (usually glass fiber), which is precoated with a thermosetting resin, onto a mandrel. This process is called filament winding.

FIGURE 68: HIGH PRESSURE LAMINATING

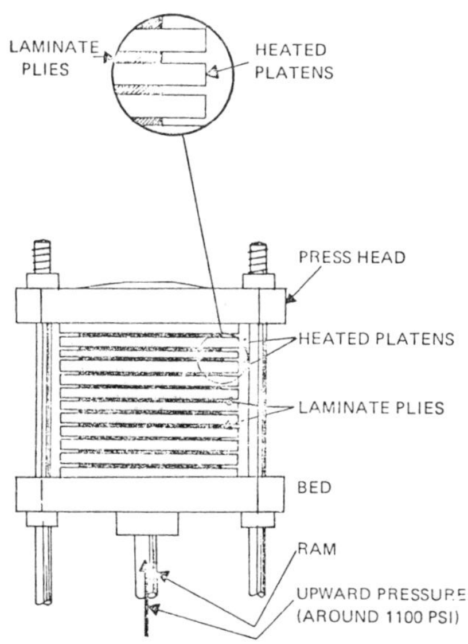

High pressure decorative laminates are produced by laying up sheets of resin, impregnated paper or other sheet material, and pressing them together in a hot multiplaten press.

Source: J. Milgrom (16)

Film lamination of one or more roll films is common for packaging end uses and is usually accomplished in conjunction with printing. Adhesive is applied to one side of the film at one printing station, and a second continuous film is brought into contact with the adhesive coated film before the films enter a set of nip rolls. The type of adhesive used in this process depends upon the individual substrate. For example, laminates of paper with plastic can use solvent based adhesives, because the vapors can pass through the paper; laminates of impervious material require the use of thermoplastic (hot melt) or thermoset adhesive formulas.

The end product of high pressure laminating may be plain flat sheets or decorative sheets, as in counter tops, rods, tubes, or formed shapes. Contour laminating is used in the manufacture of boats, industrial tanks and similar large parts; plastic, or glass fibers and resin, or glass cloth and resin mixtures are either sprayed up or laid up to produce the product. The lay up type of glass fabric and plastic construction results in products which demand the ultimate in tensile strengths from plastic materials. Extremely high tensile strengths have been obtained by continuous filament winding of glass or other materials impregnated with a thermosetting resin, and they represent some of the strongest man-made materials available, in many cases they are stronger than the metals they replaced.

Laminated plastics are used for such items as automobile timing gears, cams, silent gears, clutches, fan blades, and bearings used in machinery. In the decorative field, resin treated wood surfaces and colorful fabrics are used to produce decorative table tops which are wear and mar resisting and impervious to many solvents; in the packaging field, film laminates are used for applications already described above under Coextrusion. Coextrusion and lamination can produce equivalent multiplastic films.

Scrap rates are high in high pressure lamination, and the scrap is not salvageable. Therefore, NP generation averages about 8%. Scrap rates are also high both in the production of film laminates and in their subsequent conversion into labels, bags, or other packaging material. Little, if any, side trim or scrap cuttings from laminates is suitable for recycling. Most of it ends up as NP. The lamination process itself generates 8 to 10% NP. Laminating processes could be used to produce the multilayer products previously mentioned, where NP is used as the core material overlayed with virgin or high quality recycled plastic.

Coating

Both thermosetting and thermoplastic resins may be used as coatings. They can be applied as a melt, a thermosetting liquid, a solution, or as a finely divided powder. A wide variety of substrates such as metal, wood, paper, fabric, leather, glass, concrete, ceramics, or other plastics can be coated. Methods of coating include knife or spread coatings, spraying, roller coating, dipping, and brushing. Calendering is also a form of coating and was described earlier. The method of applying plastic coatings to metals was described under Dip and Slush Molding.

In spread (or knife) coatings (see diagram on following page), the material to be coated passes over a roller and under a long blade or knife. The plastic coating compound is placed on the substrate in front of the knife. The thickness of the coating is regulated by both the position of the knife and the speed at which the substrate is drawn under the knife.

In roller coating, two horizontal rollers are used. One roller picks up the plastic coating solution and deposits it on the second roller which, in turn, deposits the coating solution on the substrate. A plastic coating also may be applied by spraying it through a spray gun or by brushing it over the material to be coated, as in silk screen work.

Vinyl plastisol coatings are used for coating fabrics to obtain materials not unlike old-fashioned oil cloth, waterproof tarpaulins, tenting materials, etc. Plastic coatings are also applied to the back of rugs to give them nonskid properties. Foamed or cellular vinyls are also used to coat fabrics. For example, vinyl foam with a nonfoamed vinyl skin is used to fabricate winter jackets, upholstery material for furniture and automobiles and in ladies'

Knife Coating

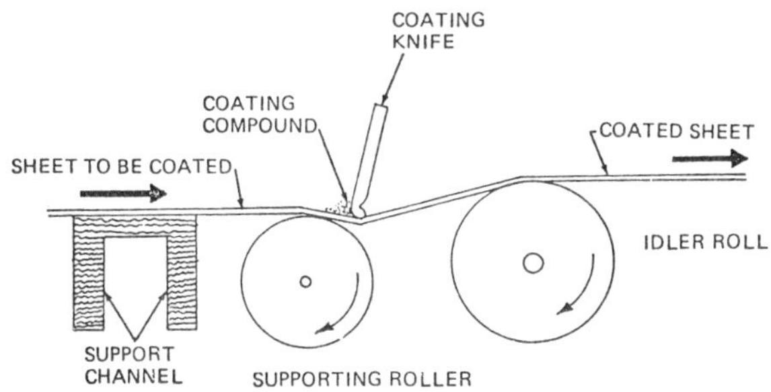

In spread (or knife coating) the material to be coated (substrate) passes over a roller and under a blade (or knife). The plastic coating compound is fed onto the substrate in front of the knife. The thickness of the coating is regulated by the amount of clearance between the knife and the substrate as well as the speed of travel of the substrate.

Source: J. Milgrom (16)

accessories such as hats, handbags, shoes and belts. In the packaging area, Saran coated films are widely used, as mentioned above, because Saran has good oxygen barrier characteristics. Because polyethylene has good heat sealability, it is often used as a coating on aluminum foil, which cannot be heat sealed.

Coating obviously involves thin layers of plastic, which makes resin requirements very demanding. Thus, molecular weight, and molecular weight distribution of the polymer are critical properties. The resins sold for this application are special formulas that command high prices. Stringent quality control is required by the resin manufacturer.

Because the end product in coating operations is typically a composite or, in the case of coatings on a plastic substrate, a multiplastic, little of the scrap can be recycled. Thus, scrap from trim, start ups, rejects, etc., becomes NP. Levels of NP can vary from 6 to 10%. Due to the high quality of resins required for coatings, use of NP within the current processing technology would be difficult. However, a high quality plastic coating could perhaps be used to upgrade the surface of a plastic item produced from scrap plastic of poor quality.

Reprocessing

In the broad sense, reprocessing is compounding. Not only are different materials mixed together in this process, but usually contaminants are removed. Reprocessing requires equipment such as Banbury mixers, two roll mills, extruders, blenders, grinders, choppers, and a variety of specialized equipment.

As indicated in the previous section, reprocessing is often an integral part of fabrication. For example, in injection molding, sprues, runners, and rejects are often fed directly into the molding equipment or into granulators that automatically remix and recycle the scrap with virgin resin. Reprocessing also is carried out by a small segment of the plastics industry, the reprocessor. Reprocessors usually function as compounders and jobbers as well. The reprocessor purchases a wide variety of scrap plastic from resin producers, fabricators, and some converters. If he reprocesses PVC, for example, his raw material may be granulated rigid or flexible PVC with or without contaminants such as dirt or burnt pieces of polymer. He also may accept a shipment of plastic bottles that did not meet specifications, or he might purchase an array of pieces of colored PVC sheet, or large chunks from an extruding operation. Reprocessors who buy polyolefin scrap also purchase strands from the extruder and batches of polyolefin resin that are still wet from the washing step.

The following are examples of operations involved in reprocessing. A chopper or a guillotine is usually used to reduce large chunks into small pieces that can be fed into a granulating machine. The granulating scrap is then blended with other additives, or fed directly to an extruder, where it is pelletized or diced. Film scrap is often made into confetti and similarly fed directly into an extruder.

A common practice is to blend scrap with virgin resin. The reprocessor uses multiple screen packs in the extruder to remove contaminants such as wood, paper, garbage, dirt, and metal. Ferrous metal is sometimes removed by placing magnets on conveyors and/or the hopper filled with scrap. Processing contaminated material through the extruders using a number of screen packs reduces production rates considerably. For example, running virgin material, one can obtain in a given piece of equipment rates of 2,000 lb/hr, whereas rates as low as 500 lb/hr are obtained with some of the heavily contaminated scrap plastic.

As a general rule, reprocessing involves blending to obtain a homogeneous product. Usually a number of small lots of scrap are mixed in a blender before being extruded and pelletized. Because the reprocessor deals with a variety of colored scrap or discolored products, he often adds carbon black to the blend and then sells a black secondary resin.

Reprocessors sometimes wash the scrap plastic to remove contaminants. As an example, the chopped material can be spread out and agitated on a large trough and cleaned with hot water. Afterward, the washed material can be dumped into a device where dirt and other contaminants are pulled down, and the plastic skimmed off as a wet slurry. This process is used with polyolefins, which have a density less than one. The plastic-water slurry is then centrifuged and the recovered plastic dried prior to extrusion.

Under certain circumstances, for example, when reprocessing colored sheet scrap, if there is sufficient amount of each color, the reprocessor will accept the scrap from the fabricator or converter and, for a toll charge, reprocess and return it to the original user in a granulated or pelletized form ready for normal processing. Because the fabricator usually must know his raw material, this method of reprocessing a company's scrap is popular. Reprocessing often involves altering the flow characteristics of the polymer by blending different lots, or adding virgin resin. Other times it may be necessary to process the scrap through the extruder more than once to remove all of the contaminants and/or alter its physical properties.

The principal difficulty in recycling or reprocessing plastics is that they are generally not compatible with each other. If a mixture of different plastic types is used to fabricate an end item, the resulting product can appear striated, and it often fibrillates. Not only does a mixture give rise to a product with a poor appearance, but the physical properties of the product also suffer. Nevertheless, some of the major thermoplastics can be mixed to a limited extent, depending on the product desired. For example, low and high density polyethylenes are miscible in all proportions but, due to the difference in melting points, difficulties can arise in certain processing.

Polypropylene also can be mixed with polyethylenes at levels up to 15% without causing serious problems and, similarly, a mixture containing up to 15% polyethylenes and the remainder polypropylene is a usable combination. Depending on the concentration of vinyl acetate in the commercial ethylene-vinyl acetate copolymer, these polymers also are partially miscible with polyethylenes. Polymers vary in their ease of being reprocessed. Polystyrene reprocesses well, but loses some physical properties. Generally, opacity increases with reprocessing, and tinted crystal polystyrene can become opaque.

Polystyrene is very sensitive to contamination with other polymers. The polyethylenes, on the other hand, tend to oxidize with each heat cycle and the melt flow decreases. Polypropylene does the opposite. Its melt flow increases as it is subjected to multiple cycles. Oftentimes contamination with other polymers tends to embrittle polypropylene.

Furthermore, due to the ease of oxidation of polypropylene, it tends to brown with recycling. PVC, because of its poor heat stability, is recycled with difficulty; however, by adding auxiliary stabilizer PVC can be recycled. ABS resins tend to lose impact as the heat cycles increase, and it also has a tendency to discolor. Light colors are difficult to maintain, and other physical properties are reduced on recycling.

Reprocessing involves considerable plastic losses due to the high state of contamination of the raw material. In some instances the reprocessor has to hand pick his way through the raw material to insure that grossly contaminated products or gross contaminants are removed. Losses in this selection process can easily go up to 10%. On extrusion and compounding, the levels of NP generation vary from 1 to 3%. Generally, short runs have high losses.

RUBBER PRODUCTS INDUSTRY

The rubber products industry consists of two major divisions:

–the tire and tube industry and
–the fabricated and reclaimed rubber products industry.

The fabricated and reclaimed rubber products industry in turn consists of the following segments:

–molded products
–extruded products
–fabricated products
–reclaimed rubber products
–latex-based products

The EPA, for convenience in designation of various water pollution-producing categories has further made designations as follows:

Subcategory E–molded, extruded, fabricated–small plants
Subcategory F–molded, extruded, fabricated–medium plants
Subcategory G–molded, extruded, fabricated–large plants
Subcategory H–wet digestion rubber reclaiming
Subcategory I–pan (heater), mechanical and dry digestion reclaiming
Subcategory J–latex-dipped, latex thread and latex molded products
Subcategory K–latex foam

AIR POLLUTION

Sources of air pollution from rubber mills (5) are [1] finely ground dusts introduced as additives, [2] fumes generated by mechanical working of the batch by the mill rollers, [3] oil mists from liquid additives, and [4] odors. A major source of air pollution from rubber mills occurs when the finely divided dusts are introduced into the batch. Opacity of the resultant dust cloud depends upon the character, density, and particle size of the additive. Opacity generally ranges from 5 to 50%, persisting from a few seconds to several minutes.

Carbon black and oil are added to the rubber in the compounding operation. To avoid many of the housekeeping problems created by both carbon black and oil, these ingredients are added automatically. Carbon black is a finely divided amorphous material that has the consistency of dust and is easily airborne. The compounding area is equipped with air pollution equipment to control this problem. Bag house particulate collectors are normally used which can produce removal efficiencies of essentially 100% when designed, maintained, and operated correctly.

Uncontrolled emissions vary from a negligible amount to about 1 pound per hour, depending upon the size of the mill, the size of the batch, and the composition of the mix. Emissions average approximately 0.5 pound per hour. Solvent vapors emanating from the mixare ordinarily uncontrolled and enter the atmosphere.

Introduction of ingredients into a Banbury mixer is effected through the feed hopper. It is at this point, during charging, that air contaminants may enter the atmosphere. Emissions are similar to those from the mills. In general, most of the dry ingredients are added at the Banbury mixer, carbon black being the most troublesome.

Generally, rubber mills are provided with hoods. The primary purpose of a hood is to carry away heat generated by the mechanical mixing action. As a secondary consideration, the exhaust hood removes dust, fumes, and mists emitted from the rolls. Sufficient volume should be exhausted to give an indraft velocity of 100 fpm through the open face of the enclosure. If an unusual dust problem is encountered, supplementary hooding can be added. The minimum required exhaust volume is equal to 200 cfm per square foot of mixer charging opening.

In general, emissions from Banbury mixers and rubber mills are in a finely divided form and smaller than 15 microns. Inertial separators are not, therefore, effective control devices for this service. The most common control device employed is the baghouse; a well-designed baghouse can be operated with 98 to 99.5% efficiency. Standard cotton sateen bags are adequate at a filtering velocity of 3 fpm. In some cases scrubbers have also proved satisfactory and advantageous in scrubbing out some oil vapors and oil mists that may be present in some blends.

As regards an air pollution from tire manufacture (13), before molding and curing, the green tire is sprayed with release agents. These agents aid in the release of air from the tire during molding and of the tire from the mold after curing. Both water- and solvent-based sprays are used. Excess spray is released to the atmosphere. In most plants the tires are placed in a hood during spraying to reduce atmospheric contamination. Wastewater is generated by wet scrubbers where used to scrub the excess spray from the air.

The potential for wastewater streams exist due to the possibility of solvent spills within this area. If wet scrubbers are used to scrub the excess spray from the air, another wastewater stream will exist.

After the molding and curing operations, the tire proceeds to the grinding operation where the excess rubber which escaped through the weepholes is ground off. If the tire is designated to be a whitewall, additional grinding is performed to remove a black protective strip. Most tires receive further grinding of the tread in order to balance the tire.

The weepholes which are ground off are relatively large particles of rubber which fall to the floor and are swept up. Their final destination is a landfill. The grindings from the white sidewall operation are relatively small and will stay airborne for long periods of time. The industry generally uses a particulate collection device such as cyclone or wet scrubber to control these emissions. The discharge from a wet scrubber will have a high solids content and will therefore be a wastewater problem. The balancing operation suffers the same problems as the white sidewall grinding operation.

After the grinding operations, the whitewall portion of a tire receives a protective coat of paint. The paint is generally water based. This operation usually occurs in a hooded area. Again, any wet air pollution equipment or runoff due to over spraying of the paint will create pollution problems. After inspection and possibly some final repairs, the tire is ready to be shipped.

The type of air pollution control equipment employed by the molded, extruded or fabricated rubber product facility can have an effect on the quantity and quality of a plants' overall process wastewater flow (17). The use of dry scrubbing equipment produces no wastewater problem. If wet scrubbing devices are used solid-laden blowdown can be settled and filtered. By recycling the water from the settled and filtered discharge, this wastewater can be minimized. Wet scrubbing devices are used in the rubber compounding and product buffing areas. Although wet scrubbing devices are used to trap rubber-buffing and metal sand-blasting fines, dry air-pollution control devices are also applicable and are used in many plants within the industry. In some plants where wastewater discharge requirements are stringent and the use of bag collector devices is also inappropriate, the wastewaters from wet scrubbing devices are settled and filtered before discharge, or a municipal treatment system is used to accept the wastewaters.

Rubber reclaiming plants are infamous for the odor problems they create in the neighborhood of the plants although details of emission and controls are conspicuously absent from the literature. Wet air pollution control devices are common in the industry; however, similar control measures are required for all plants using the same reclaiming process. Therefore, subcategorization of the rubber reclaiming industry according to the extent and type of air pollution control equipment employed is not necessary. Few air pollution controls and no wet scrubbing devices are used by either the latex dipped goods or foam industry sectors.

WATER POLLUTION

Tire and Inner Tube Manufacture

One aspect of the rubber products industry which has important pollution control considerations is the Tire and Inner Tube area (SIC*3011). For the purpose of establishing limitations, the tire and inner tube industry has been subcategorized according to the age of the production facility. Waste loads and costs of control technologies substantiate this. Factors such as the manufacturing process, final product, raw materials, plant size, geographic location, air pollution equipment, and the nature and treatability of wastewaters are similar and further substantiate the subcategorization of the tire and inner tube plants by age.

Process wastewaters for both subcategories of the tire and inner tube industry include discharges of solutions used in the manufacturing process, washdown of processing areas, runoff from raw material storage areas, and spills and leakage of cooling water, steam, processing solutions, organic solvents and lubricating oils. Primary pollutants in these wastewaters are oil and grease, suspended solids, and acidity and alkalinity (pH).

In the tire and inner tube industry, the emphasis of environmental quality control and treatment technologies is placed on the control of particulate emission and the reduction of pollutants in nonprocess wastewaters. Control and treatment of many process wastewaters has been given secondary priority. As a result, no adequate overall control and treatment technology is employed by plants within the industry. A treatment system, practicable and available to the industry, was therefore proposed for both subcategories. It encompassed a combination of the various technologies employed by the different segments of the industry to control one or more constituents in the process wastewaters.

Proposed effluent limitations and standards for the best practicable control technology currently available are:

Suspended Solids	0.064 kg/kkg (lb/1,000 lb) raw material
Oil and Grease	0.016 kg/kkg (lb/1,000 lb) raw material
pH	6.0 to 9.0

No additional reduction is proposed for the limitations and standards represented by best available technology economically achievable or for sources coming on stream after the guidelines are put into effect.

A wide variety of synthetic rubbers are used including stryrene-butadiene rubber (SBR), polybutadiene, butyl, polyisoprene, and ethylene-propylene diene rubber (EPDM). Of the three categories of compounding materials used, the fillers, extenders and reinforcers are the most important. These are used: [1] to dilute the raw crumb rubber in order to produce a greater weight or volume, and [2] to increase the strength, hardness, and abrasion resistance to the final product. Of these, carbon black and oil are the most common. A typical rubber compound might be described as follows: 100 parts rubber; 50 parts fillers, extenders, and reinforcers; 3.5 parts curing and accelerator agents; and, 8.0 parts antioxidants and pigments.

The typical tire manufacturing process consists of [1] preparation or compounding of the raw materials; [2] transformation of these compounded materials into the five tire components; and, [3] the building, molding, and curing of the final product. A flow diagram for the typical plant is shown in Figure 69.

The basic machinery units used in the compounding operation are the Banbury mixer and the roller mill. A Banbury mixer is a batch-type internal mixing device and is the hub of this compounding operation. The Banbury is used for two operations. In the first, the fillers, extenders and reinforcing agents, and the pigments and antioxidant agents are added and mixed into the raw rubber stock. The resulting mixture is known as nonproductive or nonreactive rubber stock. Because no curing agents have been added, this material will have a long shelf life, thus allowing large quantities of a particular recipe to be made and stored for later use. In the second operation, the curing and accelerator agents, in addition to a small quantity of the original list of elements, are added. This mixture, known as productive or reactive rubber stock, meets the particular compounding requirements of its final destination. Since it contains the curing agents, this mixture has a short shelf life and will be used almost immediately.

After mixing, the compound is sheeted out in a roller mill, extruded into sheets, or pelletized. The process depends on the type or batch (reactive or nonreactive) and the manufacturer. Pelletizing of a nonreactive batch enables the weighing and mixing of the reactive stock to be done automatically. The reactive compounded rubber is always sheeted out. The sheeted material is tacky and must be coated with a soapstone solution. This solution is a slurry which, when allowed to dry on the sheeted materials, prevents them from sticking together during storage.

Because it is a slurry the soapstone solution is usually recirculated. Releasing the material into a wastewater stream would create a difficult solids problem. Spills in the soapstone area are common and do create a maintenance and wastewater problem.

If a manufacturer wishes to exclude soapstone in the final effluent, a method must be provided for coping with these spills. Current techniques include the blocking of all drains in the area, diking of the area, and the use of steel grates on the floors. The diking and sealing of drains prevents the slurry from entering the drainage system. The use of steel grates helps decrease the risk of workers slipping on spilled soapstone.

Maintenance and housekeeping problems in this area are further complicated by the leakages of oil and water from the oil seals in the mills, and oil and dust from the dust ring seals of the Banburys. Each has the potential to become a wastewater pollutant if allowed to mix with the cooling water discharges or to be washed down and discharged without treatment.

The rubber stock once compounded and mixed must be molded or transformed into the form of one of the final parts of the tire. This consists of several parallel processes by which the sheeted rubber and other raw materials, such as cord and fabric, are made into the following basic tire components: tire beads, tire treads, tire cords, and tire belt (fabric). Tire beads are coated wires inserted in the pneumatic tire at the point where the tire meets the steel wheel rim (on which it is mounted); they insure a seal between the rim and the tire (2). The tire treads are the part of the tire that meets the road surface; their design and composition depend on the use of the tire. Tire cords are woven synthetic fabrics

FIGURE 69: GENERAL WATER FLOW DIAGRAM FOR TYPICAL TIRE AND CAMELBACK PRODUCTION FACILITY

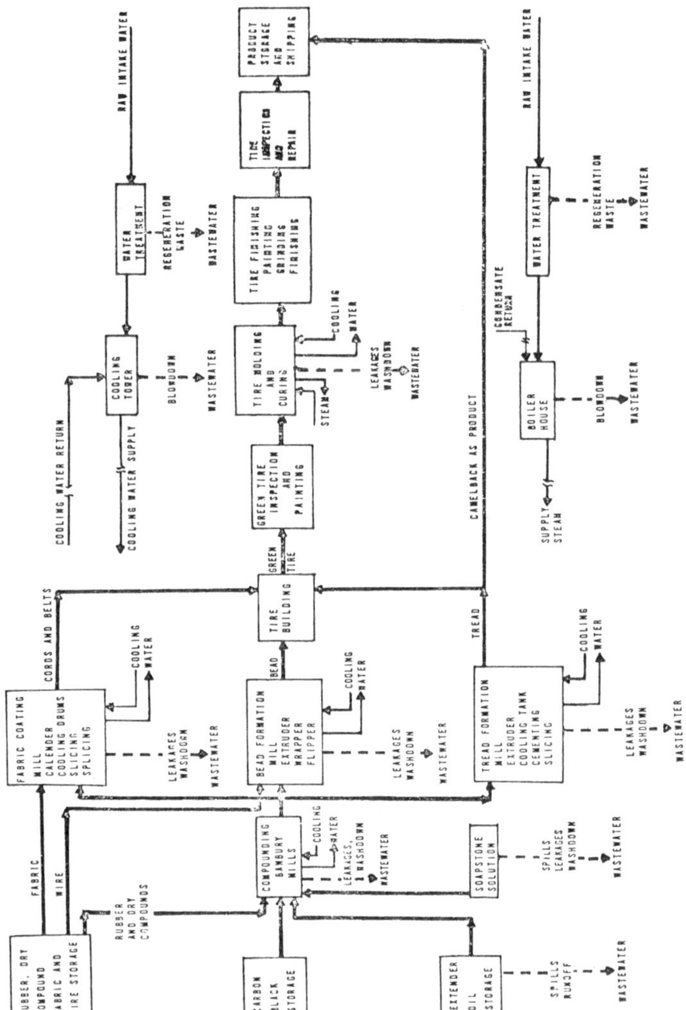

Source: EPA (13)

(rayon, nylon, polyester) impregnated with rubber; they are the body of the tire and supply it with most of its strength. Tire belts stabilize the tires and prevent the lateral scrubbing or wiping action that causes tread wear. In the formation of tire treads, the rubber stock as it is received from the compounding section is manually fed to a warm-up roller mill. Here the rubber is heated and further mixed. Heat is provided by the conversion of mechanical energy. Temperature control is provided by the use of cooling water within the rolls of the mill.

The heated stock passes from the warm-up mill to a strip-feed mill where it receives its final mixing. This mill is also cooled to control the temperature of the stock. The stock is peeled off the rollers of the mill in a thin strip which is fed continuously to an extruder. The mixing of the stock in these mills insures that the final tread will have homogeneous properties. The heating or temperature control of the stock is necessary to insure a proper extrusion with a minimum consumption of power.

At the extruder, two types of rubber stocks originating from two different strip mills are joined together to form the tire tread and sidewalls. The tread leaves the extruder as a continuous strip while still hot and therefore tacky. Next a cushioning layer is attached to the under side of the tread. The tread is then cut to the proper width, cooled in a water trough, labeled, and then cut to the proper length. Trimmings are either manually or automatically transferred back to the proper strip-feed mill and reprocessed. The ends are coated with rubber cement and the tread is then placed in a tread book and sent to the tire building machines.

Wastewater problems in this area arise from the spillage of the solvent base cements, from oil and water leakages from the various mills, and from accidental overflows from the cooling water system. The cooling water overflow would not normally be a problem since the rubber tread is relatively inert and therefore does not contaminate the water. However, it does serve as a washdown agent for an area contaminated with the cements and oils.

To produce tire cords and belts, rubber stock must be impregnated onto a pretreated fabric. The fabric is let off a roll, spliced onto the tail of the previous roll (either adhesively or by a high-speed sewing machine), and fed under controlled tension (via a festooner) to a latex dip tank. After dipping and while still under tension, the fabric is fed past vacuum suction lines or rotating beater bars to remove the excess dip before the fabric rises through a drying and baking oven.

After pretreatment, and still under tension, the fabric is passed through a calendering machine where rubber is impregnated into the fabric. The rubber fabric is next cooled by large water or refrigerant cooled drums; after cooling, the tension can be released. This treated fabric is still not ready for the tire building operation. To achieve the proper bias it must be cut to the proper angle and length, and then spliced together again. The angle and length will vary depending on the size of the tire for which it is used and whether it is a cord or belt. Once spliced, the fabric is rolled in cloth and sent to the tire building.

The rubber used to impregnate the fabric proceeds through an operation similar to that of the tread process. It passes through both a warm-up mill and a strip-feed mill prior to impregnation onto the fabric. Wastewater problems in this area arise due to the latex dripping operation and to problems with oil and water leaks and spillages which are similar to those of the tread process.

Many tire manufacturers are transferring their latex dip operations from individual plants to one large central facility. In most cases, the reasons behind such a decision are as follows:

[1] A minimal dipping operation requires a large capital expenditure.

[2] The fabric dipping and coating operation is one of the fastest operations in the plant, and, as such, is readily capable of over-supplying the plant with fabric.

[3] Dipped fabric is not that more expensive to ship than undipped fabric.

[4] The maintenance and housekeeping requirements of the dip operation are limited to one facility.

In the processing of rubber stock to tire beads, the rubber is extruded onto a series of copper-plated steel wires, which are then cemented, wrapped, and cut. The rubber stock is pretreated, as before, in a warm-up mill and strip-feed mill. Excess rubber is trimmed from the bead before it leaves the extruder and is fed back to the strip feed mills. To apply cement the coated wire is passed through a trough or set of brushes. The cement is necessary to insure the proper adhesion of the bead when it is wrapped. Wastewater problems can arise due to the use of the mills or from the spillage or overflow of the cement. They will be similar in nature to those found in this tread formation process.

The inner lining for the tire is formed by calendering or extruding the rubber stock in a manner similar to either the formation of cord fabric or tread rubber. It is this inner liner that enables a tire to be tubeless since it is light and air impervious. The tire is built up as a cylinder on a collapsible, round rotating drum. First the inner liner is applied to the drum. Then layers of cord are applied, one layer tying the beads together in one direction and another layer in the other direction. The beads are attached to the tire by folding over the ends of the cord fabric. Next the tire belt fabric is laid onto the cord. Finally the tire tread is placed over the cords and fabric and wrapped around the beads. The cylinder is removed. These green tires (uncured tires) are now ready for final processing.

The tire is molded and cured in an automatic press. Here an inflatable rubber bladder bag is inflated inside the tire, causing the tire to take its characteristic doughnut shape. The mold is simultaneously closed over the shaped tire. Heat is applied by steam via the mold and bladder bag. Excess rubber and trapped air escape through weepholes. After a timed, temperature-controlled cure, the press is cooled, the bladder is deflated via a vacuum, and the tire is removed. The tire is next inflated with air and left to cool in the atmosphere. This last inflation insures product quality and uniformity by allowing the tire to set up or achieve the final limits of its cure under controlled conditions.

Because of the large number of presses in the typical plant, there is always the potential for a mold to leak or for a bladder to break. This water is released and scavenges some of the large amounts of lubricating oil used in this area. This oily water creates a water contamination problem if it is discharged. Table 29 presents a review of the potential sources of wastewater streams as discussed above.

The discussion thus far has described a typical tire plant, and applies most readily to the production of the passenger tire. There are several variations. The first of these is due to the production of truck and industrial tires. Truck tires tend to have a greater amount of natural rubber in their treads. Natural rubber, as received in the plant, is much harder to handle than synthetic rubber. Additional roller mills are needed to break up and soften the rubber before it enters the Banbury mixer. There are also major differences in the building and molding of the tires as the larger sizes are approached. The building of a giant off-the-road tire requires the services of two men each for a half a day, whereas the passenger tire can be built in less than 5 minutes.

Larger tires are cured in giant molds which are not automatically operated. Cranes or hoists are required to open and close these molds. Curing can take up to 24 hours. Hot water, instead of steam, is used in the curing operation. The process variations associated with truck and industrial tire production do not have a significant effect on the quantity and quality of the wastewaters generated when compared to those from automobile tire production.

Another variation in the typical tire production is the manufacture of camelback. Camelback is tread used for tire retreading (2). It is produced in the same manner as tread used for newer tires, (see flow diagram, Figure 69).

TABLE 29: SUMMARY OF POTENTIAL PROCESS-ASSOCIATED WASTEWATER SOURCES FROM THE TIRE AND INNER TUBE INDUSTRY

Plant Area	Source	Nature and Origin of Wastewater Contaminants
Oil Storage	Run off	Oil
Compounding	Washdown, spills, leaks, discharges from wet air pollution equipment	Solids from soapstone dip tank Oil from seals in roller mills Oil and solids from Banbury seals Solids from air pollution equipment discharge
Bead, Tread, Tube Formation	Washdown, spills, leaks	Oil and solvent-based cements from the cementing operation Oil from seals in roller mills
Cord and Belt Formation	Washdown, spills, leaks	Organics and solids from dipping operation Oil from seals, in roller mills, calenders, etc.
Green Tire Painting	Washdown, spills,air pollution equipment	Organics and solids from spray painting operation Soluble organics and solids from air pollution equipment discharge
Molding and Curing	Washdown, leaks	Oil from hydraulic system Oil from presses
Tire Finishing	Washdown, spills, air pollution equipment	Solids and soluble organics from painting operation Solids from air pollution equipment discharge

Source: EPA (13)

FIGURE 70: GENERAL WATER FLOW DIAGRAM FOR A TYPICAL INNER TUBE PRODUCTION FACILITY

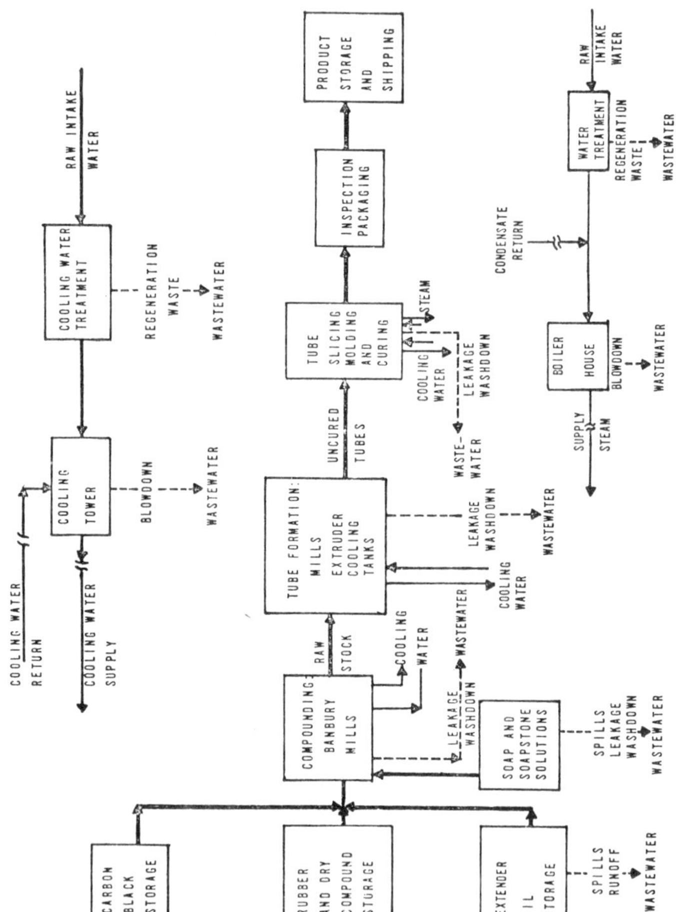

Source: EPA (13)

Since camelback production operations are usually part of a tire production facility feeding off the same machinery, wastewater problems will be similar to those already discussed. Radial tire production offers another variation to the overall process. Radial tires, like truck tires, contain more natural rubber, thus requiring more machinery in the compounding area. Whereas bias-ply tires are built in the form of a hollow cylinder, radial tires are built in the doughnut shape of the final product. Like truck tires, radial tires are cured using hot water instead of steam. Again wastewater problems will be very similar to those of the typical passenger tire manufacture.

Inner tube manufacture is very similar to tire manufacture in that the process consists of the following steps: [1] preparation or compounding of the raw materials; [2] the extension of these compounded materials to form a tube; and [3] the building, molding, and curing to form the final product. A flow diagram for the typical process is shown in Figure 70.

The basic machinery used in the compounding operation is similar to that used in tire manufacture; namely, Banbury mixers and roller mills. Both nonreactive and reactive stocks are prepared. One minor distinction of inner tube manufacture is the high usage of butyl rubbers. In addition, a soap rather than a soapstone solution is sometimes used to coat the nonreactive stock. The soap solution is not discharged and is used in a completely closed-loop system with solution make-up. In general, wastewater problems arising from this section are similar to those of the typical compounding area of a tire plant; that is, leakages and drippings of oily and particulate material.

The process by which the tube is formed is similar to the extrusion of the tread. The compounded rubber is fed to an extruder via a warm-up mill and strip-feed mill. Here the rubber is extruded into a continuous cylinder. To keep the inside of the tube walls from sticking to each other, a dry soapstone powder is sprayed inside the tube as it is formed in the extruder. The tube is labeled and passed through a water cooling tank. After cooling, the water is blown off the tube and soapstone powder is sprayed on the outside of the tube. Excess powder must be collected in either a dry or wet collection device. If a wet collection device is used, the discharge will be heavily laden with solids. Other wastewater problems are similar to those found in the tread formation process of tire manufacture.

Once extruded, the tube must be cut to length and the ends spliced together. A valve must also be attached. There is no potential wastewater problem arising from this area of operation. Once formed, the tube must be molded and cured. Again, this operation is very similar to that of the tire manufacture. Wastewater problems include only water leakage and spills. After curing, the tube is inspected for defects, packaged and sent to warehousing and shipping. Table 29 summarizes the potential sources of wastewater streams as discussed above. The primary water usage in a tire and inner tube facility is for noncontact cooling and heating. Discharges from service utilities supplying cooling water and steam are the major source of contaminants in the final effluent. Characteristics of these wastewaters are COD, BOD, suspended solids, and dissolved solids.

Table 30 presents the raw waste loading for the combined process and recirculating cooling water. Flow variations are due mainly to the use of once-through cooling water in certain plants as opposed to recirculated cooling water in others. In order to adequately estimate the wastewater discharge flow rates, the plant effluent was divided into process wastewaters and nonprocess wastewaters. The process wastewaters consist of mill area oily waters, soapstone slurry and latex dip wastes, area washdown waters, emission scrubber waters, contaminated storm waters from raw material storage areas, etc. The nonprocess wastewaters are sanitary and clean storm waters, utility wastewaters such as once-through cooling water, boiler blowdown, cooling tower blowdown, water treatment wastes and uncontaiminated contact cooling water like tread cooling waters.

Plants A and B are new plants using totally recirculated cooling water. Plants E and G are old facilities also using recirculated water. A comparison of these four plants indicates that no significant variations in flow exist due to age of the plant. Plant F typifies a plant using once-through water as its primary source of process cooling. COD and BOD loadings vary

TABLE 30: RAW WASTE LOADS OF TOTAL EFFLUENT FROM TIRE AND INNER TUBE FACILITIES[1]

Plant	Category	FLOW L/kkg (gal/1000lb) of raw material	COD kg/kkg (lb/1000lb) of raw material	BOD kg/kkg (lb/1000lb) of raw material	SS kg/kkg (lb/1000lb) of raw material	TDS kg/kkg (lb/1000lb) of raw material	OIL kg/kkg (lb/1000lb) of raw material
A	New	6344 (762)	1.890	0.067	0.960	4.800	0.248
B	New	3430 (412)	0.184	0.002	0.047	0.159	0.075
C[2]	New	8251 (991)	NA	NA	1.155	NA	0.794
D	New	10883 (1,306)	0.142	0.012	0.092	0.879	0.009
E	Old	5453 (655)	0.100	0.001	0.440	0.001	0.027[3]
F	Old	123480 (14,824)	3.398	0.296	1.358	0.001	0.650
G	Old	3220 (387)	0.645	0.093	3.429	0.000	0.267
H	Old	72427 (8,695)	0.001	0.036	0.676	0.000	0.167
I	Old	10610 (1,274)	0.615	0.148	2.812	1.810	0.172

[1] Includes utility wastewaters.

[2] Estimated, raw material consumption not known.

[3] Includes treatment by in-plant sumps.

NA - Data not available

Source: EPA (13)

to a great degree by the type and amount of chemicals used in the treatment of boiler and cooling tower makeup waters. Larger loadings for older plants indicate an increased amount of process wastewater pollutants in the effluent. Loadings measured in Plant A are high due to the practice of discharging washdowns of soapstone and latex dip areas noticed during the sampling period. This plant uses holding lagoons. Because all wastes are contained within the plant's boundaries, Plant A discharges contaminants which other exemplary plants (using different technologies) cannot accomplish. These contaminants lead to a correspondingly higher loading.

Suspended solid loadings evolve primarily due to water treatment blowdowns, wastes, and boiler blowdowns. In addition, the suspended solid loadings in process wastewater can increase due to spills, leakage, and soapstone discharge. Loadings for old plants tend to be higher than those for new plants. This is due in part to the use of older water treatment techniques and the larger volumes of process wastewater containing solids discharged by older facilities. The quantity of dissolved solids discharged is related to the amount of recirculated nonprocess water and the water supply source. Plants using well water typically have higher dissolved solid loadings, than those using municipal or river water sources.

Table 30 also shows that the plant's final end product has no significant effect upon the raw waste loading in the final effluent. Data from Plant H, which produces primarily truck and industrial tires, is not substantially very different from Plants E, F, or G, which, while producing a combination of products, produce mainly passenger tires. Loadings from Plant I are similar to the others, even though this plant's primary product is the manufacture of inner tubes.

Molded Products

Rubber molding processes typically consist of the following (17): [1] compounding of the rubber stock; [2] preparation of the mold preforms or blanks; [3] molding; and [4] deflashing. A flow diagram for a typical molding operation involving compression, transfer, and injection molding processes is shown in Figure 71.

Compression Molding: Compression molding is the oldest method of making molded parts. The uncured rubber is formed to the approximate shape, referred to as a preform, and placed in the individual cavities of the mold. As the mold is closed under pressure, the compound conforms to the shape of the cavity and the excess material is forced out into a flash groove. Larger molding facilities, or plants using special rubber compound recipes, compound their own rubber stock from basic ingredients such as rubber, carbon black, colorants, extender oils, antioxidants, and accelerators. Compounding is generally carried out in either a Banbury mixer or compounding mill. These pieces of equipment require cooling water. Leakages of lubricating oil and grease are common. In some plants, airborne particles generated during the compounding operation are controlled by wet scrubbing devices.

After compounding, the rubber stock is worked on a warm-up mill and formed to approximately the required shape ready for molding by either a calender or extruder. The formed rubber is cooled and generally dipped in an antitack agent. In many cases, the formed rubber is cooled in an open tank which produces a cooling water overflow. The antitack liquid is generally a zinc stearate solution or its equivalent. Soapstone slurry is not normally used because its antitack properties are persistent and adversely affect the quality of the subsequent molding operation.

The preforms are prepared from the calendered or extruded stock rubber by cutting, slicing or stamping out. Cutting can be accomplished by a machine or by hand. Slicing is generally carried out on a meat slicing machine or guillotine. In the case of the slicing machine, lubricant water flows over the cutting surface to facilitate the operation. The waste lubricant water is noncontaminated. The preform stamping machine can be equipped with cooling water. Although the exact shape of preform is not crucial, it is necessary to ensure that there is sufficient rubber in the preform to fill the mold. The preforms are placed in the open mold, usually by hand. In some cases release agent powders or liquids are spread on

FIGURE 71: FLOW DIAGRAM FOR THE PRODUCTION OF A TYPICAL MOLDED ITEM

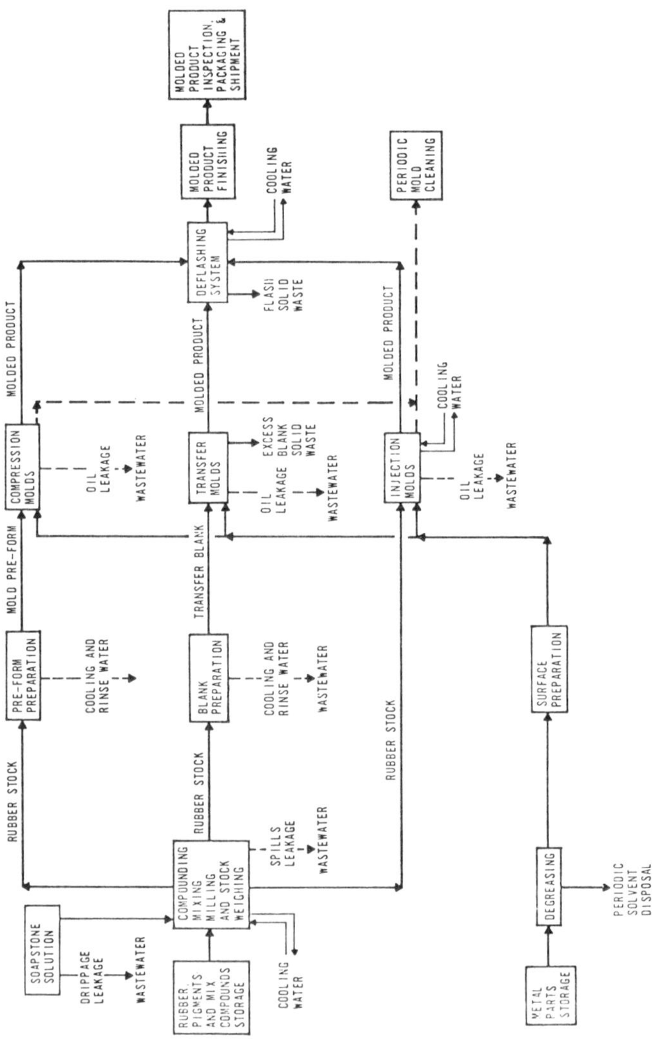

Source: EPA (17)

the mold surfaces. For small molded items, each mold generally has multiple receptacles which enable several items to be molded simultaneously. The mold is closed and held together, normally by hydraulic oil pressure, during the curing cycle. The molds are generally heated with steam flowing through channels in the mold plates. Steam condensate is recycled to the boiler. Some older mold systems are electrically heated. Oil leakage frequently occurs from the hydraulic mold closing systems and the hydraulic pump itself. The molding cycle can vary considerably depending on the curing properties of the rubber and the size of the molded item. At the conclusion of the molding cycle, the items are removed from the mold and sent to the deflashing operation.

The rubber overflow or flash must be removed from the part before it is shipped. Deflashing usually is carried out by hand, grinding wheel, or press-operated dies. In some cases, the rubber parts are tumbled in Dry Ice (solid carbon dioxide) using machines that are similar to cement mixers. The thin rubber flash becomes brittle while the main body of the part is not cooled sufficiently and remains flexible. The thinner frozen flash thus breaks off in tumbling while the heavier main part is not harmed. Blasting frozen rubber parts with fine shot also removes flash. Rubbers which are freeze-resistant are not used in this dry-ice process.

Metal-bonded items, which consist of a molded rubber component bonded to the metal part, are manufactured in a manner similar to that for all rubber products. In most cases the metal parts enter the molding plant contaminated by grease. The grease is picked up during the manufacture of the metal part or is applied for shipping and storage purposes. The metal part is first degreased. Usually the degreasing system consists of a rotating drum in which the part is brought in contact with a suitable degreasing solvent, frequently trichloroethylene. The solvent is drummed and hauled from the plant when saturated with grease. In a few cases the metal part is pickled to prepare the surface for bonding with rubber. The waste acid pickling liquor contains metal ions, and frequently heavy metal ions, and requires containerization or treatment before it is discharged.

In some molding plants, poor quality molded items are recycled to reclaim the metal component for reuse. This is generally practiced in cases where the metal part is large or valuable, or where the molding operation is particularly difficult, producing a high proportion of rejected molded items. To reclaim the metal, the reject rubber is ground and buffed from the metal; the metal item is then sandblasted clean. The grinding and buffing operations generally create airborne rubber-buffing particles, which are most effectively removed from the air with wet scrubbing devices. These devices produce a wastewater discharge which, if a water recycle system is used, can be reduced to a low daily volume.

The metal surface to which the rubber is to be molded is normally prepared to provide good adhesion between the metal and the rubber. The mating surface of the metal part is first sandblasted to roughen it and then coated with rubber cement to improve the adhesion of metal to rubber. The metal surface is painted with cement by hand for small items; larger metal surfaces are often sprayed with cement.

The prepared metal part and its mating rubber component are placed in the mold cavity and molded in a similar manner to an all-rubber product. Deflashing is generally carried out by hand or with a grinding wheel. One of the disadvantages of compression moldings is that the flash tends to be largest at the thick side of the molded item, making removal by tumbling difficult or impossible. The cost of preparing individual preforms and the placing of each in the mold cavity is another disadvantage of compression molding.

Transfer Molding: Transfer molding involves the transfer of the uncured rubber stock from one part of the mold to another. The stock, in the form of blanks, is placed in a recess called the pot or transfer cavity. The pot is fitted with a ram or piston which is inserted over the stock. The force of the press when applied to the ram plus the heat from the mold causes the stock to be softened and flow through runners into the previously empty molding cavities, where the stock is cured in the desired form. The rubber for transfer molding is compounded in the same way as for compression molding. The rubber stock blanks, which are fed into the transfer pot of the mold, are generally cut from extruded or sheeted-out rubber stock and take the form of slabs. Frequently, the weight of the rubber blank is brought within a certain weight tolerance by trimming overweight blanks. Underweight blanks and the trimmings are recycled to the sheet-out mill. The weight of the blank is regulated to ensure that sufficient rubber is available in the transfer pot to fill the mold cavities. Transfer molds are normally heated by steam and are operated by hydraulic oil systems similar to compression mold hydraulic systems. Oil leaks and spillages are frequent. Curing times in the mold are similar to those of compression molds and likewise are based on product dimension and rubber stock properties.

When the mold is opened the item is pulled or cut from the runners. The runners and the residual rubber in the transfer pots are discarded as waste. The molded item is deflashed by methods similar to those described above for compression molded products. Articles containing metal inserts are generally manufactured by transfer molding and the overall processes are very comparable to those used in compression molding including the preparation of the metal component.

Transfer molding permits closer dimensional control and generally reduces flash so that the parts can be easily finished. Complex shapes can be readily manufactured by this method. Small parts can be made more economically by this technique because of the labor cost savings on preforms and finishing. The disadvantages of this method are that the formulations and control of stocks are much more critical than with compression molds. Stocks must flow well and knit properly, and still cure relatively fast. As explained above, the small amount of compound remaining in the transfer pot, as well as the material in the runners, is wasted. Therefore, this molding method requires careful cost calculations for scrap loss when making items requiring premium priced polymers such as the fluororubbers, silicones, and polyacrylates. Transfer molds are more expensive and in general require more maintenance than compression molds.

Injection Molding: Injection molding is the newest method of molding and requires the greatest degree of sophistication both from the standpoint of materials and mold design. Basically, it is the same as transfer molding with the exception that the stock is injected into the cavities. There are essentially three different types of injection-molding machines. One machine uses a ram to force the stock through runners into the cavities; another uses a screw; the third is a combination of the first two and is a reciprocating screw. From the point of view of water use, there is little difference between the three types of machines. All require cooling water. In some cases the injection molding machines are equipped with their own closed-loop cooling systems.

The molds are often mounted on a revolving turret which takes the molds through a cyclic process. If required, the mold surfaces are treated with release agents followed by closing the mold before injection. Rubber is then injected into the closed mold after which the mold is opened and the molded item removed. Deflashing of the molded item can be carried out by either hand or machine techniques similar to those methods used for the deflashing of compression and transfer molded products.

In order to make injection molding profitable, very short cycles are required which are generally in the 45 to 90 second range. This requires curing temperatures of approximately 400°F. Parts must be readily removed from the molds to keep the heat loss and cycle time to a minimum. All the advantages mentioned for transfer molding apply to injection molding. Efficiency is greatest for a large volume of small items with relatively thin walls and of complicated shape.

Table 31 presents a review of potential process-associated wastewater streams produced in the manufacture of molded items as described in the foregoing text.

TABLE 31: PROCESS-ASSOCIATED WASTEWATER SOURCES FROM THE PRODUCTION OF MOLDED RUBBER ITEMS

Plant Unit or Area	Source	Nature and Origin of Wastewater Contaminants
Oil Storage	Spills and leaks	Oil pick-up by storm run-off
Compounding	Washdown, spills, leaks and discharges from wet air pollution control equipment	Solids from soapstone dip tank Oil and water leaks from mixers and mills Solids from wet air pollution control equipment discharges
Blank and Preform Extrusion and Preparation	Rinse waters, spills and leakages	Rubber fines in lubricant and rinse water, anti-tack agent in cooling tank overflow oil from machinery
Curing	Spills and leaks	Oil from hydraulically operated curing presses

Source: EPA (17)

Extruded Products

As described above, rubber extrusion is used to prepare the preforms and blanks used in compression and transfer molding processes. The extrusion operation is a minor element of the overall production process for molded items. Rubber extrusion, however, plays a more significant role in the manufacture of such items as rubber belting and sheeting. Extruded rubber products can be grouped into two principal classes based on the nature of the manufacturing process. The simpler manufacture, such as sheeting production, involves essentially compounding, extrusion, and curing. On the other hand, items such as belting require a more involved manufacture consisting of all the above processes plus building. However, the building operation, as used in belting manufacture is a relatively straight forward process with little wastewater impact. Therefore, the manufacturing processes used to produce rubber sheeting and belting will be described concurrently.

The majority of the processes used to manufacture sheeting and belting are very similar and serve as good examples of the production methods used to manufacture extruded items. A flow diagram for a typical extrusion production facility is presented in Figure 72. Table 32 presents the sources of wastewater from extrusion facilities. The rubber stock is compounded from the basic ingredients on a compounding mill or Banbury mixer. These pieces of machinery require cooling water. Leakage of lubricating oil and grease can occur. Wet scrubbers are sometimes used to control air polluiton by airborne particles produced in the mixing area. In some cases the rubber is sheeted out on a sheeting mill and dipped in soapstone slurry. Soapstone leakage can occur in this area. After compounding, the rubber is worked on a warm-up mill and fed to the extruder. The extruded rubber is produced as a sheet. In cases where the dimensions of the extruded rubber sheet are critical, the extruded rubber is calendered to the desired thickness. The extruded or calendered rubber is cooled in a cooling tank before storage for further processing. In some cases the extruded or calendered rubber is dipped in soapstone slurry for storage.

Belting is manufactured by extruding the rubber onto the wire reinforcement in the extruder, or calendering the rubber sheeting onto reinforcement fabric that has been frictioned with rubber. Calenders require cooling water. Oil and grease leakage can be produced by the calendering machinery. Belting or extruded and calendered sheeting is cured using a rotacure or press curing technique. A rotacure is an air heated drum. The sheeting and belting pass around the rotating drum and are cured. The press curing technique consists of two heated belts which hold the rubber belting or sheeting between them under pressure to facilitate the curing process. The heated belts turn and drag the sheeting or belting through the press. Cured belting or sheeting is inspected, cut to length, and stored before shipment. In some plants a certain amount of the sheeting is shipped in an uncured state for use in the manufacture of tank linings and other large rubber items. Such sheeting is supplied from the extrusion or calendering line before the belting-formation or curing operations.

FIGURE 72: FLOW DIAGRAM FOR THE PRODUCTION OF TYPICAL EXTRUDED ITEMS SUCH AS BELTING AND SHEETING

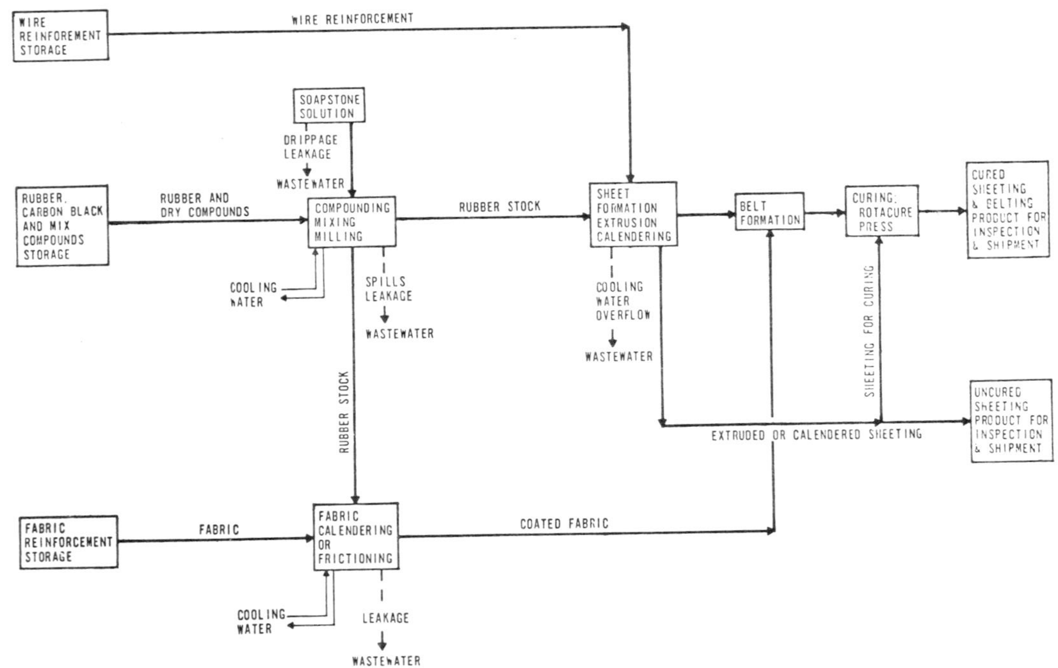

Source: EPA (17)

TABLE 32: PROCESS-ASSOCIATED WASTEWATER SOURCES FROM THE PRODUCTION OF EXTRUDED RUBBER PRODUCTS INCLUDING RUBBER HOSE AND BELTING (SIC 3041)

Plant Unit or Area	Source	Nature and Origin of Wastewater Contaminants
Oil Storage	Spills and leaks	Oil pick-up by storm run-off.
Compounding	Washdown, spills, leaks, and discharges from wet air pollution control equipment	Solids from soapstone dip tank. Oil and water leaks from mixers and mills. Solids from wet air pollution control equipment discharges.
Extrusion	Cooling waters, spoils, and leakages	Anti-tack agent in cooling tank overflows, oil from machinery.
Calendering	Spills and leaks	Solids from soapstone dip tank. Oil and water leaks from mixers and mills.
Curing[1]	Condensate	Organics and lead leached by steam vulcanizer condensate.
Testing[2]	Spills and leaks	Oil pick-up hydraulic testing water.

[1]Waste waters generated by curing operations are essentially limited to hose manufacture.
[2]Testing waters are used only in hose manufacture.

Source: EPA (17)

Fabricated Products

The types of products and processes covered by this manufacturing description are varied. It can be said that this industry sector is a catch-all for manufacturing types not included in the molded or extruded product groups. Processes employed in this industry sector are compounding, milling, fabrication, molding, and vulcanization. Figure 73 illustrates the production steps of typical hose items.

Rubber hose generally consists of three components. They are the tube (lining), the reinforcement, and the outer cover. The tube is the innermost rubber element. Some hose, such as a vacuum cleaner hose, does not have an inner lining or tube inside the reinforcement (usually a wire spiral). The primary function of the tube is to retain the transported material. The type and thickness of rubber used depends upon the intended service of the hose and the type of hose fitting or connecting device to be used.

The reinforcement is the fabric, cord, or metal elements built into the body of the hose to withstand internal pressure or external forces. The type of reinforcing material depends upon the method of manufacture and the service conditions. The rubber used to bond together the individual elements of the reinforcing material is considered a part of the hose reinforcement. The cover is the outermost element. It is normally made of rubber and its prime function is to protect the reinforcement from outside damage or abuse.

While most hose is used for pressure service, there are many applications where an essential property of the hose is its resistance to collapse under suction and vacuum service. The usual method of preventing hose carcass collapse is to build a metal reinforcement, generally a steel wire spiralled in the form of a helix, into the body of the hose.

Hose manufacture can be classified according to the manner in which the hose is produced. Such factors as service, size, production volume, and cost usually determine the method by which the hose is made. The greatest proportion of all hose manufactured today is produced by highly mechanized equipment specially designed for the purpose. Three fundamental methods of hose manufacture exist, producing the following types of hose: [1] machine-wrapped ply hose; [2] hand built hose; and [3] braided and spiralled hose.

Coated Materials: Materials coated with rubber compounds are generally an essential ingredient in the manufacture of fabricated products. Many plants, starting with a rubber compounding operation, coat the fabric material and ship the coated fabric to another plant where it is fabricated into the finished article. Other plants, however, have an integrated facility where the rubber stock is compounded, the fabric is coated, and finally the article is built and vulcanized; all in the one plant.

Rubber-coated materials generally consist of woven or nonwoven fabrics to which a rubber compound or composition has been applied either by impregnation of the fabric or by application to one or both sides. Synthetic rubber materials such as acrylic rubber, butadiene-acrylonitrile, butadiene-styrene, chloroprene, chlorosulfonated polyethylene, fluorinated polymeric compositions, polyisobutylene, polysulfide and silicone polymers are used where particular physical properties such as water and solvent resistance, gas impermeability, flame resistance, surface-release characteristics, abrasion resistance, and good aging properties are required.

Rubber-coated fabrics are generally used for industrial applications where their characteristic odor and color limitations are not objectionable. Such products are frequently specified for usage requiring low-temperature flexibility. Typical uses for rubber-coated textiles include service raincoats, balloon bags, diaphragms, gaskets, inflatable life rafts, pontoons, etc. Rubber coatings are generally applied by calendering techniques.

Rubber-coated cotton sheeting is used to produce friction tape, raincoats, gasketing, and diaphragms. Rubberized knit cotton fabric is used to manufacture rubber overshoes, boots, and diaphragms. Rubber-coated duck and canvas is fabricated into pontoons, lifeboats, and

FIGURE 73: FLOW DIAGRAM FOR THE PRODUCTION OF TYPICAL HOSE ITEMS (INCLUDING REINFORCED TYPES)

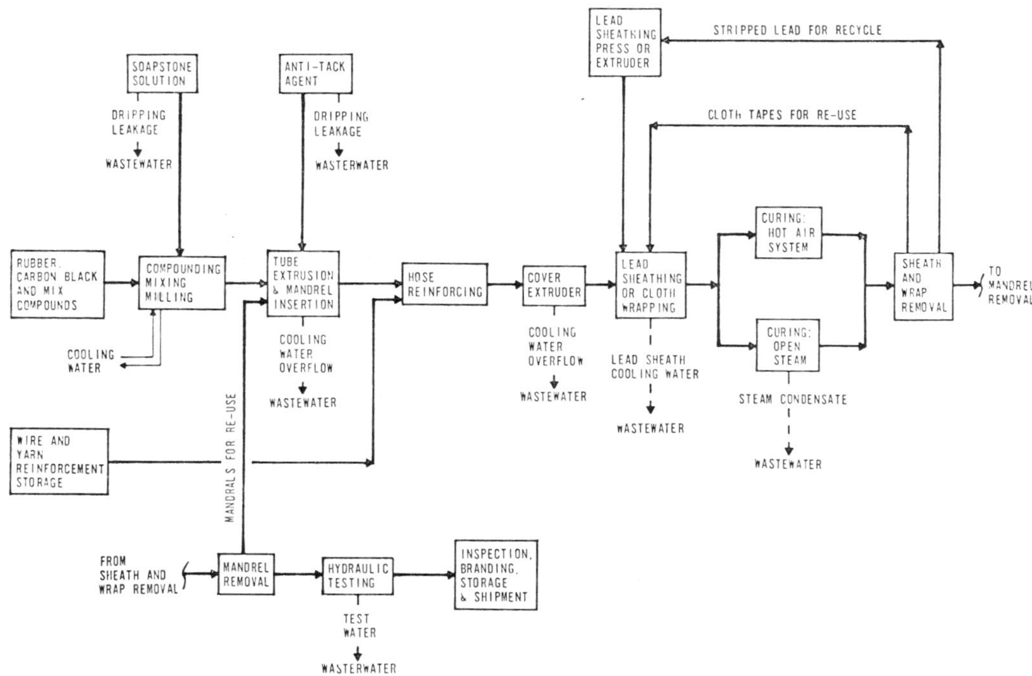

Source: EPA (17)

tarpaulins. The origin of coated fabrics has been traced to the application of preservative resins on Egyptian mummy wrappings. More recent use of coated fabrics during the early nineteenth century evolved from the linseed oil-coated fabrics (oilskins) used by seafarers and the rubber-coated protective garments developed by Charles Macintosh, still known as macintoshes. These early products were accompanied by objectionable odors and poor aging properties resulting either in tackiness or in embrittlement. Improvement in rubber coatings followed Goodyear's discovery of the vulcanization technique in 1839.

In early usage of coated fabrics, the base material was required to give strength and tear resistance to the finished product. In such constructions heavy cotton sheetings, drill weaves, sateens, broken twills, and canvas were required to give proper service. The fabric is often treated before coating. Fabric treatments can include desizing to produce pliability, surface shearing or brushing to remove knots and flaws, dyeing to match the coating, matting to improve softness or coating adhesion, and flame proofing.

Most recent constructions of coated fabrics employ knitted textiles or nonwoven textile webs to achieve maximum softness and tear resistance. These fabrics depend on the toughness of the coating to furnish abrasion resistance and long service life. The use of very lightweight and very strong synthetic fiber fabrics has permitted design of lightweight air-holding products for inflated structures and similar uses. Saturated nonwoven cotton textiles have found extensive applications as garment linings and interfacings.

Before the coating process, the rubber stock is compounded from basic rubber ingredients such as rubber crumb, extenders, pigments, accelerators, and antioxidants. The wastewaters arising in a typical compounding area result from leaks and spillage and the principal contaminants are oil and grease, and suspended solids. The fabric to be coated is received from the textile mill either already dipped in latex or the dipping can be carried out in the rubber coating plant. If a fabric dipping operation is employed, latex spillage and washdown are potential wastewater sources.

Rubber coating is performed in either three or four-roll calenders. The compounded rubber stock is generally prepared for the calender on a warm-up mill. The three-roll calender applies the coat to one side of the fabric and the four-roll calenders coat both sides of the fabric. The top roll of the three-roll calender or the bottom and offset rolls of the four-roll calender are run at a different speed to the center roll, usually two-thirds as fast, to friction the rubber coating onto the fabric.

Rubber coated fabrics need to be cured at elevated temperatures for periods of time varying from ten minutes to several hours. The curing ovens can be 30 feet high and hundreds of feet long, or 6 to 8 feet in height and 8 to 20 feet in length for products requiring a shorter curing cycle. The principal requirement for the ovens is that of uniform temperature distribution to obtain uniform product quality. After curing, the coated fabric is cooled and rolled.

Fabricated products such as rainwear, rafts and pontoons are built using dies or jigs to cut the coated material and rubber cements to join the various sections. In general the building areas are dry and no wastewaters should arise. In the event of a spillage of rubber cement, the spill or leak would most effectively be wiped away.

The types and characteristics of wastewaters produced by a rubber coating or fabrication manufacturing facility are similar to those of a rubber footwear facility and are presented in Table 33 (page 187).

Rubber Footwear: The rubber footwear industry (SIC 3021) is the second largest subsector of the general fabricated products group and within this subsector canvas footwear constitutes the major product type. The process description presented below pertains to canvas shoe production which utilizes all the major processing technologies commonly found in the manufacture of general fabricated products. A schematic flow diagram for a typical canvas shoe production facility is shown in Figure 74. The various rubber stocks consumed

FIGURE 74: FLOW DIAGRAM FOR THE PRODUCTION OF TYPICAL CANVAS FOOTWEAR ITEMS

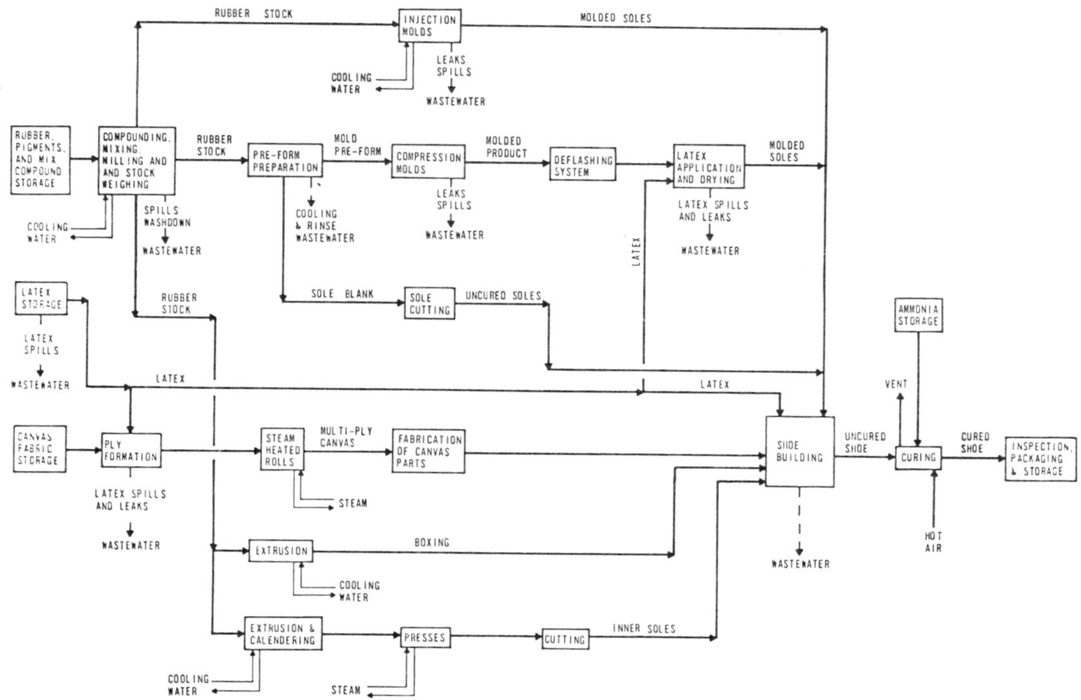

Source: EPA (17)

in a canvas shoe plant are compounded in Banbury mixers or compounding roll mills and then sheeted out. The sheeted rubber is dipped in an antitack solution to prevent sticking during storage. The canvas shoe is built from four major components: soles, canvas uppers, boxing, and inner soles. These components are made separately by varying operations before being brought together in the fabrication operation.

The soles are generally molded using injection, compression, and transfer molding techniques. All molding processes can produce oil spills and leaks; however, compression and transfer molding equipment generally produce more oil spillage than the injection molding machines.

The molded soles are deflashed, usually in a buffing machine, before coating with latex adhesive. The latex coating is dried in an oven. The canvas components for footwear are made from two or three-ply fabric. The fabric is received at the plant as single sheets. Latex is applied to the plies, which are pulled together and passed over a heated drum. The sheets are stacked and the multilayer canvas is stamped to shape. The different canvas components making up the shoe uppers are stitched together on sewing machines. The boxing, or edging, which protects the join between the sole and the canvas uppers is extruded as a long strip from rubber stock.

The inner sole is extruded in a flat sheet from a special rubber stock. The extruded sheet is passed through heated presses. Blowing agents, such as sodium bicarbonate and azodicarbonamide, which are mixed into the rubber stock in the compounding area, decompose and release gases which blow the extruded sheet into cellular sponge. The inner soles are then stamped out of the cellular sheet.

The shoe is built from the various components on a last. Firstly the canvas upper is cemented at its edges, and placed over the last. The inner sole is attached to the bottom of the last. The outer sole, toe and heel pieces, and boxing are placed on the shoe using latex as an adhesive. The complete, uncured shoe is usually inspected and placed in an autoclave to cure.

The autoclave is air heated. Anhydrous ammonia is injected into the autoclave to complete the cure. Curing with ammonia produces a good surface texture on the rubber and eliminates the residual tackiness associated with rubber that is cured conventionally. Some shoes are cured without ammonia. This is done where the tackiness of the product is not very important or where the compounding recipe can be modified to eliminate the tackiness often associated with regular air curing. Steam is not used for curing because in many cases the steam would stain the canvas parts of the shoe.

The curing cycle can last about one hour and approximately two to five pounds of ammonia are used for every thousand pairs of shoes cured. At the end of the curing cycle the ammonia/air mixture is vented to the atmosphere. No air pollution problems or requirements appear to be associated with this practice and air pollution control devices are not envisaged.

As described above, latex is used in several applications as an adhesive. The latex is received at the footwear plant either in bulk via tank truck or in 55-gallon drums. Spills, leaks, and cleanout wastewaters, laden in uncoagulated latex solids, are frequently produced. The nature of the wastewaters produced by a typical footwear production facility are listed in Table 33.

Tire Retreading: Tire retreading is an industry dominated by independents. Approximately 5,000 retreading shops or plants are currently registered with the Federal Department of Transportation and the number of tires retreaded by the industry as a whole approaches 32 million each year. There are a few large retreading plants which are operated by the major tire companies; in most aspects these are very comparable to a plant manufacturing new tires. An average retreading requires approximately 10 pounds of rubber per passenger tire and 35 pounds per truck tire. Very few tire retreading facilities compound their own rubber stock. Those that do mix stock will have potential wastewater contamination prob-

TABLE 33: PROCESS-ASSOCIATED WASTEWATER SOURCES FROM RUBBER FOOTWEAR PRODUCTION

Plant Unit or Area	Source	Nature and Origin of Wastewater Contaminants
Oil Storage	Spills and leaks	Oil pick-up by storm run-off.
Latex Storage	Spills, leakage, washdown, and cleanout waters	Dissolved organics, suspended and dissolved solids.
Compounding	Washdown, spills, leaks and discharges from wet air pollution control equipment	Solids from soapstone dip tank. Oil and water leaks from mixers and mills. Solids from wet air pollution control equipment discharges.
Preform Extrusion and Preparation	Rinse waters, spills and leaks	Rubber fines in lubricant and rinse water anti-tack agents in cooling tank overflows, oil from machinery.
Ply Formation	Spills, leaks and washdown	Dissolved organics, suspended and dissolved solids.
Shoe Building	Spills, leaks and washdown	Dissolved organics, suspended and dissolved solids.
Curing	Discharges from air pollution equipment	Ammonia used in curing.

Source: EPA (17)

lems with oil from machinery and suspended solids from soapstone dip equipment and wet scrubbing devices, although well-designed curbing should contain such spillages. After compounding, the stock is extruded to produce the camelback tread rubber. The extruded camelback is usually passed through a water cooling tank which has an overflow. However, this cooling water overflow is uncontaminated and does not require treatment.

The majority of retreading shops purchase the rubber stock from an outside supplier in the form of camelback tread or extruded rubber. The worn tire is first visually inspected to ascertain its suitability of retreading. Those failing the inspection are removed from the retreading shop as solid waste. The satisfactory worn tire is buffed with a grinding wheel to remove the old, worn tread rubber. Rubber buffings collected in the buffing area are solid waste and are periodically containerized and removed from the plant. Few plants have air pollution control devices to remove the fine buffings from the air.

The buffed casing is coated with rubber cement and the camelback tread or extruded rubber is applied around the tire and cut to length. The tire with tread rubber is placed in the curing mold and the mold is closed. Most of the curing molds are steam heated and the steam condensate is recycled to the boiler. Some molds are heated with electricity; these are generally older than the steam heated molds.

After curing, the tire is removed from the mold. The rubber flash is buffed off the tire before it is inspected and shipped. The process wastewater types generated in a typical retreading shop are minor and arise from spillage and washdowns. These wastewater types are adequately covered by those wastewaters listed in Table 29 for general molded products. The contaminants of these wastewaters are oil and suspended solids. However, most plants do not have any process wastewaters at all, and the wastewater discharges are limited to boiler blowdown, cooling water, and sanitary wastewaters.

Reclaimed Rubber Production

The quantity of scrap rubber being reclaimed and reused and the number of rubber reclaiming plants operating in the United States have both steadily declined over the past decade. This decline has occurred despite increased production and use of rubber products. As a percent of new rubber produced, reclaimed rubber fell from 19 percent in 1958 to 10 percent in 1968. Some of this reduction is probably due to development of new rubbers not compatible with reclaimed rubber, but, undoubtedly, the major decreases were caused by cost, quality, and environmental reasons.

Table 34 indicates the usage of reclaimed rubber during the 1960's and it can be seen that substantial reductions have occurred in some applications. Competitive materials, such as rugs and colored plastics, have reduced usage in automotive mats and mechanical parts from 105 million pounds in 1960 to approximately 55 million pounds. Similar reductions are noted for other mechanical goods, hose, shoe heels and soles, and hard rubber products.

Reclaimed rubber is the product resulting from the treatment of ground scrap tires, tubes, and miscellaneous waste rubber articles with heat and chemical agents whereby a substantial devulcanization or regeneration of the rubber compound to its original plastic state is effected, thus permitting the product to be reprocessed, compounded, and revulcanized. The term devulcanization which is frequently associated with reclaiming is a misnomer. Actually all the commercial reclaiming processes employed are based on depolymerization of the rubbers. This depolymerization can occur either by promoting thermal scission or breaking of the polymer chain or by oxidation at points other than at cross-linking sites. Some scission of the existing cross-links may also occur.

There are two fundamental factors which determine the type of reclaim. The first, and the most important, is the type of scrap from which the reclaim is made. The second is the process by which the scrap is reclaimed. By far the most important source of raw material is tire scrap. The supply is plentiful and well distributed so that it is relatively easy to collect. The quality of rubber in tires is high, giving an unusually high percentage

TABLE 34: CONSUMPTION OF RECLAIMED RUBBER BY PRODUCT

	Million Pounds of Product			
Automotive Products	1960	1967	1968	1969
Tires, Inner Tubes, and Tire Repair Material	380.6	379.5	415.3	396.0
Automobile Mats and Mechanical Parts	104.8	55.8	57.6	55.3
Hose and Belt	36.1	26.2	24.9	31.8
Mechanical Goods	52.6	31.8	23.5	32.9
Non-Automotive Products				
Cements and Dispersions	15.7	16.4	18.4	19.7
Heels and Soles	17.0	8.7	11.4	5.6
Hard Rubber	26.2	14.3	7.4	6.9
Rubber Surface	--	4.9	5.1	3.3
Other	22.9	11.0	12.1	8.3
Total	655.9	548.6	575.7	559.6

Source: EPA (17)

of rubber hydrocarbon at low cost. The whole tire creates problems due to the tire-cord fiber contained in the carcass portion. This fiber has to be removed either by mechanical means or by chemical methods such as those used in the digester process.

Three basic techniques are used at existing plants to produce reclaimed rubber: the digester process, the pan process, or the mechanical process. A generalized material flow diagram for the three processes is shown in Figure 75. Broadly, the reclaiming process can be divided into three major parts, two of which are mostly mechanical and the other predominantly chemical. The rubber scrap is first separated and ground, then given heat treatment for depolymerization, and finally processed by intensive friction milling. All three processes employ similar rubber-scrap separation and size-reduction methods. They differ in the depolymerization and the final processing steps.

Rubber Separation and Size Reduction: The rubber scrap is first sorted, and then reduced by mechanical chopping or cracking on a very heavy cracker mill to a suitable size for the particular depolymerization step being used. A cracker mill consists of two horizontal heavy steel rolls revolving at different speeds. The roll surfaces are corrugated, giving a scissor-cutting action. In the case of tires, the wire beads are broken in the cracker mill and removed with magnetic separators along with any other ferrous material which may have been picked up during their use. The cracked ground stock is conveyed to a vibrating screen of a given mesh size. The oversized material is returned to the crackers for further grinding. The stock which passes through the screen is conveyed to storage bins to await use in the depolymerization process.

Most reclaiming plants require fiber-free scrap in the depolymerization process. A series of screens, air separators, and sizing equipment are used to remove fiber from ground rubber scrap. First the rubber scrap is passed through hammermills and beaters which remove fiber from the rubber. This screened mass then goes to an air-flotation table where the final separation of fiber and rubber takes place. After passing through magnetic separators, the rubber crumb which results goes to storage bins. Rubber scrap separation and size reduction is followed by the appropriate depolymerization process.

Depolymerization: Digester Process— The digester process consists of placing the ground scrap, water, and reclaiming agents into a steam-jacketed agitator-equipped autoclave (digester). The batch is then cooked for 5 to 24 hours at 370° to 450°F. During this time the rubber becomes partially depolymerized, and attains the consistency of soft granules. Reclaiming agents that are used include petroleum and coal tar-base oils and resins, as well as various chemical softeners, such as phenol alkyl sulfides and disulfides, thiols (mercaptans), and amino compounds.

Reclaiming agents or oils are used to speed the depolymerization, to impart desirable processing properties. Natural rubber can be reclaimed in the absence of any reclaiming agents. SBR types, however, require the assistance of these reclaiming agents or catalysts to produce a reclaim. The reclaiming agents generally function by catalyzing the oxidative breakdown of the polymer chain and oxidatively disrupting sulfur cross-links.

Rubber scrap which has not been defibered mechanically requires chemical degradation of the fibers in the digestion process. Defibering agents, such as caustic soda or the chlorides of zinc and calcium, and plasticizing oils are added to the digester to complete the charge. The fiber from tires and other rubber scrap products is hydrolyzed and goes into solution. The presence of synthetic rubber such as SBR in the tire scrap necessitates the use of the metallic chlorides in place of the caustic soda since the caustic solution produces a heat-hardening effect with SBR instead of softening.

At the end of the digestion period the contents of the digester are blown down under internal pressure into a blowdown tank. Water is added to the soupy mass to facilitate the subsequent washing operations. After thorough agitation, the mixture is discharged onto continuous vibrating screens where a series of spray nozzle showers wash the rubber free from the digester liquor. The washed rubber, which is in the form of a slurry, is then

FIGURE 75: FLOW DIAGRAM OF TYPICAL MECHANICAL, PAN (HEATER), AND WET DIGESTER RECLAIM PROCESSES

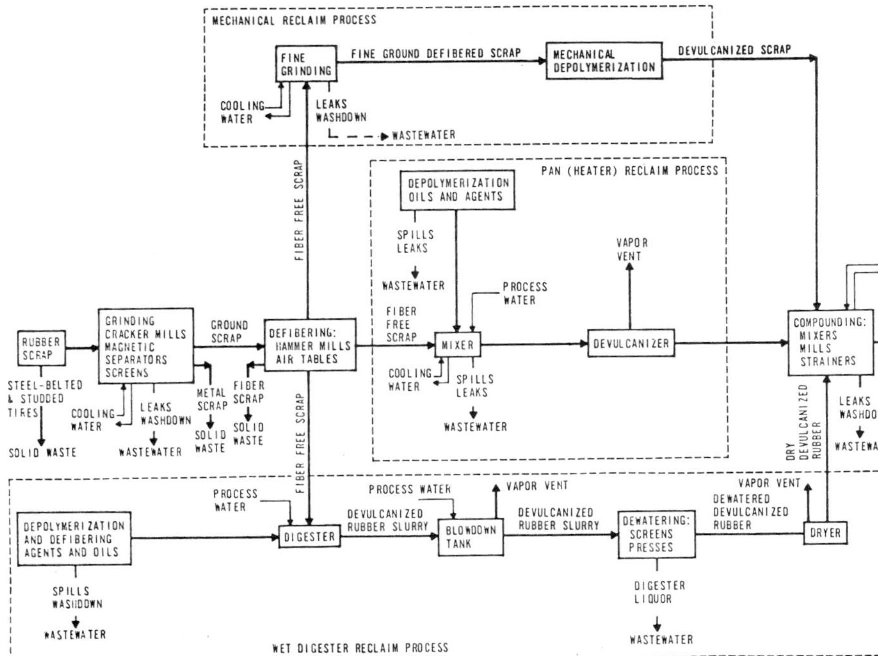

Note: Some reclaiming facilities operate more than one type of process

Source: EPA (17)

passed through a dewatering press which forces out much of the occluded water. The dewatered rubber is dried in a hot-oven or tray dryer prior to further processing. In cases where chemical defibering is carried out in the digestion process, the rubber has to be washed free from the decomposed fiber as well as the digestion liquor. This washing procedure generates wastewater which can be alkaline or laden with metal chlorides and fine sludge-like particles of hydrolyzed fiber and rubber. Chemical defibering and the subsequent washing process create an effluent problem. At present, mechanical fiber separation is used to reduce the wastewater problem inherent in the digester process.

Pan Process: The finely cracked scrap, which is usually free from fiber, is reduced to a finer particle size by grinding on smooth steel rolls. The finely ground scrap is blended in an open mixer with the correct amount of reclaiming oils and is then placed in open pans which are stacked on a carriage and placed in a large horizontal heater. The heater is a single-shell pressure vessel into which live steam is passed. Depolymerization is carried out at about 365°F for 2 to 18 hours. After this treatment the heater is vented, the pans discharged, and the cakes of rubber sent on for further processing.

Since the condensate from this operation is highly contaminated with gums, resins, etc., it cannot be returned to the boilers and, therefore, must be treated as a waste. This waste is similar in composition to the digester alkali and chloride wastes. The pan process is relatively inexpensive because the equipment is simple, and the washing and drying steps are eliminated. The supply of cheap fiber-free scrap rubber can be a limiting factor in the operation of the pan reclaiming process.

Mechanical Process: The mechanical reclaiming process, unlike the other two preceding processes, is continuous. The fine ground, fabric-free rubber scrap is fed continuously into a high-temperature, high-shear machine. The machine is a horizontal cylinder containing a screw for forcing and working the material along the chamber wall at 350° to 400°F in the presence of reclaiming agents and catalysts for depolymerization. The rate of depolymerization is controlled by the speed of screw while the compression and temperature are maintained constant. The discharged reclaimed rubber needs no drying and is ready for further processing.

Final Processing: The final stage of the reclaiming operation, namely milling, first involves the mixing and blending of the material from any of the various depolymerization processes in a Banbury internal mixer with small amounts of reinforcing materials such as clay, carbon black, and softeners. This aids in smoothing dried stock and obtaining uniformity. The reclaim is then given a preliminary refining on a short two-roll mill having a high-friction ratio between the roll surfaces. The sheet thickness after the first refiner pass is about 0.01 inch. The reclaim is then strained to remove foreign matter before going to the final thickness of 0.002 to 0.005 inch. The strainer is an extruder which contains a wire mesh screen held between two strong perforated steel plates in the head of the machine. The strainer not only removes foreign matter, but also plasticizes and blends the reclaim.

Each reclaimer may complete his reclaiming operations in either of two ways, by sending his product to the customer in the form of slabs stacked on pallets or on bales. Slabbed reclaim is made on a mill and the discharged sheet is wrapped on a rotating drum of a specified diameter, until the proper thickness is obtained. The wrapped layers or sheet are then cut off the drums, forming a solid slab of a certain length, width, and weight. The slabs are then dusted with talc to prevent sticking to each other, tested, and shipped to the customer.

Baled reclaim, is also made on a mill, except the thin milled sheet is conveyed to a baler, where the rubber is compacted to form a bale. The bale is then encased in a bag, stacked on a pallet, tested, and sent to the customer. The effluent wastewaters occurring during the various reclaiming processes are identified in Table 35.

TABLE 35: PROCESS-ASSOCIATED WASTEWATER SOURCES FROM RUBBER RECLAIMING

Plant Unit or Area	Source	Nature and Origin of Wastewater Contaminants
Wet Digester Reclaim Process		
Grinding	Washdown, spills, leaks	Bearing oil from machine drives. Oil from seals on milling equipment.
Depolymerization, Defibering and Oil Storage	Spills, runoff	Oil, solids, caustic and organics.
Blowdown Tank	Air pollution equipment	Oils and organics.
Dewatering	Digester liquor	Oil, solids, caustic and organics from spent depolymerization and defibering agents and excess oil. High concentrations of fibrous material removed from tires.
Dryers	Air pollution equipment	Oils and organics.
Compounding	Washdown, spills, leaks, air pollution equipment	Bearing oil from machine drives. Oils from seals on milling equipment. Solids from soapstone dip tanks. Solids from air pollution equipment.
Pan (Heater) Reclaim Process		
Grinding	Washdown, spills, leaks	Bearing oil from machine drives. Oil from seals on milling equipment.
Depolymerization Agent and Oil Storage	Spills, runoff	Oil, solids, caustic and organics.
Devulcanizer	Air pollution equipment	Oil and organics.
Compounding	Washdown, spills, leaks, air pollution equipment	Bearing oil from machine drives. Oils from seals on milling equipment. Solids from soapstone dip tanks. Solids from air pollution equipment.

Source: EPA (17)

Latex-Based Products

To manufacture sundry rubber goods from latex compounds, it is necessary to convert the compounds into solids of the desired form. Latex compounds are generally stabilized by the latex producer before shipping to the rubber goods manufacturing facility. Here, the stabilized latex compound is transformed into the final rubber goods. Prior to forming the rubber goods, the latex is compounded (mixed) with various ingredients, such as antioxidants, in accordance with a specific recipe. The mixing of latex compounds is a simple operation consisting of weighing out the proper amounts of the various solutions and emulsions required, and then stirring these materials into the latex, usually in a large tank equipped with a mechanical agitator.

In some cases, such as the compounding of latex for foam sponge, some of the ingredients must be added just prior to or during the foaming operation, and in these cases complicated automatic proportioning equipment is sometimes used.

Several types of rubber goods are fabricated from latex mixtures. The basic manufacturing processes and their wastewater generations for these product types are described below. The major classes of latex-based goods are dipped goods, such as finger cots and surgical gloves, and foam-backing materials.

Latex-Based Dipped Goods: There are two principal techniques used in the manufacture of dipped rubber goods. One is the straight-dip method employed in the production of very thin-walled dipped goods from which water can readily and quickly be removed by evaporation. The second technique is coagulation-dipping where the rubber goods are produced by coagulating a film of rubber from a rubber latex onto shaped forms. Thicker-walled items are made by coagulation-dipping rather than those manufactured by the straight-dip method. The thicker rubber deposit cannot be dried as readily and must be assisted by coagulation in order that it does not disintegrate during subsequent processing.

Figure 76 is a flow diagram for a typical coagulation dipping operation. Such a facility might be engaged in the manufacture of rubber gloves. A closed-loop conveyor transports the forms through various processing units. The forms can be made of glazed porcelain or polished metal.

The forms are first dried and heated to 100° to 120°F in a conditioning oven prior to dipping in the coagulant solution. The coagulant solution is usually a mixture of organic solvents and coagulants. Combinations of ethanol and acetone are generally used as solvents. Typical coagulants are calcium nitrate, calcium chloride, and zinc nitrate. A surfactant is sometimes added to the mixture to ensure good wetting of the forms, and release agents are added in cases where the form has a complicated shape and removal of the dipped goods from the forms is difficult.

After coating with coagulants, the forms are dipped in the rubber latex. The rubber latex and ingredients are compounded prior to the dipping operation. In some cases, the latex storage and compounding tanks are cooled with cooling coils or jackets to prevent degradation of the latex. The coagulant film on the surface of the form causes the rubber emulsion to break. The latex solids coalesce to produce a film of rubber that covers and adheres to the form.

The coated forms are passed through a preliminary drying oven to dry the film sufficiently so that it does not disintegrate and wash away in the subsequent washing step. In the washing operation the soluble constituents of the rubber film are leached out and rinsed away. Important constituents of the leachate are the emulsifiers used originally in the production of the latex and metal ions from the coagulant mixture.

The coated forms are sent through a drying oven to dry the goods. In some applications, such as rubber gloves manufacture, the goods are not only dried but heated sufficiently so that the rubber coating can be rolled downward on itself to form a reinforced cuff bead.

FIGURE 76: FLOW DIAGRAM FOR THE PRODUCTION OF TYPICAL LATEX-BASED DIPPED ITEMS

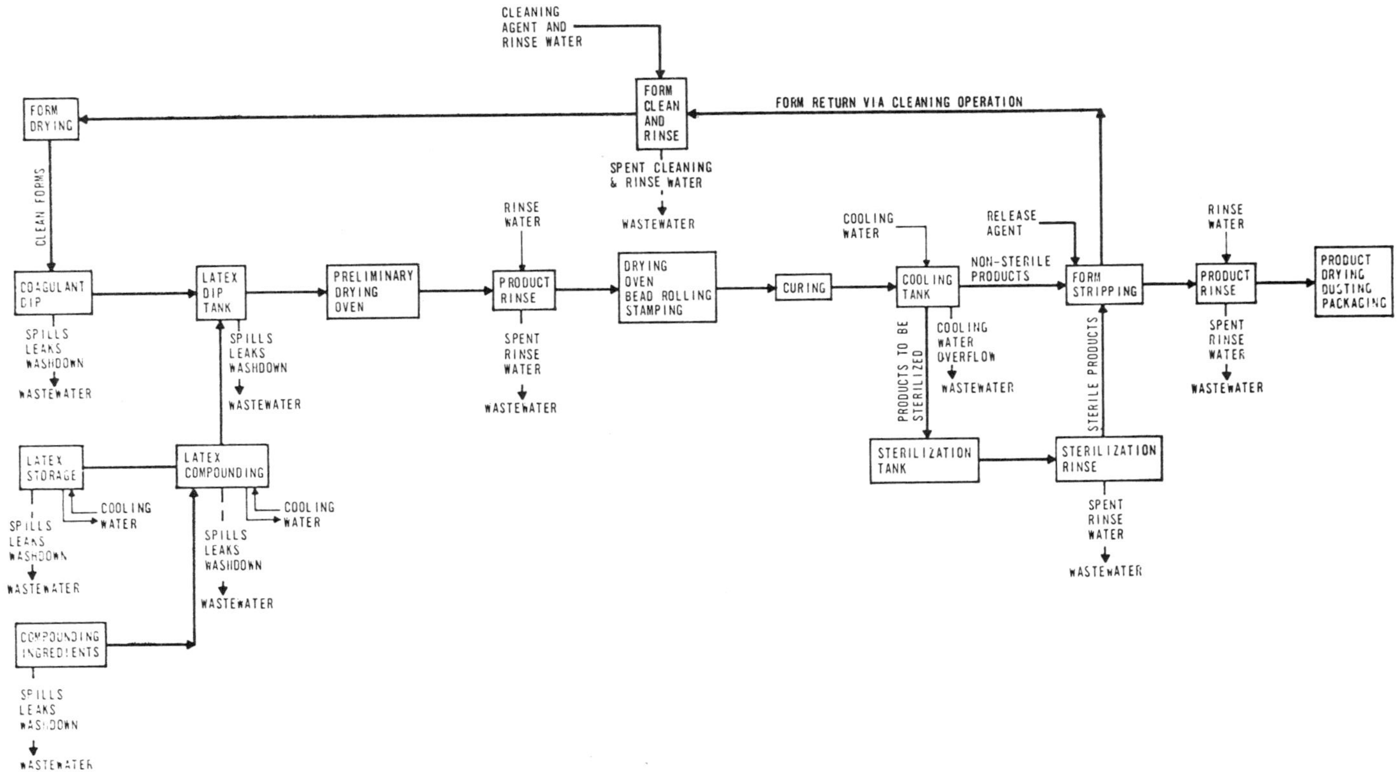

Source: EPA (17)

In most applications the rubber goods are stamped with the proprietary brands and other information such as size in a stamping unit after the drying process. The rubber goods are cured in an oven at approximately 200°F. After curing the items are cooled in a water cooling tank and mechanically stripped from the forms usually with the aid of a lubricating detergent. The detergent is subsequently washed from the goods in a rinse tank. The final manufacturing operation consists of drying the goods, dusting them inside and outside with talc to prevent sticking and packaging.

In cases where sterilized products are required, such as surgical rubber gloves, the goods can be immersed in a chlorine dip tank (free chlorine concentration typically 1,000 mg/l) to provide disinfection and improve the surface finish of the glove. After disinfection the goods are dipped in a hot water tank (approximately 170°F) to remove the residual chlorine from the rubber product. These two operations generally occur between the post-curing cooling tank and the final drying and packaging operation. In many cases the gloves are sterilized by dipping in a hot water bath (approximately 200°F).

Periodically, it is necessary to clean the form upon which the goods are deposited. When this is necessary, the forms are passed through a bath containing a cleaning agent. In the case of porcelain forms the agent used can be chromic acid (mixture of potassium dichromate, sulfuric acid, and water). The cleaned forms are rinsed of residual chromic acid in a subsequent rinse tank. The tank is equipped with a fresh water makeup and overflow to blow down the accumulation of cleaning agent. The cleaning frequency is generally in the order of once a week. Other methods of cleaning involve simply scrubbing the forms with cleaning agents followed by rinsing.

The wastewater sources and characteristics of a typical dipping operation are presented in Table 36. The straight-dip method is the simplest of any used in making articles from latex. The forms are dipped directly into the latex and slowly removed. After dipping, the form is slowly rotated while the liquid film is drying to ensure a uniform film thickness. The films are dried at room temperature or in warm air at 120° to 140°F. Thicker articles can be made by a multiple-dipping process with drying between dips. Latex deposits vary from 0.005 to 0.10 inch per dip, depending on the viscosity of the latex compound.

Cement-Based Dipped Goods: It is appropriate to discuss the manufacture of dipped goods from rubber cement here because the production process has similarities to the manufacture of latex-based dipped goods as described above (see Figure 77).

Various products are made via cement dipping processes. The following process description is oriented towards the manufacture of gloves having a high electrical resistance. Cement dipping results in a product which has good electrical resistance since no water or ionic species are trapped in the cement as would be the case with latex dipped gloves.

The solid rubber required for the cement recipe is compounded in a small Banbury mixer or compounding mill. The recipe ingredients include antioxidants, curing agents, and pigments. The compounded stock is cut in small pieces to facilitate dissolution in the solvent. These pieces of stock are separated by weight into predetermined quantities and placed in a bin. The wastes generated in the rubber compounding and weighing areas result from spills and leakages from machinery, powders from compounding, and washdown wastewaters.

The rubber cement is prepared in blend tanks using fixed quantities of rubber stock and solvent. The solvent used is generally aliphatic in nature, for example, hexane. The blended cement is pumped to a storage tank prior to its use in the dipping operation. Several rubber cements of different colors and properties are stored simultaneously awaiting the dipping operation. Solvent and rubber cement leaks can occur in this area. The gloves are formed by dipping on glazed porcelain forms. In the case of linesmen gloves the rubber layer is built up with about twenty to thirty dips. The dipped products are allowed to drip-dry between dips. The temperature and humidity of the air in the drying room is controlled to ensure good drying conditions. It is possible that the quality of the exhaust drying air

TABLE 36: PROCESS-ASSOCIATED WASTEWATER SOURCES FROM LATEX-BASED DIPPED GOODS PRODUCTION

Plant Unit or Area	Source	Nature and Origin of Wastewater Contaminants
Latex Storage and Compounding	Spills, leaks, and cleanout rinse waters	Dissolved organics, suspended and dissolved solids. High quantities of uncoagulated latex.
Coagulant Dip	Spills, leaks, and cleanout rinse water	Dissolved organics, dissolved solids.
Product Wash, Cooling, and Rinse Tanks	Spills, leaks and overflow waters	Dissolved organics, suspended and dissolved solids.
Sterilization Process	Spills, leaks, and rinse overflow waters	Dissolved solids, sterilzation agent.
Form Cleaning	Form wash and rinse waters	Dissolved organics, suspended and dissolved solids.
All Plant Areas	Area washdown and storm runoff	Organics, suspended and dissolved solids.

Source: EPA (17)

FIGURE 77: FLOW DIAGRAM FOR THE PRODUCTION OF TYPICAL CEMENT DIPPED ITEMS

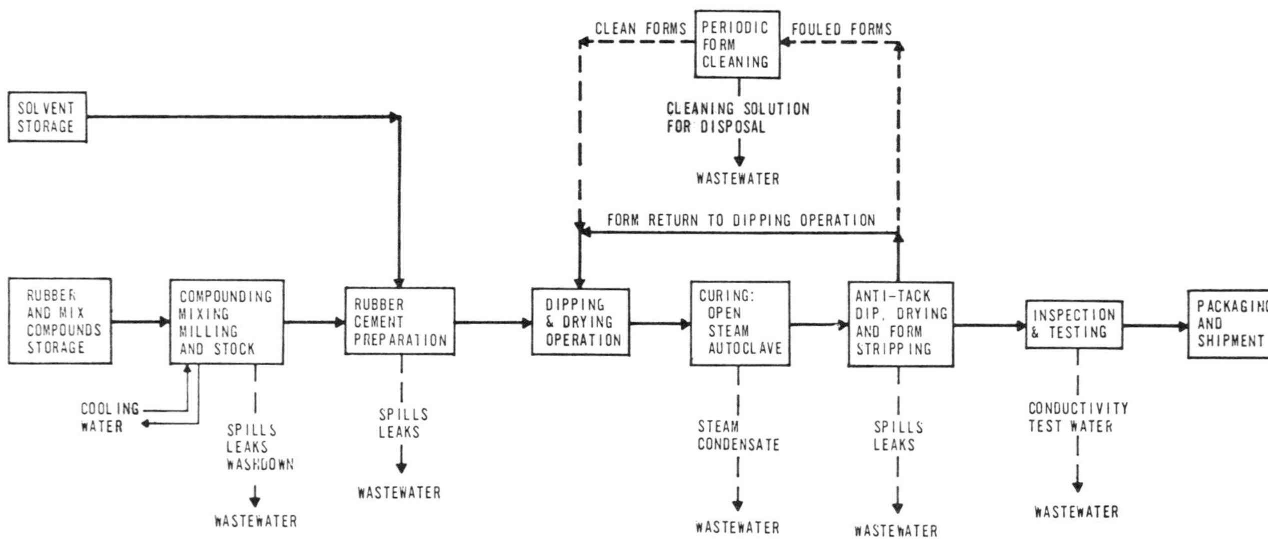

Source: EPA (17)

which is solvent laden, will be subject to control under air quality legislation in the future. When the dipping and drying operation is completed, the gloves are stamped with size and brand information and the cuff bead is formed by rolling the cuff back on itself.

The gloves are cured in an open steam autoclave vulcanizer. The temperature and length of curing depend on the type of glove and properties of the rubber. The steam condensate leaches organics from the rubber. The condensate is extremely low in volume and is discharged to the plant drain. At the end of the curing cycle the gloves are removed from the vulcanizer and left to air cool.

When partly cool the gloves are dipped in a soapstone slurry or equivalent antitack agent prior to final cooling. The soapstone slurry dries leaving a powder on the gloves which are then stripped from the forms. The gloves are dusted in a rotating drum with talc powder and sent to product inspection.

Gloves which pass a visual inspection are tested for electrical resistance. This operation involves filling the gloves with water and placing the gloves in a tank of water. A high voltage is applied between two electrodes; one electrode inside the water-filled glove and the other outside in the water-filled tank. At a given voltage, a satisfactory glove limits the current flow. Gloves removed from the tank are dried and packaged prior to shipment.

Periodically the forms require cleaning. This is carried out with a mild scouring slurry followed by rinsing of the forms. Spent scouring slurry and the rinse waters are low in volume. The types of wastewater generated by a typical cement dipping operation are listed in Table 37.

Latex Molded Products: Porous molds prepared from plaster of Paris or unglazed porcelain with pore sizes smaller than the smallest rubber particles, are used in the rubber sundry industry. The latex compound is poured through a funnel-shaped opening into the mold. The latex compound is allowed to dwell in the mold until a deposit of the desired thickness has developed on the mold wall. The mold is emptied of excess compound and placed in an oven to dry at 140°F for one hour.

With some articles, to prevent pour lines, the mold is rotated on all planes for 15 to 30 minutes to give the latex an opportunity to flow to all extremities of the mold interior before setting. This technique is used for dolls and squeeze toys. The interior rubber surfaces are dusted with talc to prevent sticking when being removed from the mold. The article, after stripping from the plaster mold, may be returned to the 140°F oven for 30 minutes or a shorter period at a higher temperature to facilitate drying.

The types of wastewater generated by this sector of the industry are similar to those produced by a latex-dipping facility. They are characterized by uncoagulated latex solids, and are evolved by spills and leaks and tank washing operations. The wastewater types are similar to those produced by the manufacture of latex-dipped goods (Table 36).

Latex Thread: The manufacture of thread from latex (referred to as latex thread) makes use of some of the general principles and methods described above. The most widely used method is extrusion of the latex compound through fine orifices into a coagulant bath which gels the thread, followed by mechanical handling of the thread during toughening, washing, drying, and curing operations. The coagulant bath is usually dilute acetic acid.

Latex Foam: Although the number of plants and rubber companies involved in this sector of the industry has decreased over the past several years, foamed-latex sponge rubber still constitutes one of the more important applications for latex, both natural and synthetic. (See Figure 78).

The latex may consist entirely of natural latex or synthetic SBR latex or it can be a mixture of natural and an SBR latex. The load-bearing capacity of the foams at a given density falls significantly as SBR is used in place of natural rubber. Latex rubber foams are generally prepared in slab or molded forms in the density range of 4 to 8 pounds per cubic foot.

TABLE 37: PROCESS-ASSOCIATED WASTEWATER SOURCES FROM CEMENT DIPPED GOODS PRODUCTION

Plant Unit or Area	Source	Nature and Source of Wastewater Contaminants
Oil and Solvent Storage	Spills, leaks, and storm runoff	Oil and organics pick-up by storm water.
Compounding and Weighing	Spills, leaks, and washdown	Rubber ingredient solids, oil and water leaks from mixers and mills.
Cement Preparation	Spills, leaks and washdown	Soluble and insoluble organics from solvent spills and leaks.
Curing	Condensate	Soluble and insoluble organics leached from product.
Form Stripping	Spills, leaks, and washdown	Solids from anti-tack agent.
Form Cleaning	Spills, leaks, rinse and cleaning agent discharges	Solids and surfactants from the cleaning agent.
Product Inspection	Conductivity Test Water Overflow	Water is uncontaminated.

Source: EPA (17)

FIGURE 78: FLOW DIAGRAM FOR THE PRODUCTION OF TYPICAL LATEX FOAM ITEMS

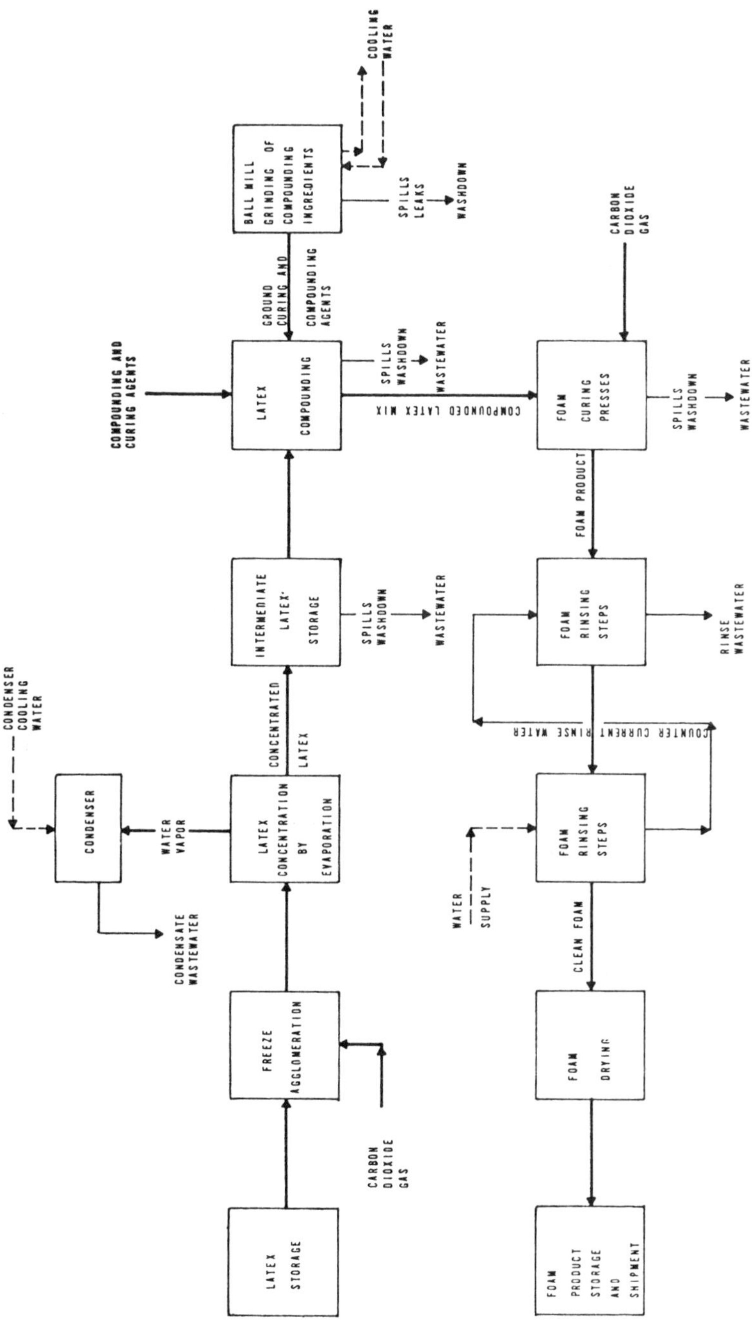

Source: EPA (17)

Many different processes are patented for preparing this type of product, but there are two of prime commercial interest for manufacturing such articles as molded-foam cushioning stock. These are the Dunlop (the most widely used), and the Talalay processes. Some producers have developed variations which in effect are a combination of the two processes.

The basic aspects of the Dunlop processes are: [1] mechanically whipping the latex to a froth; [2] settling the frothed mass with a coagulant or gelling agent; and [3] vulcanizing the rubber so that the foam is permanent.

The latex is first whipped on a batch or continuous basis to produce the foam. The Oakes continuous mixer is the standard piece of equipment used by the industry to prepare the foamed latex. The gelling agents are added to this foam.

Proper coagulation of the latex to give a stable foam (commonly referred to as gelation) is the key to the process. The gelling agent is one that can be mixed into the frothed latex, then remain dormant long enough to allow the froth to be poured into molds before producing the gelling effect. The gelling system usually consists of sodium silicofluoride in conjunction with zinc oxide.

The foam is poured into molds and cured. The molds are usually steam heated. When the curing cycle is completed, the product is removed from the mold and washed with water to remove those ingredients of the latex recipe which are not held permanently in the foam matrix. The foam is dried in a hot air dryer prior to inspection, storage, and shipment.

In the Talalay process the froth is produced by chemical rather than mechanical means. Hydrogen peroxide and enzymatic catalysts are mixed into latex and the mixture is placed in the mold. Decomposition of the peroxide by the added enzyme results in the liberation of oxygen, which causes the latex mix to foam up and fill the mold. The foam is rapidly chilled and carbon dioxide is then introduced to gel the latex. The gelled foam is then handled in a manner similar to that used in the Dunlop process.

The wastewaters generated by the manufacture of latex foam products are similar to those produced at a latex-based dipped goods facility with the exception that there is an additional zinc-laden rinse water generated by washing the foam product. These wastewaters are identified in Table 36. An important characteristic of this type of wastewater is the presence of uncoagulated latex solids.

For supported flat-stock foam, that is, foam backing on various fabrics such as carpets, scatter mats, upholstery fabrics, etc., a different type of gelatin agent is employed in place of the sodium silicofluoride gelling agent used in latex foam production. Either ammonium acetate or ammonium sulfate is employed in combination with zinc oxide.

The froth is prepared with an Oakes machine, the gelling agent is added at the machine, and the foam is applied to the fabric by spreading directly on the fabric or spreading on a belt and transferring the gel to the fabric via the belt. The gelling is carried out at elevated temperatures, usually with the aid of infrared lamps. To prevent uneven shrinkage, the fabric is carried through the high-temperature zone and drying ovens on tenter frames. For this application the foam is poured in narrow thicknesses, from ⅛ inch up to a maxium of ½ inch. The wastewaters produced by the manufacture of foam backing are comparable to those generated by a typical latex-dipped goods manufacturing facility (Table 36).

AQUEOUS WASTE CHARACTERIZATION

Tire and Inner Tube Area

To substantiate the data and conclusions on total final effluent, Corps of Engineers water discharge permit applications were obtained by EPA (13) for a large segment of the tire and inner tube industry. Comparison of Corps permits for plants considered old and new revealed that the above findings and conclusions are substantially correct. Table 38 lists the main characteristics and the loadings corresponding to a typical old and typical newer tire production facility.

TABLE 38: INDUSTRY-WIDE AVERAGE VALUES OF FINAL EFFLUENT LOADS FOR TIRE AND INNER TUBE PLANTS[1]

	FLOW - L/kkg. (gal./1000 lb.) of raw material	COD - kg/kkg (lb./1000 lb.) of raw material	SS - kg/kkg (lb./1000 lb.) of raw material	TDS - kg/kkg (lb./1000 lb.) of raw material	OIL - kg/kkg (lb./1000 lb.) of raw material
Old Tire Facilities (to 1958)					
Minimum	6,267 (752)	0.088	0.001	0.349	0.003
Maximum	386,478 (46,396)	7.294	5.854	105.660	3.363
Average	37,000 (4,442)	0.780	1.000	12.000	0.120
New Tire Facilities (1959 to Present)					
Minimum	2,303 (276)	0.083	0.032	0.387	0.011
Maximum	48,070 (5,771)	2.020	1.397	12.980	0.187
Average	10,500 (1,261)	0.580	0.310	5.535	0.042

[1]Includes utility wastewaters.

Source: EPA (13)

Raw waste loads in the process wastewaters leaving the production facility are presented in Table 39.

TABLE 39: RAW WASTE LOADS OF PROCESS WASTEWATERS FROM TIRE AND INNER TUBE FACILITIES

Plant	Category	FLOW L/kkg(gal/1000lb) of raw material	COD kg/kkg(lb/1000lb) of raw material	BOD kg/kkg(lb/1000lb) of raw material	SS kg/kkg(lb/1000lb) of raw material	TDS kg/kkg(lb/1000lb) of raw material	OIL kg/kkg(lb/1000lb) of raw material
A	New	290 (35)	1.57	0.067	0.882	5.12	0.248
B	New	165 (198)	NA	0.001	0.013	0.3739	0.075
C[2]	New	NA	NA	NA	NA	NA	NA
D	New	NA	0.193	0.010	0.064	3.210	0.009
E	Old	110 (13)	0.046	0.000	0.017	0.019	0.027
F	Old	700 (84)	NA	NA	0.57	NA	0.650
G	Old	590 (71)	NA	NA	0.001	NA	0.260
H	Old	4300 (516)	2.199	0.356	0.520	19.765	0.163
I	Old	769 (92)	NA	NA	2.610	NA	0.172

[1]Estimated

[2]No data available from Plant C

NA - Data not available

Source: EPA (13)

Flow rates are estimates only, mainly due to the intermittent nature of the waste discharges. Although there appears to be no significant difference in the measured flow rates as shown by data, the composition of the flows originating from old and new plants differs greatly. New Plant A uses large amounts of washdown water which comprises the bulk of their process wastewaters. New Plant B process wastewaters consist largely of discharges from an extensive wet air pollution train. The discharges from this equipment are the primary constituent of the process wastewaters.

The process wastewater flow rates leaving older plants are due to other factors such as spills, leakage, runoff from storage areas and inherent plant practices of older facilities. Therefore the data indicate that, given the same housekeeping policies and the same degree of wet air pollution equipment and controls, the process wastewater flow rates from older plants will be higher than from newer plants.

Two important characteristics of the process wastewaters are suspended solids and oil. The suspended solids are generally higher from older plants due to greater maintenance and poorer housekeeping and control practices. The same can be said for the oil. Suspended solids evolve from the powdered substances used in the compounding area and from the collection of particulates by wet air pollution control equipment.

Molded, Extruded and Fabricated Area

Wastewater characterization data was obtained from literature, EPA documents and company data by EPA (17). Plant visits were made to confirm this data. The generalized process descriptions and diagrams of typical molded, extruded, hose, footwear, and cement-dipped production facilities, respectively, in the preceding text indicate the location of water supply and wastewater generation streams.

Table 40 summarizes total effluent quantities and contaminant loadings. All data is normalized to a unit of raw material consumption. Indicated flow rate values include both process and nonprocess wastewaters. Nonprocess wastewater can include once-through cooling water, cooling tower blowdown, boiler blowdown, water treatment wastes, domestic wastes, and steam condensate. The largest portion of the total effluent flow is cooling water discharge. Fluctuations in the flow rates generally reflect water management practices. For instance, plants G and C use recirculated cooling water, whereas plants A and D have once-through cooling systems.

Values tabulated for raw waste loads include both process and nonprocess wastewaters. In addition to the values listed, tests were conducted for other contaminants, such as phenols, chromium, and zinc. Phenol content in all the plants visited was minimal and it can be assumed that phenols in Subcategories E, F, and G total raw effluents will not be significant.

Chromium and zinc can be present in the total plant effluent from nonprocess wastewaters evolved from cooling tower blowdowns, but concentrations are usually below the measurable level. As with flow rates, COD and BOD loadings reflect water management techniques. Typically, the use of cooling water treatment chemicals in recirculated cooling water systems will raise the COD and BOD loadings in the final effluent.

Suspended solids and oil loadings are attributable to process wastewater discharges. These loadings in the total effluent appear minor when compared to COD and BOD loadings. There are two major reasons for this: first, process flow rates are minor relative to nonprocess discharges and dilution by nonprocess wastewaters reduces their concentrations in final raw effluents; secondly, the plants listed are exemplary. Sound housekeeping and maintenance practices substantially reduce suspended solids and oil loadings in raw effluents. Dissolved solids loadings in the raw effluents is a function of both the water management techniques, particularly with utility services, and the quality of the water supply source. Typically, the use of a cooling water recirculation system or the use of an underground raw water source will increase the dissolved solids loadings in the final effluent.

TABLE 40: RAW WASTE LOADS OF TOTAL EFFLUENT FROM EXEMPLARY SUBCATEGORIES E, F, AND G FACILITIES

Plant	Product Type	Size[1]	Flow[2]		COD[3]	BOD[3]	SS[3]	TDS[3]	Oil[3]	Lead[3]
A	Seals, General Molded Items, Rubber-Metal Bonded Items	1	69,771	(8,362)	1.395	1.645	0.0676	45.839	1.785	–
B	Cement Dipped Gloves	1	11,340	(1,359)	0.028	0.037	0.102	4.354	0.010	–
C	Molded Sport Grips, Rubber-Metal Bonded Items	2	10,560	(1,265)	0.528	0.249	0.137	8.342	0.074	–
D	Hose, Various Sizes	2	120,042	(14,405)	2.404	1.214	0.120	72.841	0.094	0.001
E	Belting and Sheet Rubber	3	22,122	(2,651)	1.349	0.194	0.221	68.598	0.103	–
F	Large Seals and Weather Stripping	3	29,683	(3,557)	1.703	0.161	1.828	25.189	0.138	–
G	Hose, Various Sizes	3	7,290	(875)	1.903	0.205	0.28	2.753	0.224	0.059
H	Canvas and Cement Dipped Footwear	3	60,369	(7,235)	4.601	0.385	1.734	12.886	0.394	–

[1]Size 1 is less than 3,720 kg/day (8,200 lb/day) raw materials consumption
Size 2 is 3,720 to 10,430 kg/day (8,200 to 23,000 lb/day) raw materials consumption
Size 3 is greater than 10,430 kg/day (23,000 lb/day) raw materials consumption
[2]l/kkg (gal/1,000 lb) of raw material
[3]kg/kkg (lb/1,000 lb) of raw material

Note: Raw waste loads include utility wastewaters

Source: EPA (17)

The raw wastewaters from plants D and G contain lead and, in general, have a higher COD content. This is attributable to vulcanization techniques employed in hose manufacturing. Raw wastewater loadings of plant B are lower than the other plants. This plant produces cement dipped goods. Their manufacture requires less heavy machinery and a correspondingly smaller amount of nonprocess and process wastewaters is evolved. The primary source of process wastewaters within the whole industrial category is related to the use of heavy machinery, and various antitack solutions. Leakage of bearing, gear, and seal oil can readily pass into nearby drains and be carried into the wastewater effluents. Oil powered hydraulic systems provide additional potential for oil contamination of wastewaters.

Antitack agents which are allowed to spill onto the floor can, when not properly handled, contaminate plant effluents, contributing a potentially high suspended solid loading. Washdown of the dipping areas where antitack agents are in use will create an additional suspended solids loading in the effluents. In addition, uncontrolled or untreated runoff from outdoor oil storage areas will contribute to additional oil loadings in the effluent. Flow, oil, and suspended solids contributions from these sources are the major components of the total process effluent. Smaller plants, with a typically less efficient operation (smaller throughput per machine) and older machinery, normally have higher normalized oil and suspended solids loadings. Larger storage areas in relation to production level are also typical of smaller plants and potential contamination from these areas is relatively greater.

The discharge from wet air emission control devices is another process wastewater which, although of less significance, is common to the entire industry. Particulate air collection equipment is necessary in the compounding areas and those areas where grinding or buffing of rubber occurs. This would include such processes as buffing of products to remove flash and the grinding of rubber from metal parts. Air pollution control devices in the compounding areas are typically the dry bag-type collectors. Devices in other areas can be either wet or dry. If wet, there does exist the potential for wastewaters when discharges are not recycled.

Other process wastewaters of minor importance include the discharge of contact cooling water, product testing waters, and spillage of mold release solution. Product testing and contact cooling waters are not contaminated and in some cases are recycled. Mold release solutions are usually applied manually. Spillage from this area is minute and easily controlled.

Process wastewaters specific to particular products within this industry include vulcanizer condensate from the curing of both lead-sheathed and cloth-wrapped hoses, vulcanizer condensate from the curing of cement dipped items, and latex discharges from fabricated rubber production facilities.

Discharges of condensate from the curing of lead sheathed hose are characterized by a high lead concentration (approximately 60 mg/l). However, a relatively small lead loading results, due to the extremely low flow rates of the condensate.

Vulcanizer condensate from the curing of cement dipped goods is characterized by high COD concentrations (approximately 800 mg/l) caused by solvents evaporated from the product by the steam condensate. Flow rates, however, are small, resulting in minor COD loading (0.4 kg/kkg of raw material). The vulcanizer condensate produced during the curing of cloth-wrapped hose has a lower COD concentration and loading than the cement-dipped condensate.

Latex discharges are characterized by COD, BOD, and suspended solids loadings, but with proper handling and controls they are relatively minor. In summary, the major process streams are: [1] spills, leakage, washdown, and runoff from processing and storage areas; [2] vulcanizer condensate from the curing of lead-sheathed and cloth-wrapped hoses; and, [3] vulcanizer condensate from the cure of cement dipped items. Flow rates and loadings for these streams are listed in Table 41.

TABLE 41: RAW WASTE LOADS OF PROCESS EFFLUENT FROM TYPICAL SUBCATEGORIES E, F, AND G FACILITIES

Wastewater Types	Size[1]	Product Types	Flow[2]		- - -COD[3]- - -		- - -SS[3]- - -		- - -Oil[3]- - -		- - -Lead[3]- - -	
General Process Wastes												
Spills, Leaks, Washdown and Runoff	1	All Product Types	16,200	(1,944)	2.960	182	3.500	216	1.000	62	–	–
Spills, Leaks, Washdown and Runoff	2	All Product Types	9,810	(1,177)	1.770	180	1.220	122	0.600	60	–	–
Spills, Leaks, Washdown and Runoff	3	All Product Types	6,210	(745)	1.120	180	0.900	145	0.380	61	–	–
Specific Process Wastes												
Vulcanizer Condensate	All Sizes	Hose manufactured by the lead sheathed cure process	450	(54)	–	–	0.030	63	–	–	0.030	63
Vulcanizer Condensate	All Sizes	Cement Dipped Goods and Hose manufactured by the cloth wrapped cure process	528	(63)	0.435	823	0.005	10	–	–	–	–

[1]Size 1 is less than 3,720 kg/day (8,200 lb/day) raw materials consumption
Size 2 is 3,720 to 10,430 kg/day (8,200 to 23,000 lb/day) raw materials consumption
Size 3 is greater than 10,430 kg/day (23,000 lb/day) raw materials consumption

[2]l/kkg (gal/1,000 lb) of raw materials

[3]In all of these categories the left-hand column represents kg/kkg (lb/1,000 lb) of raw materials; and the right-hand column represents mg/l of raw materials

Note: Data obtained from U.S. Army, Corps of Engineers - Discharge Permit and Plant Visit Sources

Source: EPA (17)

Based on the discussion above, several conclusions can be drawn about the wastewaters generated by Subcategories E, F, and G:

[1] Process wastewaters are of a low flow rate and have little impact on the total effluent flow rate.

[2] The most significant process wastewater streams occur by spillage, leakage, washdowns, and runoff. They contribute the majority of the suspended solids and oil in the final effluent. The flow rate of this type of process wastewater is dependent on plant size, and increases relative to production level as plant size decreases.

[3] Other process streams worthy of consideration are vulcanizer condensate from the lead-sheathed and cloth-wrapped hose production and cement dipped goods manufacture.

[4] Other process streams are of negligible importance, their impact being minor or undetectable.

Reclaimed Rubber Area

Data for characterizing the two reclaim subcategories were obtained primarily from EPA documents and companies (17). A data collection visit was made at a plant which used both the wet digester and pan processes in order to obtain further first-hand information. Since the total number of facilities producing reclaimed rubber is small, each existing plant was also interviewed to supplement the primary data on processing techniques, and wastewater types and treatment methods.

Table 42 summarizes the total process and nonprocess wastewater effluent quantities and contaminant loadings of the plant visited. All data are normalized to a unit of weight of reclaimed rubber product.

TABLE 42: RAW WASTE LOADS OF TOTAL EFFLUENT FROM EXEMPLARY SUBCATEGORIES H AND I PROCESSES

Plant	Product Type	Subcategory	Flow[1]	COD[2]	BOD[2]	SS[2]	TDS[2]	Oil[2]
1	Reclaimed Rubber	H	89,451 (10,720)	4.686[3]	0.896	1.890[4]	11.832	0.427[3]
1	Reclaimed Rubber	I	74,765 (8,960)	2.813	0.535	1.246	7.801	0.210

[1] l/kkg (gal/1,000 lb) of products
[2] kg/kkg (lb/1,000 lb) of products
[3] Includes reductions made through the reuse of digester liquors
[4] Value does not include any fibrous material which is removed at this plant prior to digestion

Note: Raw waste loads include utility wastewaters

Source: EPA (17)

Table 43 presents raw waste loads of process wastewater effluents. The flow rates presented include both process and nonprocess wastewaters. For this plant, nonprocess wastewaters include once-through cooling water and steam condensate. Nonprocess water contributes the bulk of the flow rate.

Rubber scrap coming to the plant is segregated. Tires are reclaimed by the wet digester process; whereas fiber-free scrap, such as inner tubes, is fed to the pan process. Since tire reclaiming requires additional grinding and associated cooling water, particularly as preparation for the mechanical defibering operation, the wet digester process wastewater flow rates are higher than those of the pan process. Incidentally, if tires were reclaimed by the pan process, the flow of nonprocess cooling water would be greater.

The nonprocess cooling water is relatively contaminant free, containing COD, BOD, and suspended solids attributable to the water source rather than the reclaiming process. Therefore, the contaminant loadings presented in Table 42 are almost entirely attributable to process wastewater streams. In the wet digester process (Subcategory H), the major constituents include COD, BOD, suspended solids, and oil. Analytical testing was made for phenols and heavy metals, in particular zinc, but the levels were negligible. The existing wet digester processes do not produce zinc-laden wastewaters as were encountered with the older or more conventional digester processes.

Values presented in Table 42 for Subcategory H represent raw waste loads after in-plant control. As mentioned earlier, fiber is removed from the scrap mechanically, thereby reducing the suspended solids loading of the effluent. In addition, dewatering liquors and vapor condensates are reused as makeup to the digester, thus reducing final oil and COD loadings. The final loadings of importance are oil, suspended solids, and COD.

In the reclaim processes for Subcategory I (including pan, mechanical and dry digestion reclaim processes) the process wastewaters have lower flow and contaminant loadings than the wet digestion process. This is due to the absence of the dewatering liquor wastewater stream and the lower usage of depolymerizing oils in the dry reclaiming processes.

The primary source of process wastewater loadings for Subcategory H is dewatering liquor. High COD and oil loadings are characteristic of this discharge. When mechanically defibered scrap is fed to the wet digester process, suspended solids are contained in the dewatering liquor owing to the carrying over of depolymerized rubber fines. If defibering is carried out chemically in the digestion step, additional suspended solids due to the fiber will be present.

A second major source of contaminant loadings for both Subcategory H and Subcategory I is spills, leaks, and washdown from processing areas. The discharge is qualitatively similar to the corresponding discharge of Subcategories E, F, and G; however, flow rates and loading on a per-day basis are substantially higher.

A third major source of contaminant loadings is air control equipment used to collect light organics which are vaporized or entrained in the vapors leaving the pan devulcanizers or the wet digester system. Flows and loadings from the wet digester process are substantially higher than those of the pan process. In the wet digestion process, the oil contained in these condensates can be recycled.

In summary, wastewaters generated by the reclaiming industry, Subcategories H and I contain the following major contaminant constituents: COD, suspended solids, and oil. Toxic materials like zinc and phenol were not present in the wastewaters from this industry sector. For Subcategory H, dewatering liquor contributes the majority of the total COD, oil and suspended solid loadings. In both subcategories, spills, leaks, washdown, and runoff from processing areas are a substantial source of effluent contamination. Additional contamination is attributable to the control of vapor emissions from both the pan and wet digester processes.

Loadings for wastewaters from Subcategory H are substantially higher than those of Subcategory I. In-plant controls such as the use of mechanically defibered scrap and the recycle of dewatering liquors do not reduce the wastewater loadings associated with the Subcategory H process to the levels resulting from the Subcategory I reclaiming process.

Latex-Based Products Area

Wastewater characterization data for Subcategories J and K were obtained from literature, EPA documents, company records, and first-hand plant data (17). Plant visits were made at two latex dipping facilities and one latex foam plant. Generalized process flow diagrams, indicating both product flow and wastewater generation points, have been presented earlier in this volume.

TABLE 43: RAW WASTE LOADS OF PROCESS EFFLUENTS FROM TYPICAL SUBCATEGORIES H AND I PROCESSES

	Subcategory	Flow[1]	COD[2]		BOD[2]		SS[2]		Oil[2]	
Wet Digester Process										
Spills, Leaks, Washdown and Runoff	H	3,130 (374)	0.345	110	0.085	27	0.190	287	0.760	240
Vapor Condensate from Air Pollution Control Devices	H	1,620 (195)	0.396	240	0.108	70	0.108	70	0.783	480
Dewatering Liquor (Chemical Defibering)[3]	H	2,391 (286)	9.010	3,910	1.800	790	256.000	106,650	25.747	10,770
Dewatering Liquor (Mechanical Defibering)	H	2,391 (286)	9.010	3,910	1.800	790	1,400	585	25.747	10,770
Pan, Dry Digester and Mechanical Process										
Spills, Leaks, Washdown and Runoff	I	3,130 (374)	0.345	110	0.085	27	0.190	287	0.483	154
Vapor Condensate from Air Pollution Control Devices	I	160 (20)	0.055	340	0.025	160	0.001	8	0.010	65

[1] l/kkg (gal/1,000 lb) of products

[2] In all of these categories the left-hand colum represents kg/kkg (lb/1,000 lb) of products; and the right-hand column represents mg/l of products

[3] The characteristics of dewatering liquor generated in a chemical defibering process are estimates based on the quantity of fiber in scrap tires. The suspended solids values represent maximum levels. It is possible that much of the suspended solids will be converted to dissolved solids during the chemical defibering process.

Source: EPA (17)

Table 44 summarizes the total process effluent quantities and contaminant loadings produced by the latex-based industry sector. All data is normalized to a unit of latex consumption. Flow rates are higher for the latex foam due to the larger amounts of processing and product washing waters required.

TABLE 44: RAW WASTE LOADS OF PROCESS EFFLUENTS FROM SUBCATEGORIES J AND K FACILITIES

Plant	Product Type	Flow[1]	COD[2]	BOD[2]	SS[2]	TDS[2]	Oil[2]	Surfactants[2]	Zinc[2]
J	Latex Dipped Prophylactics, Balloons, Finger Cots	43,200 (5,180)	7.60	5.75	3.38	49.50	0.24	0.08	–
K	Latex Dipped Gloves	102,920 (12,350)	69.73	15.66	310.70	39.58	13.30	0.66	–
L	Latex Foam	17,700 (2,124)	75.85	20.44	8.71	23.94	10.11	0.09	3.54

[1] l/kkg (gal/1,000 lb) of raw material
[2] kg/kkg (lb/1,000 lb) of raw material

Note: Raw waste loads do not include utility wastewaters

Source: EPA (17)

Effluent streams were measured for COD, BOD, suspended solids, dissolved solids, oil, surfactants, zinc, and phenols. Phenols were not found in any of the wastewaters. Zinc was found in process wastewaters from the latex foam facility. No zinc was found in wastewaters generated by latex dipped facilities. COD, BOD, suspended solids, and dissolved solids are attributable to latex in the wastewater.

Loadings for plant K are substantially higher because of the contributions of detergent-laden wash waters which also produce a higher flow. The oil in the wastewaters is more an extractable organic material than bonafide oil and grease. This characteristic of the oil analysis is based on a carbon tetrachloride extraction procedure. Surfactants are contributed by the emulsifying agents used in the latex mix and additionally in plant K by the detergent in the washing wastewaters.

The principal source of wastewaters within these subcategories is product wash waters. These wash waters are characterized by a COD, BOD, dissolved solids and suspended solids loading. The loadings of these wastewaters are highly dependent on the washing techniques employed by the plant. Plant K uses detergents in the washing operations, whereas plant J uses only high temperature water (approximately 200°F). The use of detergents adds to COD and BOD loadings in the raw effluent. Flows are also highly dependent on the technique employed. Higher flow rates are normally evolved from multiple washings.

In addition to the other loadings, discharges of product wash water from latex foam facilities can contain high concentrations of zinc. Zinc oxide is used as a curing agent during foam manufacture. Zinc components, which are not held, or fixed, in the foam matrix are removed by the wash waters.

A second source of contamination results from spills, leaks, washdown, and runoff from latex storage, compounding, and transfer areas. This wastewater will contain latex and is characterized by COD, BOD, suspended solids, dissolved solids, oil, and surfactant loadings. COD, BOD, and suspended solids are present due to the latex in the wastewater. Oil and surfactants are contributed by coagulation agents, extractable organics, and emulsifier aids.

A third source of wastewater, relevant to latex dipping operations, is from cleaning wastes. Plant J employed a biodegradable cleaning compound in hot water. The operation is continuous, the form being cleaned automatically after each complete dipping operation. Wastewaters from this operation are characterized by loadings of COD, BOD and suspended solids. Plant K cleans forms manually and only periodically. Discharges from this area, although characterized by similar contaminants, will be intermittent in nature and will be of a substantially lower flow. Literature and EPA documents indicate that certain latex dip operations use chromic acid solutions to clean forms. Potentially these wastewaters can appear in the process effluent. A summary of the individual process stream characteristics is presented in Table 45.

In summary, process wastewaters from Subcategories J and K operations are characterized by COD, BOD, suspended solids, oil and surfactant loadings. In addition, discharges from these operations can contain substantial quantities of zinc and chromium. The primary wastewater is product wash waters. The characteristics of the wash water are highly dependent on the techniques employed. In the production of latex foam, the wash water will contain zinc. Other sources of process wastewaters include spills, leaks, washdown, and runoff from latex storage, compounding and transfer areas, and discharges from form cleaning operations.

WATER POLLUTION CONTROL AND TREATMENT METHODS

Tire and Inner Tube Area

In order to review and fully evaluate the wastewater control and treatment technologies used in the rubber processing industry, selected plants were visited by EPA to conduct operation analyses, review water and wastewater management programs, and evaluate wastewater treatment facilities. The plants were selected as being exemplary or advanced in their wastewater control and treatment technologies, based on effluent and treatment data from the technical literature, EPA documents, Corps of Engineers Permit to Discharge Applications, and individual company treatment data (13).

Plants producing passenger tires (both bias and radial ply), truck tires, camelback, and inner tubes were visited and studied to determine if the type of product affected the quality and quantity of wastewater streams and/or the control and treatment technology employed. Both singleproduct and multiproduct plants were included so that the effects of combined lines on the plant wastewaters could be evaluated; likewise, plants of various sizes were studied to determine the impact of production levels.

Age was a major consideration because determination of the effect, if any, of newer processing technology and machinery on the control and treatability of process waste streams was one of the principal objectives of the investigative phase of this project. Table 46 is a summary of the products manufactured, raw material usage, and wastewater control and treatment technologies utilized at the tire and inner tube plants visited.

Two alternative approaches exist for the control and treatment of process wastewaters from both old and newer tire and inner tube production plants. The first approach is to combine process and nonprocess wastewaters and to treat the entire plant effluent. Where land is available, end-of-pipe treatment is the approach favored by many of the tire manufacturers. Generally, the reasons supporting this approach are as follows:

[1] In older plants, in-plant sewers for process and nonprocess wastewaters are usually combined, thus making combined treatment more attractive.

[2] Process flows are usually small relative to nonprocess flows.

[3] The treatment of nonprocess wastewaters has received the bulk of industry's attention. High suspended solid loadings in blowdown and water treatment wastes are the major pollutant in the combined plant effluent from tire facilities.

TABLE 45: RAW WASTE LOADS OF PROCESS EFFLUENTS FROM TYPICAL SUBCATEGORIES J AND K FACILITIES

Waste Water Type	Flow[1]		COD[2]		BOD[2]		SS[2]		Surfactants[2]		Zinc[2]	
Subcategory J												
Spills, leaks, washdown and runoff from latex storage, compounding and dip tank areas[3]	476	(57)	–	–	–	–	1.86	3,900	–	–	–	–
Single product wash and rinse waters[4]	43,200	(5,180)	7.60	176	5.75	133	3.38	78	0.08	2	–	–
Multiple product wash and rinse waters[4]	381,000	(45,710)	41.14	108	9.24	24	3.51	9	0.66	2	–	–
Subcategory K												
Foam rinse waters	3,600	(432)	47.63	13,230	17.62	4,900	25.22	7,000	0.09	25	3.51	975
Spills, leaks, washdown and runoff from latex storage, compounding and transfer areas	600	(72)	17.51	29,180	2.17	3,620	0.72	1,200	0.002	3	0.02	33

[1] l/kkg (gal /1,000 lb) of raw materials

[2] In all of these categories the left-hand column represents kg/kkg (lb/1,000 lb) of raw materials; and the right-hand column represents mg/l of raw materials

[3] Obtained from available U.S. Army, Corp of Engineers - Discharge Permit Information; not necessarily typical

[4] Both single and multiple product wash and rinse water streams do not exist at any one plant

Source: EPA (17)

TABLE 46: WASTEWATER CONTROL AND TREATMENT TECHNOLOGIES AT EXEMPLARY TIRE AND INNER TUBE PLANTS*

Plant	Product	Raw Material Usage, kg/day	Control Measures	Primary Effluent Treatment	Secondary Effluent Treatment
A	Passenger tires, implement tires, front tractor tires	120,000	Recirculation of soapstone, absence of drains in dirty areas	Sedimentation and holding lagoon	No discharge due to spray irrigation and evaporation of wastewaters
B	Passenger tires, industrial tires	310,000	Recirculation of soapstone, baffled oil separator for oil storage area, absence of drains in dirty areas	Sedimentation and lagoons	None
C	Passenger tires	349,000	Recirculation of soapstone	Primary settling basins	None
D	Passenger tires, truck tires, tractor tires, camelback	840,000	Recirculation of soapstone, oil separator, local oil sumps	Gravity separator for boiler blowdown, some water treatment wastes and washdown of soapstone area	Discharge of some water treatment wastes to municipal treatment facility
E	Passenger tires, truck tires	210,000	Blockage of drains, local oil sumps and gravity separators, curbing of soapstone area, recirculation of soapstone	None	Discharging of process waste to municipal treatment facility
F	Passenger tires, truck tires, inner tubes, camelback	681,000	Recirculation of soapstone, oil sump for oil storage area	Primary settling basin and clarifying basin	None
G	Passenger tires, truck tires	246,000	Recirculation of soapstone, oil sump and curbs for oil storage area	None	None
H	Truck tires, industrial tires	244,000	Recirculation of soapstone	Sedimentation basins for boiler blowdown, cooling tower and water treatment wastes	Discharge of treated wastes to municipal treatment facility
I	Inner tubes	75,000	Containment of soap and soapstone solutions	None	None

*Includes process and nonprocess wastewater control and treatment facilities

Source: EPA (13)

However, end-of-pipe treatment systems also have several disadvantages:

[1] The combined effluent treatment system usually requires one or two lagoons for settlement and retention. Lagooning of the wastes requires large land area, which is not readily available at many plant locations.

[2] Because of dilution, the effectivenesss of treatment for oil removal from process wastewater is reduced. In several of the systems observed, oil passed through untreated (although it was present in significant quantities), because its concentration was below the capabilities of the treatment system employed.

The second approach employed is control and treatment of a segregated and undiluted process wastewater. This approach has been followed in plants having partially or wholly segregated process and nonprocess sewers. This would, of course, include any plant using recirculated cooling water. The main advantages for this treatment scheme over combined end-of-pipe treatment are:

[1] Higher pollutant removal rates.

[2] Smaller land area required for treatment facilities.

The primary disadvantage of a segregated system approach is that separate process and nonprocess sewers are required.

Upon examining these alternatives, control and treatment of segregated process wastewaters was considered to be most applicable to the tire producing industry. End-of-pipe treatment of combined wastewaters is not feasible for pollution control because of: the ineffectiveness of such systems in removal of process wastewater contaminants; and, the large land requirements. All costs, therefore, are related to the treatment of a segregated process waste stream.

With proper in-plant control, the process streams consist of readily separable lubricating and extender oils and settleable solids. Volumetric flow rates for process wastewaters are small. Therefore, the initial treatment applicable from a cost and proven operation basis is an API-type gravity separator. The performance and efficiency of a gravity separator can be improved by addition of an absorbent filter.

Fabricated and Reclaimed Products Area

Several selected rubber processing plants were visited by EPA (17) to provide further accurate data on the performance of the wastewater control and treatment technologies used by the industry. These data collection visits encompassed analysis of process operations, review of water and wastewater management programs, and evaluation of wastewater treatment facilities. The plants visited were considered exemplary or advanced in their approach to wastewater control and treatment or the thoroughness of their housekeeping procedures. The plant selection was made based on effluent and treatment data obtained from published literature, EPA records, Corps of Engineers Permit to Discharge Applications, and company historical data on wastewater quality and treatment.

As examples of Subcategories E, F, and G industries, plants manufacturing molded, extruded, and fabricated products were sampled. Compression, transfer, and injection molding technologies were represented as well as small and large-sized molded items. The major extruded items such as belting and sheeting were included in the visits. Diverse fabricated products such as hose and rubber footwear as well as rare sectors of the industry such as cement dipped goods were studied during the plant visits. In addition all three ranges of plant size were represented by the data-collection plant visits.

As an example of the reclaimed rubber industry, the largest reclaiming plant in the U.S. was visited. Wet digester and pan (heater) reclaim processes are employed. Therefore, this plant is representative of both Subcategories H and I. The wet digester process is exemplary since physical defibering is carried out and the highly contaminated digestion-dewatering liquor is recycled. In addition vapor vents on both the wet digester and pan processes are condensed and decanted to recover what would otherwise be polluting reclaiming process oils.

Two plants manufacturing latex dipped goods were visited. The products made at the two facilities include surgical gloves, balloons, prophylactics, and finger cots. One plant employed chemical coagulation and settling before discharge to a municipal treatment plant. This primary treatment, or pretreatment for a publicly owned treatment works, produced good effluent. The second facility utilized a stabilization and settling pond system. This effluent had good quality for direct discharge.

The only significant latex foam plant in the industry was sampled. The treatment facilities used by this plant include chemical coagulation of latex solids and chemical precipitation of the zinc-laden foam rinse waters. A summary of the products, processes, production capacities, and wastewater control and treatment technologies of the exemplary rubber processing plants visited is presented in Tables 47 and 48.

A process developed by *W.H. Scott, Jr., U.S. Patent 3,726,794; April 10, 1973; assigned to Clarkson Industries, Inc.* provides an apparatus and system for the removal of latex from wastewater. When latex is used in a rug finishing process for attaching burlap or foam rubber backing to rugs, for example, the latex mixing tanks, dip troughs, pipe line, etc., must be frequently cleaned and washed out with water resulting in large amounts of wastewater with varying quantities of latex and latex foam therein which heretofore have been dumped into the sewer or river prior to or after some batch type of settling techniques used in the industry.

The present process for removing latex from latex laden wastewater includes a mixing compartment into which latex laden wastewater is supplied and control means connected to the compartment for sensing the rate of flow of the latex laden wastewater therethrough. Pump means activated by the control means upon sensing the rate of flow above a predetermined rate supplies treating chemicals from a source to the compartment upon activation of the pump means whereby the chemicals are mixed with and treat the latex laden wastewater in the compartment. A settling tank means receives the chemicals and latex laden wastewater from the compartment and settles out the major portion of the latex sludge therein, the settling tank including means for removing the settled latex sludge therefrom.

Finally the latex laden wastewater from the settling tank is supplied to a polishing tank for final clarification before being discharged to a sewer.

WATER POLLUTION CONTROL ECONOMICS

Tire and Inner Tube Area

Effluent quality data for older tire and inner tube and for newer tire facilities are presented (along with cost data) in Tables 49 and 50. The treatment technology involves the isolation of wastes with curbing, the protection of oily areas to prevent storm runoff contamination, and the separation of settleable solids and oily material from the wastewater. A flow diagram of the system used as a basis for costing is presented in Figure 79.

Data from Corps of Engineers permit applications and plant data obtained during inspection visits were used to obtain the average or typical plant size and wastewater discharge flows, and raw waste loadings. In order to adequately estimate the wastewater discharge flow rates, the plant effluent was divided into process wastewaters and nonprocess wastewaters. The process wastewaters consist of mill area oily waters, soapstone slurry and latex dip wastes, area washdown waters, and emission scrubber waters, contaminated storm waters from raw material storage areas.

TABLE 47: WASTEWATER CONTROL AND TREATMENT TECHNOLOGIES AT SUBCATEGORIES E, F, AND G PLANTS WITH EXEMPLARY FEATURES

Plant*	Product	Process	Production Capacity (kg/day, lb/day)	Control Measures	Primary Effluent Treatment	Secondary Effluent Treatment
Subcategory E - Small Plants						
A	Seals, general molded items	Compression and transfer molding	340 (750)	Most drains in milling and curing areas are blocked.	None	None**
B	Cement dipped gloves	Cement dipping	500 (1,100)	No floor drains	None	None**
Subcategory F - Medium Plants						
C	Molded sport grips, rubber-metal bonded items	Compression, transfer and injection molding	8,600 (19,000)	Most drains in compounding, milling and curing areas are blocked. Dry dust collection devices.	None	None**
D	Hose, various sizes	Extrusion and fabrication	9,000 (20,000)	Most drains in compounding and milling areas are blocked.	None	None**
Subcategory G - Large Plants						
E	Belting and sheet rubber	Extrusion	26,000 (58,000)	In-plant containment of oil leaks and spills.	None	None**
F	Large seals and weatherstripping	Molding and extrusion	26,000 (58,000)	Most drains in compounding, milling, and curing areas are blocked.	Gravity oil separators and holding pond.	None**
G	Hose, various sizes	Extrusion and fabrication	75,700 (162,500)	Blocking of some floor drains and use of dry clean-up methods.	None	None***
H	Canvas and cement dipped footwear	Molding, fabrication and cement dipping	120,000 (265,000)	In-plant containment of oil leaks and spills.	Gravity oil separation, and chemical coagulation and clarification of latex-laden wastes.	None**

*The definition of the plant sizes is based on total raw material usage: Small plants, less than 3,720 kg/day (8,200 lb/day); Medium plants, 3,720-10,430 kg/day (8,200-23,000 lb/day); and Large plants, greater than 10,430 kg/day (23,000 lb/day).
**Direct discharge to stream.
***Discharge to municipal treatment system.

Note: Subcategories E, F, and G include General Molded, Extruded, and Fabricated Products, i.e., Hose, Belting, Seals, Packing, Gaskets, Footwear, Cement Dipped Goods, and Tire Retreading.

Source: EPA (17)

TABLE 48: WASTEWATER CONTROL AND TREATMENT TECHNOLOGIES FOR SUBCATEGORIES H, I, J, AND K PLANTS WITH EXEMPLARY FEATURES

	Subcategory H Wet Digestion Rubber Reclaiming	Subcategory I Pan (Heater), Mechanical, and Dry Digestion Reclaiming	Subcategory J Latex-Dipped, Latex Thread, and Latex-Molded Products		Subcategory K Latex Foam
Plant	**I**	**I**	**J**	**K**	**L**
Product	Reclaimed rubber	Reclaimed rubber	Ballons, prophylactics, and finger cots.	Surgical gloves, hot water bottles, rubber syringes, and pharmaceutical items.	Foam mattresses and pillows.
Process	Wet digestion	Pan (Heater)	Latex dipping	Latex dipping compression molding	Talalay foam process
Production capacity kg/day (lb/day)	56,000 (123,000)	45,000 (100,000)	4,300 (9,500)	900 (2,000)	90,000 (200,000)
Control measures	Physical defibering and and return of process oils and digester liquor.	Return of process oils.	Few in-plant drains. Minimal water usage for tank cleaning.	Few in-plant drains.	Countercurrent foam rinse system.
Primary effluent treatment	Oil separation and recycle of process oils and digester liquor.	Oil separation and recycle of process oils.	Settling ponds.	Coagulation and clarification of latex solids.	Latex waste chemical coagulation and clarification. Zinc precipitation and clarification.
Secondary effluent treatment	None. Direct discharge to stream.	None. Direct discharge to stream.	Settling pond effluent discharged to stream. Evaporation pond for latex waste waters.	None. Discharge to municipal treatment system.	Proposed: equilization, carbonate precipitation, and filtration of present primary effluent.

Source: EPA (17)

FIGURE 79: HYPOTHETICAL WASTEWATER SEGREGATION AND TREATMENT FACILITY FOR TIRE AND INNER TUBE PLANTS

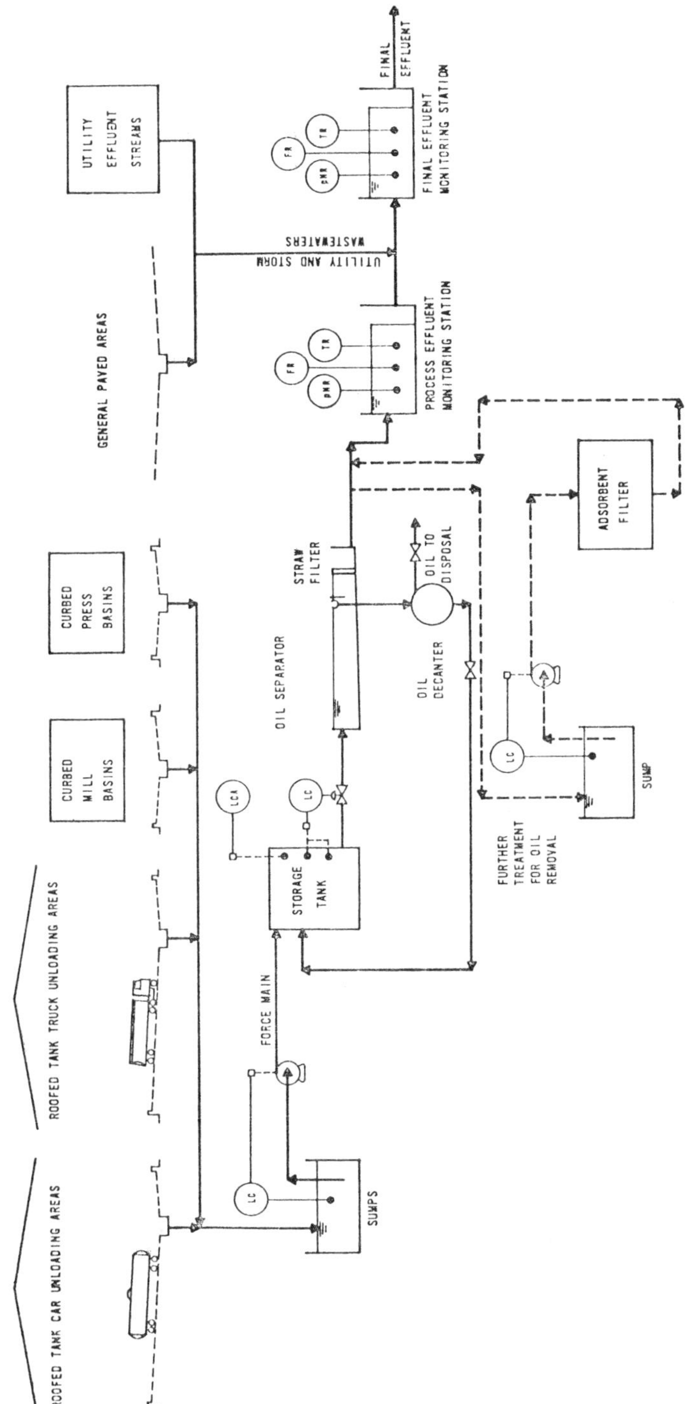

Source: EPA (13)

The nonprocess wastewaters are sanitary and clean storm waters, utility wastewaters such as once through cooling waters, boiler blowdown, cooling tower blowdown, water treatment wastes and uncontaminated contact cooling water like tread cooling waters.

From these data, a typical process wastewater flow was estimated to be 3.785 l/sec (60 gpm) for a plant consuming 205,000 kg (450,000 lb) of raw materials per day. For the older tire and inner tube plant subcategory the average oil loading is 0.246 kg/kkg (lb/1,000 lb) of raw material consumed. The suspended solids loading for this subcategory is estimated to be 0.319 kg/kkg (lb/1,000 lb) of raw material. Based on these typical profiles for old and new production facilities, treatment cost data were generated and are presented in Tables 49 and 50. The costs shown for a typical older facility are based on the worst case of having to install a completely segregated sewer system rather than isolating the process wastewater stream from the combined sewer line.

TABLE 49: OLD TIRE AND INNER TUBE PRODUCTION FACILITY SUBCATEGORY

Estimated Wastewater Treatment Costs for Different Degrees of Treatment Based on a Typical Raw Material Consumption of 205 Metric Tons/Day

	Treatment or Control Technology* A	B
Investment**	$779,000	$808,000
Annual Costs		
Capital costs	$ 78,000	$ 81,000
Depreciation	156,000	162,000
Operating and maintenance costs (excluding energy and power costs)	52,000	54,000
Energy and power costs	2,000	2,000
Total Annual Costs**	$288,000	$299,000

Parameters, kg/kkg raw material	Raw Waste Loads	Effluent Quality A	B
Suspended solids	0.319	0.064	0.064
Oil and grease	0.120	0.048	0.008

*Technology A is isolation of process wastewaters followed by API gravity separation
Technology B is Technology A followed by an absorbent filter
**August 1971 dollars

Source: EPA (13)

TABLE 50: NEW TIRE PRODUCTION FACILITY SUBCATEGORY

Estimated Wastewater Treatment Costs for Different Degrees of Treatment Based on a Typical Raw Material Consumption of 205 Metric Tons/Day

	Treatment or Control Technology* A	B
Investment**	$597,000	$628,000
Annual Costs		
Capital costs	$ 60,000	$ 63,000
Depreciation	119,000	126,000
Operating and maintenance costs (excluding energy and power costs)	41,000	43,000
Energy and power costs	2,000	2,000
Total Annual Costs**	$222,000	$234,000

(continued)

TABLE 50: (continued)

Parameters, kg/kkg raw material	Raw Waste Loads	Effluent Quality A	Effluent Quality B
Suspended solids	0.319	0.064	0.064
Oil and grease	0.083	0.048	0.008

*Technology A is isolation of process wastewaters followed by API gravity separation
Technology B is Technology A followed by an absorbent filter
**August 1971 dollars

Source: EPA (13)

Molded, Extruded and Fabricated Area

Two fundamental approaches can be applied for the control and treatment of process wastewaters produced by Subcategories E, F, and G production facilities. The first approach is to combine process and nonprocess wastewaters and to treat the entire plant effluent. Where land is available, on-site end-of-pipe treatment is practiced at some of the plants. At other plants, the total effluent is discharged to the local municipal treatment facilities. In either case, the reasons for treating or discharging a combined effluent are as follows:

[1] In older plants, in-plant sewers for process and nonprocess wastewaters are usually combined, thus making combined treatment more attractive.

[2] Process flows are usually small relative to nonprocess flows.

[3] The treatment of nonprocess wastewaters has received the bulk of industry's attention. For example, high suspended solid loadings contained in blowdown and water treatment wastes are a major pollutant in the combined plant effluent from rubber processing facilities.

However, end-of-pipe treatment systems also have several disadvantages:

[1] The combined effluent treatment system usually requires oil removal equipment and/or a holding pond or lagoon system to separate the oil. Lagooning of the wastes generally requires considerable land area.

[2] Because of dilution, the effectiveness of treatment for oil and solids removal from process wastewater is reduced. In several of the systems observed, oil passed through the system with low removal efficiency, because its concentration was below the capabilities of the treatment system employed. This phenomenom occurs even though the oil loading is significant, because of the dilution afforded by nonprocess wastewaters.

The second approach employed is control and treatment of an isolated and undiluted process wastewater. This approach has been followed in plants having partially or wholly-segregated process and nonprocess sewers. The key to this treatment approach is the reduction, containment, and isolation of the contaminated process wastewaters prior to treatment. The principal advantages for this treatment approach over the end-of-pipe treatment of the combined plant effluent are: higher pollutant (primarily oil and suspended solids) removal rates; and, smaller treatment units and land area requirements.

The primary disadvantage of a segregated system approach is that separate process and nonprocess sewers are required. This will necessitate in-plant modifications which, if not carefully planned, can lead to significant production disruptions. Evaluation of these alternatives tives indicated that control and treatment of segregated process wastewaters is the most effective and generally applicable treatment approach for this industry. End-of-pipe treatment of combined wastewaters is not feasible for pollution control because of the ineffectiveness of such systems in removal of diluted process wastewater contaminants, and the larger land requirements. Incidentally, a viable alternative treatment for wastewaters (process as well as nonprocess) generated by this industry is discharge to the local publicly owned treatment works. Such discharge is contingent on adequate pretreatment.

All costs presented are related to the control, followed by treatment, of segregated process wastewater streams. It can be further stated that this type of treatment approach is considered to be the best practicable control and treatment technology currently available to the industry.

With proper in-plant control, the process streams consist of readily separable lubricating and extender oils, and settleable solids. Volumetric flow rates for process wastewaters are small. Therefore, the initial treatment applicable from a proven operation basis is an API-type gravity separator. The performance and efficiency of a gravity separator can be improved by addition of an absorbent filter. The corrugated plate interceptor (CPI) type of device is also applicable generally to this industry. The choice between the API or CPI separators will depend on land availability and the type of wastewater control, handling, and treatment equipment already present at the plant. Since the cost and treatment effectiveness of the API and CPI type separators are comparable, the effluent limitations treatment and costs presented are based on the use of API type separators.

Effluent quality data for Subcategories E, F, and G are presented in Tables 51, 52, and 53. The treatment basis includes isolation of wastes with curbing, protection of uncontaminated waste streams (such as cooling water and storm runoff), blocking unnecessary floor drains, the covering of oil-handling areas to prevent contamination of storm runoff, and separation of settleable solids and oily material from the wastewater. A flow diagram of a typical system, used as a basis for costing, is presented in Figure 80.

TABLE 51: ESTIMATED WASTEWATER TREATMENT COSTS AT DIFFERENT DEGREES OF TREATMENT FOR A SUBCATEGORY E PLANT

		Treatment or Control Technology*	
		A	B
Investment**		$79,000	$84,000
Annual Costs			
Capital costs		$ 7,900	$ 8,400
Depreciation		15,800	16,800
Operating and maintenance costs (excluding energy and power costs)		11,400	12,900
Energy and power costs		600	600
Total Annual Costs**		$35,700	$38,700
Parameters, kg/kkg (lb/1,000 lb) raw materials	**Raw Waste Loads**	**Effluent Quality**	
		A	B
Suspended solids	3.500	0.640	0.640
Oil and grease	1.000	0.480	0.160

*Technology A is isolation of process wastewaters followed by API gravity separation
Technology B is Technology A followed by an absorbent filter
**August 1973 dollars

Source: EPA (17)

TABLE 52: ESTIMATED WASTEWATER TREATMENT COSTS AT DIFFERENT DEGREES OF TREATMENT FOR A SUBCATEGORY F PLANT

		Treatment or Control Technology*	
		A	B
Investment**		$163,000	$171,000
Annual Costs			
Capital costs		$ 16,300	$ 17,100
Depreciation		32,600	34,200
Operating and maintenance costs (excluding energy and power costs)		18,200	23,200
Energy and power costs		800	800
Total Annual Costs**		$ 67,900	$ 75,300
Parameters, kg/kkg (lb/1,000 lb) raw materials	**Raw Waste Loads**	**Effluency Quality**	
		A	B
Suspended solids	1.220	0.400	0.400
Oil and grease	0.600	0.300	0.100

*Technology A is isolation of process wastewaters followed by API gravity separation
Technology B is Technology A followed by an absorbent filter
**August 1973 dollars

Source: EPA (17)

TABLE 53: ESTIMATED WASTEWATER TREATMENT COSTS AT DIFFERENT DEGREES OF TREATMENT FOR A SUBCATEGORY G PLANT

		Treatment or Control Technology*	
		A	B
Investment**		$212,000	$223,000
Annual Costs			
Capital costs		21,200	22,300
Depreciation		42,400	44,600
Operating and maintenance costs (excluding energy and power costs)		20,800	27,800
Energy and power costs		1,200	1,200
Total Annual Costs**		$ 85,600	$ 95,900
Parameters, kg/kkg (lb/1,000 lb) raw materials	**Raw Waste Loads**	**Effluency Quality**	
		A	B
Suspended solids	0.900	0.250	0.250
Oil and grease	0.380	0.189	0.063

*Technology A is isolation of process wastewaters followed by API gravity separation
Technology B is Technology A followed by an absorbent filter
**August 1973 dollars

Source: EPA (17)

FIGURE 80: HYPOTHETICAL WASTEWATER SEGREGATION AND TREATMENT FACILITY FOR SUBCATEGORIES E, G, AND I PLANTS

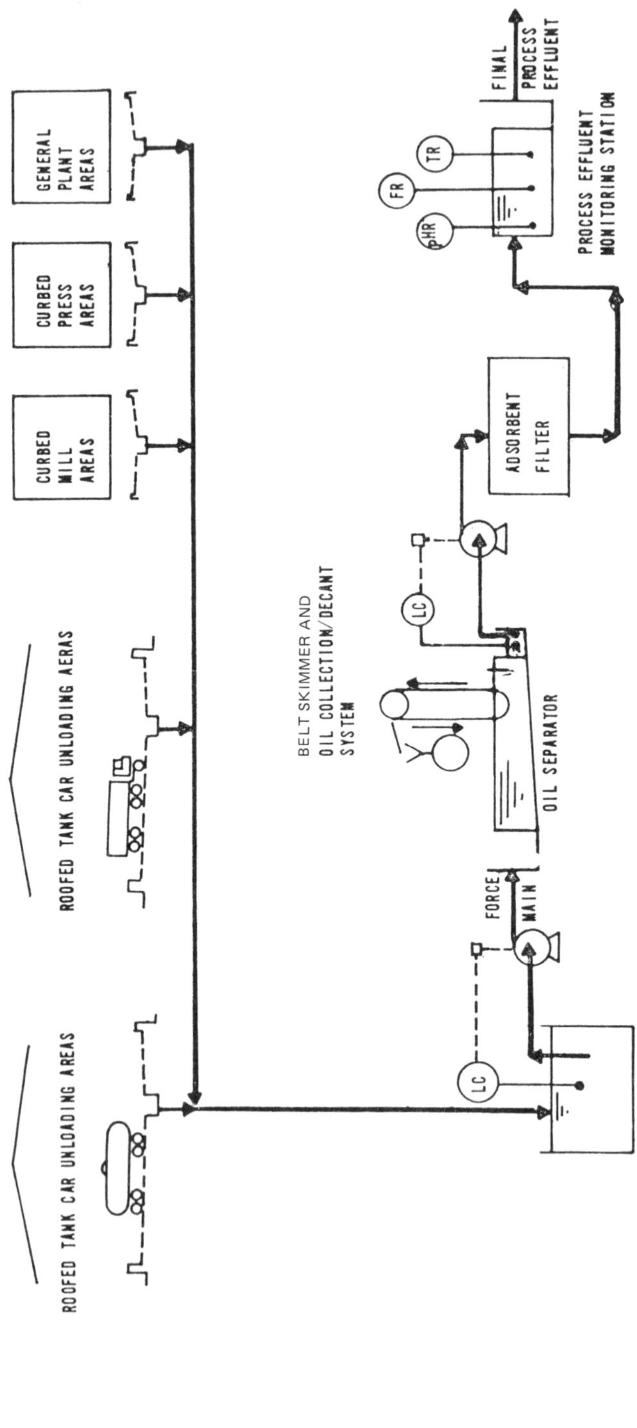

Source: EPA (17)

Reclaimed Rubber Area

Wet Digestion: The rubber reclaiming industry is presently undergoing a decline in both the number of operating plants and the quantity of reclaimed rubber produced. The wet digester reclaiming process has borne the brunt of this decline, and it must be assumed that the financial resources of wet digester reclaim plants and their ability to shoulder further operating costs such as wastewater treatment costs are limited. In view of this and since all of the existing wet digester plants discharge their wastewaters directly or are connected to the local publicly owned sewage treatment works, it is not possible to propose an end-of-pipe treatment technology which has been proven to be operable and successful while in service, as well as economically practicable.

With this in mind, attention has been duly given to in-plant modifications to the wet digestion process which have been implemented and proven successful in at least one plant. Such modifications, while reducing wastewater pollution and end-of-pipe treatment needs, also have the potential to reduce raw material wastage and consumption. However, based on the preliminary findings of this guideline study, it cannot be said that the yearly reductions in process operating costs afforded by the in-plant modifications will offset the annual operating and maintenance costs directly attributable to the modifications. Such benefits can be assessed only by personnel intimately familiar with the wet digestion process and the economics of its operation.

The principal wastewater streams in a wet digestion reclaim plant are the dewatering liquor and the vapor condensates from the blowdown tank and dryer. The type of treatment proposed for Subcategories E, F, G, and I, namely, the isolation of wastes loaded with oil and suspended solids followed by oil and solids removal, is not appropriate as BPCTCA control and treatment for Subcategory H since the costs benefits are not favorable. This argument is particularly valid when the minimal benefits to be achieved by this type of alterations are compared to the greater pollution reduction that can be attained by the application of the recommended recycle. It should be remembered at this point that the wet digester process is only marginally profitable and, at present, cannot bear large wastewater treatment costs. In addition, most wet digester plants discharge to municipal systems and are subject to pretreatment guidelines.

A fundamental change that can be made to the wet digestion reclaiming process is the conversion from chemical defibering to mechanical defibering. With chemical defibering, defibering chemicals are added to the digestion mix. The fibers are solubilized in the digestion step and leave the process system in the dewatering liquor which is discharged. The defibering chemicals, which can be of a toxic nature, are an additional constituent of the dewatering liquor. In the mechanical defibering variation, the scrap rubber is finely ground prior to digestion. The freed fiber is then elutriated from the rubber scrap on air separation tables.

With mechanical defibering the dewatering liquor is free of high levels of solubilized fiber as well as the chemical defibering agents themselves. Costs incurred by a conversion from chemical to mechanical defibering have not been fully developed for this document since the required technology for the conversion is specific to the reclaiming industry and outside the scope and technology boundaries of the guideline study. However, the costs associated with a recent conversion from chemical defibering to mechanical fiber separation illustrate the order of magnitude of the capital costs. The conversion of a 130,000-pound per day wet digestion plant from chemical defibering to mechanical fiber separation cost $611,000 in 1969-70. This modification included $31,000 for fiber lint air emission controls.

A major reduction in the volume and loadings of the process wastewaters can be achieved by adding a recovery and recycle system to the wet digestion process. In essence the dewatered liquor can be recycled back to the digestion step with a blowdown of accumulating contaminants and a makeup of the digestion liquor. At the same time the vapor condensates from the blowdown tank and dryer are decanted to recover the insoluble oils and organics. The oils and organics are recycled to the digestion liquor makeup operation, and the water underflow from the decantation is returned with the dewatered liquor. As a result of this recycle system, the volume and loadings of the dewatering liquor are considerably reduced. It should also be noted that mechanical, opposed to chemical, defibering is required in order for this system to be feasible. The dewatering digester liquor must be fiber-free in order that part of it can be recycled. The quality of the effluent achieved by this recycle system is presented in Table 54 together with the associated control and treatment costs. The costs do not include the costs incurred by the necessary conversion from chemical to mechanical defibering.

TABLE 54: ESTIMATED WASTEWATER CONTROL COSTS FOR A WET DIGESTION RECLAIM PLANT (SUBCATEGORY H)

		Control Technology*
Investment		$126,000
Annual Costs		
Capital costs		12,600
Depreciation		25,200
Operating and maintenance costs (excluding energy and power costs)		19,300
Energy and power costs		4,700
Total Annual Costs		$ 61,800
Parameters, kg/kkg (lb/1,000 lb) of product	**Raw Waste Loads**	**Effluent Quality**
COD	9.75	6.11
SS	256.10	2.31
Oil	27.29	0.58

*The control technology includes recycle of dewatering liquor and decantation of vapor condensates followed by recycle of the oils and water underflow.

Source: EPA (17)

Pan (Heater), Mechanical, and Dry Digestion Rubber Reclaiming: The extent of the wastewater problems and treatment types for all dry reclaiming processes can best be represented by those of the pan, or heater, reclaim process.

The wastewater types are essentially similar to those produced by Subcategories E, F, and G plants with the additional contribution of the oven vapor condensates. The majority of the process wastes are created by spills and leaks in the cracking, grinding, and milling buildings as well as around material and fuel oil storage areas. The contaminants are principally oil and suspended solids. The vapor condensates contain organics vaporized from the rubber mix in the depolymerization oven.

The control and treatment approach proposed for this subcategory is isolation of the process wastewaters preventing dilution of the contaminants and a decrease in their treatability. As for Subcategories E, F, and G wastes, the key to this type of control and treatment is the reduction and collection of the contaminated process wastewaters prior to treatment for oil and suspended solids removal in an API-type separator. The separable organics in the depolymerization oven condensates can also be separated in the separator.

The anticipated effluent quality achieved by the proposed treatment is presented in Table 55 together with the associated cost data.

TABLE 55: ESTIMATED WASTEWATER TREATMENT COSTS AT DIFFERENT DEGREES OF TREATMENT FOR A PAN, DRY DIGESTER, OR MECHANICAL RECLAIM PLANT (SUBCATEGORY I)

		Treatment or Control Technology[1]	
		A	B
Investment[2]		$265,000	$277,000
Annual Costs			
Capital Costs		26,500	27,700
Depreciation		53,000	55,400
Operating and Maintenance Costs (excluding energy and power costs)		25,500	28,200
Energy and Power Costs		1,200	1,200
Total Annual Costs[2]		$106,200	$119,700
Parameters kg/kkg (lb/1000 lb) product	Raw Waste Loads	Effluent Quality A	Effluent Quality B
Suspended Solids	0.192	0.192	0.192
Oil and Grease	0.493	0.240	0.144

[1]Technology A is isolation of process waste waters followed by API gravity separation.
Technology B is Technology A followed by an absorbent filter.
[2]August 1973 dollars.

Source: EPA (17)

Latex-Based Products Area

The latex-based products can be effectively separated into two subcategories. The first subcategory (Subcategory J) includes latex dipping, latex thread, and products manufactured in porous molds. The Subcategory K subcategory consists of latex foam manufacture. Since the required treatment is different for the two subcategories, separate discussions of the cost data are presented.

Subcategory J: Only one level of control and treatment has been considered in the evaluation of treatment effectiveness versus cost data. This recommended treatment includes chemical coagulation and primary clarification of latex-laden wastes followed by biological oxidation. The biological treatment cost data have been based on an aerated lagoon and settling pond system. The reason for this selection is that the BOD concentration in the process wastewaters is typically too low to support an activated sludge type biomass. Since the process wastes can be separated from the utility wastewaters, the proposed treatment system is limited to the treatment of the process wastes.

A profile was made of the latex dipping industry to determine the typical size of a production facility. The average, or typical, plant has a daily consumption of 2,100 kg (4,700 lb) of latex solids. The associated process wastewater flow, derived from plant visit data, is 153,000 liters (40,000 gallons) per day. The model treatment plant, consisting of chemical coagulation, clarification, and biooxidation is illustrated in Figure 81. This plant is equivalent to BPCTCA. The treatment designs upon which the cost data are based correspond to the highest expected raw waste load within the industry.

FIGURE 81: HYPOTHETICAL END-OF-PIPE SECONDARY WASTEWATER TREATMENT FACILITY FOR SUBCATEGORY J PLANTS

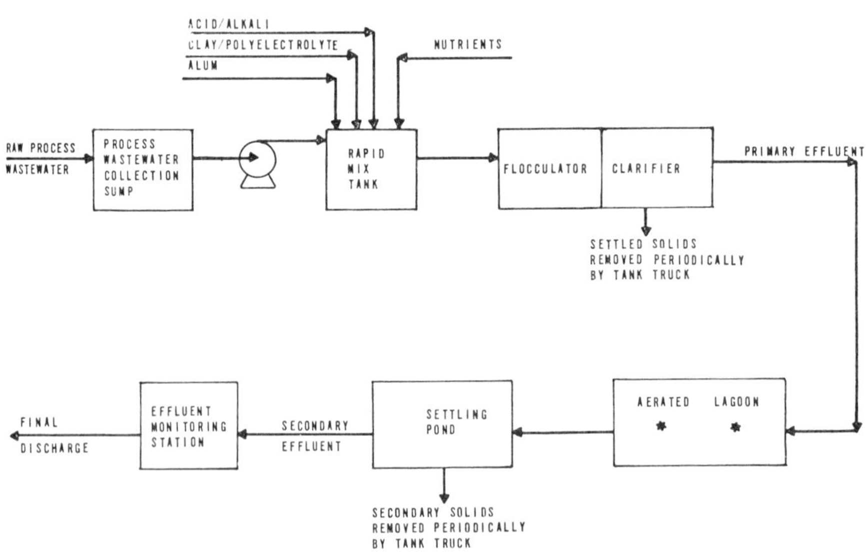

Source: EPA (17)

Relatively conservative cost figures are preferred for this type of general economic analysis. The capital costs were generated on a unit process basis, with the following percent add on figures applied to the total unit process costs in order to develop the total installed capital cost requirements:

Item	Percent of Unit Process Capital Cost
Electrical	20
Piping	15
Instrumentation	8
Site work	5
Engineering design and supervision construction	10
Construction contingency	15

The treatment costs incurred by the associated technology can be represented in terms of incremental costs per unit of production. Treatment cost data for a typical latex dipping plant consuming 2,100 kg (4,700 lb) of latex solids per day indicate that the BPCTCA and BATEA treatment will cost $0.042 per kg ($0.019/lb) of latex solids usage.

Since land costs vary appreciably between plant locations, it was decided to exclude land cost from the total capital cost estimates. Land costs must be added on an individual case basis. Annual costs were computed using the following cost basis:

Item	Cost Allocation
Capitalization	10 percent of investment
Depreciation	5-year straight line with zero salvage value
Operations and maintenance	Includes labor and supervision, chemicals, sludge hauling and disposals, insurance and taxes* and maintenance**
Power	Based on $0.015 kwh for electrical power

*Computed at 1.6 percent of the total capital costs
**Computed at 3.2 percent of the total capital costs

The short-term capitalization and depreciation write-off period is what is currently acceptable under current Internal Revenue Service Regulations pertaining to industrial pollution control equipment. All costs were computed in terms of August 1973 dollars, which correspond to an Engineering News Record Index (ENR) value of 1920. The total capital and annual costs for the model treatment technologies are presented for a typical latex-dipping plant in Table 56, together with raw waste load and treated effluent quality.

TABLE 56: ESTIMATED WASTEWATER CONTROL COSTS FOR A LATEX DIPPED PLANT (SUBCATEGORY J)

	Control and Treatment Technology*
Investment	$236,700
Annual Costs	
Capital costs	23,700
Depreciation	47,300
Operating and maintenance costs (excluding energy and power costs)	25,100
Energy and power costs	1,600
Total Annual Costs	$ 96,700

(continued)

TABLE 56: (continued)

Parameters, kg/kkg of raw material	Raw Waste Loads	Effluent Quality
BOD	18.2	2.20
Suspended solids	10.90	2.90
Oil	0.90	0.73
Chromium	0.0533	0.0036

*The control and treatment technology includes chemical coagulation, clarification, biological oxidation, and secondary clarification

Source: EPA (17)

Subcategory K: Two levels of control and treatment have been studied in the evaluation of treatment cost and effectiveness. The first level of the proposed technology is chemical coagulation and clarification of latex-based wastes, as well as chemical precipitation and clarification of zinc-laden foam rinse waters. The second technology level proposed involves biological oxidation of the combined primary effluents. The biological treatment cost data are based on an activated sludge treatment system. This type of system was selected because the BOD loading of the combined effluent is high (approximately 400 mg/l) and would be able to support an active mixed liquor.

The treatment cost is also based on a combined process effluent which includes the slightly contaminated barometric condenser flows. This waste stream has been included in the cost evaluation because the residual zinc concentration after chemical precipitation requires dilution by the condenser discharge in order to avoid biological inhibition.

The only significant latex foam plant in the United States was used as a basis for the industry. The plant has a daily consumption of 68,000 kg (150,000 lb) of latex solids; the average process wastewater flow is 1,608,000 liters (425,000 gallons) per day. The model treatment plant, illustrating both BPCTCA and BATEA, consists of chemical coagulation of latex solids, chemical precipitation of zinc, and biological oxidations. (See Figure 82.)

The influent raw waste loads, upon which the treatment designs are based, represent the raw effluent conditions at the sole U.S. latex foam plant. This results in the generation of cost data which are the most appropriate for the industry as it is known to exist. The total capital and annual costs for the model treatment techniques for the sole latex foam plant are presented in Table 57 together with the raw waste loads and treated effluent qualities. The treatment costs for a latex foam production facility can be expressed as an incremental cost per unit of production. The cost data for a plant consuming 68,000 kg (150,000 lb) of latex solids daily indicate that the BPCTCA treatment will cost $0.003 per kg ($0.003/lb) of latex consumption and that the additional costs of the BATEA treatment will approximate $0.003 per kg ($0.003/lb) of latex consumption.

TABLE 57: ESTIMATED WASTEWATER TREATMENT COSTS AT DIFFERENT DEGREES OF TREATMENT FOR A LATEX FOAM PLANT (SUBCATEGORY K)

	Treatment or Control Technology*	
	A	B
Investment**	$348,000	$649,000
Annual Costs		
Capital costs	34,800	65,000
Depreciation	69,600	129,800
Operating and maintenance costs (excluding energy and power costs)	55,300	98,400
Energy and power costs	3,500	13,100
Total Annual Costs**	$163,200	$306,300

(continued)

TABLE 57: (continued)

Parameters, kg/kkg of raw material	Raw Waste Loads	Effluent Quality A	Effluent Quality B
BOD	20.44	9.43	1.41
Suspended solids	8.71	1.60	0.94
Zinc	3.54	0.083	0.083

*Technology A is chemical coagulation and clarification of latex wastewaters and chemical precipitate and clarification of zinc-laden wastewaters

Technology B is Technology A followed by biological oxidation treatment

**August 1973

Source: EPA (17)

SOLID WASTE GENERATION FROM MANUFACTURE OF SPECIFIC PRODUCTS

A report by the U.S. Environmental Protection Agency (21) is concerned only with the solid waste problem which originates from the fabrication of rubber products and which must be disposed of at the manufacturing site. It is not concerned with ultimate disposal by the consumer of the product or the product package.

In the manufacture of the 10.7 billion pounds of salable rubber products in 1968, it is estimated that 1,058 million pounds of solid wastes originated from the fabricating facilities and had to be disposed of at those locations. These solid wastes had no present value and represented a disposal cost of about $9.5 million for collection, haulage and dumping. This sum does not include the cost to the fabricator of processing these semifinished goods, but this internal cost is a constant stimulus to the producer to keep wastes to the lowest level possible.

The kinds and relative quantities of this manufacturer's solid waste will vary according to the product or products made at a particular location. The complexity of the industry has necessitated grouping the products made into six general categories and the types of waste into five categories. Table 58 gives the estimated breakdown of fabricator's solid wastes for the entire industry in 1968.

TABLE 58: SOLID WASTE GENERATED BY THE FABRICATED RUBBER PRODUCTS (MILLIONS OF POUNDS)

Type of Waste	Type of Rubber Product: Tires and Tire Products	Foot-wear	Belts	Hose	Foam and Sponge	Mech. Goods	Total
Paper, cardboard, and wood	100.5	8.5	6.8	11.7	9.6	65.5	202.6
Rubber compound	89.3	45.0	10.2	19.2	54.0	140.7	358.4
Textile materials	44.2	33.0	6.3	6.3	0.1	42.8	132.7
Metals	72.7	00.5	0.6	3.6	---	21.3	98.7
Other Materials	105.2	17.5	7.6	17.2	16.2	102.0	265.7
Total Wastes	411.9	104.5	31.5	58.0	79.9	372.3	1,058.1

Source: EPA (21)

FIGURE 82: HYPOTHETICAL END-OF-PIPE SECONDARY WASTEWATER TREATMENT FACILITY FOR SUBCATEGORY K

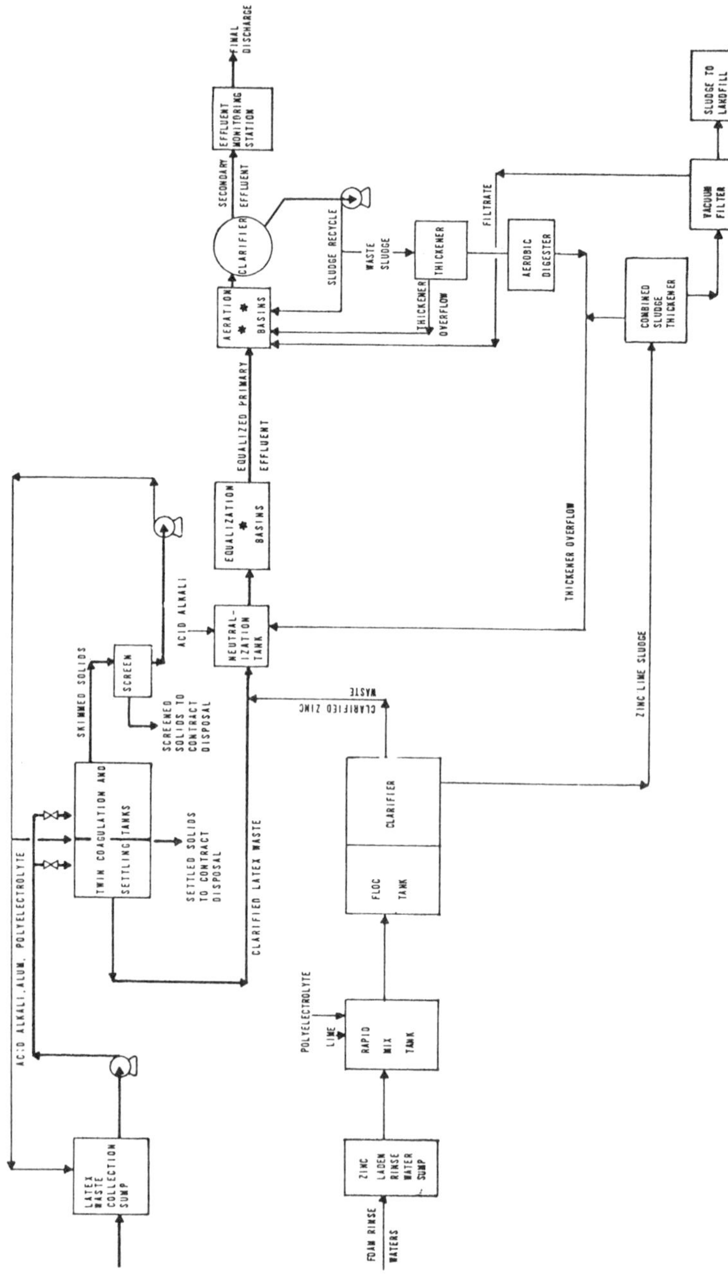

Source: EPA (17)

The tire and tire products segment of the industry accounts for about three-quarters of the total weight of finished products but generates only about 40 percent of the total solid waste. Because of the large volume of a single type of product, tires are generally made by semiautomated procedures in modern or modernized factories. This permits the use of bulk raw materials reducing the amount of input paper bags to be disposed of. It permits standardized assembly procedures and better waste cost control and provides the possibility of utilizing process waste in less critical products such as flaps and other tire accessories. An important part of the rubber compound waste is cure-bags which are relatively short-lived accessories to the tire curing operation and which when discarded cannot always be sent to reclaimers.

For the footwear segment the paper input is small because much of the raw material enters in a preprocessed condition to avoid soil. Because of the domination of hand work on a great variety of small units, the wastage in rubber compound and textile material trimmings is too high to use up entirely in lower quality components such as inner soles and counter stocks. There is a limited market for these wastes to others, but the net disposable waste is high.

For the belting, hose, and mechanical goods segments, which are usually combined in a single plant, the big waste item is rubber compound, usually in the cured state so it cannot be recycled. This segment is made up of a very large number of items largely hand-built to rigid specifications. As a result there is a sizable amount of spoilage and quality rejects which cannot be safely sold as off-grade products but must be scrapped and disposed of as solid waste.

Most foam and sponge product raw materials are received as bulk water suspensions and so the input paper is relatively small. The foam process in itself produces wastage of rubber compound as trimmings and this makes up a large part of the solid waste. Many attempts have been made to reconstitute this foam scrap into useful products but they have met with little success as they lack the special properties of new foam. The waste is not at all attractive to rubber reclaimers because of its bulkiness.

The problem of solid wastes is no novelty to the fabricated rubber products industry forced upon them by the great number of units of an endless variety that they must produce to rigid specification by largely hand assembly methods. Rubber industry management is well aware of the internal cost of producing solid waste. This cost is reflected in high reject rates, wasted raw material, and most of all, wasted labor, a high price commodity. Short of further automation and other capital investment, there is little likelihood that production wastes can be reduced further in the near future.

Most top management is not aware of the actual cost of solid waste disposal which includes in-plant collection and baling, and outside hauling and dump fees or investments. In many cases only lower management levels are aware of the actual disposal costs. Several plants have salvage operations which sell all the waste they can and hope that this revenue will offset part or all of the solid waste disposal costs.

Certainly more careful segregation of waste by kind would make disposal more effective and possibly reduce the total amount by allowing more rework. Powder spillage and trim scrap could be kept out of general waste and returned to process. Reworking of plastic separator sheets was reported by one major company, cutting that particular waste item to one-sixth of what it was. Such investments for reprocessing can reduce the disposal costs and amounts of solid waste to be disposed.

The segregation of solid waste will allow for the incineration of the nonrubber portion which accounts for sixty-percent of the material now being disposed of in land fill sites. Present techniques waste valuable land fill sites by discarding many noncontaminating combustibles in them. Further, incinerator technology should be developed that will enable industry or municipality to burn rubber products without polluting the atmosphere and thus eliminate much of the need for land fill operations in or near the crowded metropolitan areas. Incineration is a short term solution to the problem. Ultimately reuse and recycle are the long term solutions to the total solid waste problem.

Tire Manufacturing

The tire and tire product segment of the industry produces 70 percent of finished product weight of the entire fabricated rubber products industry and six major companies produce 83 percent of the tires at 42 locations. Ten other companies produce the remaining 17 percent at 15 locations. Usually inner tubes are produced in plants separate from the tire casing plant and some tread stock for retreading is produced in detached small facilities. In general all other solid tires and auxiliary materials are made in the larger plants which produce tire casings. Passenger car tire sizes account for 65 percent by weight and 80 percent by number of all tires manufactured in the United States.

The amount of solid waste generated in the manufacture of one million pounds of tires and tire products is shown below. These are average values for 1968 but because of the similarity of operation from plant to plant, they are representative of most operations in this category. There is little reason to believe that the situation was much different in the past ten years and no major improvement is to be expected in the next few years because this wastage has been recognized as a serious manufacturing cost for many years and is under constant attack. The total wastes amount to about 5.5 percent of the total weight of product made. One million pounds of tire industry product is equivalent to 40,000 passenger tire casings.

Type of Solid Waste	Pounds of Waste per Million Pounds of Product Shipped
Paper, cardboard and wood	13,400
Rubber compound	11,900
Textile material	5,900
Metal	9,700
Other	14,000
Total Waste	54,900

The paper, cardboard and wood input waste is largely paper bags in which carbon black, mineral pigments, and chemicals are received. These are of little value for reuse in the paper industry because of severe contamination by the powders they contained. At least in the larger facilities carbon black is received in dustless bulk shipments in cars or large returnable containers so the raw material bags for this item are eliminated. There is a sizable amount of in-process paper and wood waste from separator paper, temporary wrappings, and card and paper stock used to identify and control in-process materials. All of this paper and wood waste could be handled in suitable incinerators if properly segregated. In most cases no attempt is made at segregation, proper incinerators are not available, and the solid waste is handled by general collection, baling or semicompression, and hauling to dumps.

The rubber compound waste includes solids containing appreciable amounts of rubber which makes them unsuitable for combustion in any ordinarily satisfactory incinerator. Not only is the rubber hydrocarbon content such that it burns with an obnoxiously sooty flame, but the gases contain a large concentration of acid and sulfurous gases. The waste is made up of in-process trimmings and molding waste, in-process spoiled stock, curing bags, quality rejects and experimental and test products.

Some of the in-process waste is used up within the plant itself or related operations as raw material in low quality goods such as flaps and light service solid tires and rollers, but this means of disposal is limited by the product mix. Curing bags are a tire-like structure of rubber with some fabric which is inserted in the tire when it is cured to exert internal pressure. For passenger tires, these units weigh 12 pounds apiece and will serve to cure 250 to 300 tires before they must be scrapped. It is estimated that 7 to 8 million pounds of these bags are scrapped each year, but since they can be handled by reclaimers with some difficulty only a part of the curing bags appear in the solid waste figures above. Finished tire rejects for quality reasons average one percent for passenger tires and less than one percent for heavier duty tires. In addition, there is a fraction of one percent of finished tires which are partially or completely destroyed in testing in the laboratory, on tracks, or in road tests. These cannot be sold as off-quality goods and they are not readily disposed of for reclaiming as new tires are much more difficult to process than are service-aged tires which have undergone desirable physical changes. All of these new unsalable tires are included in the waste numbers above, broken down into their known components.

Textile fabric and tire cord scrap (ends, clippings, defects) which have not been rubberized find a ready market with scrap dealers. Even the uncured tire carcass scrap (trimmings and spoilage) which is 60 percent rubber compound and 40 percent textile material usually can be sold to scrap dealers for resale to the mechanical rubber goods industry. Most of the amount under textile materials would be the textile content of finished goods which cannot be sold.

The metal waste is largely the high carbon steel wire from the tire bead. This consists of defective material, in-process waste, trimmings and the wire component from finished goods rejects. There is some wastage of metal inner tube valves but this material finds a ready resale as scrap metal and does not appear in the above figures. At one time scrap bead wire could be compressed, baled and sold as scrap steel for use in the steel industry but it was never too desirable because of its bulkiness and of recent years has found no market at all because of changing steel production technology.

Changing tire construction designs may lead to changes in both the quantity and kind of waste from the industry in the future. The changes probably have not had too much effect through 1973 but may be quite important thereafter. Most of the pneumatic tires made through 1968 were of the conventional bias construction in which the carcass was built of diagonal layers of tire cord made of conventional cotton or man-made fibers. In 1969, large numbers of belted bias tires were made in which the carcass is reinforced with circumferential belts of cords of glass fiber, synthetic fiber or even fine steel wire. It is estimated that these tires will give 15 to 40 percent more service life but they will weigh about 18 percent more than the conventional bias tire.

There is no reason to believe that the net solid waste ratio will change materially and although fewer units may be marketed the weight of product will remain the same. The only change should be that there will be glass fiber in the scrap which will make it less attractive for resale and more difficult to handle in incinerators.

Another type of construction which is an important market factor is the radial tire in which the cords lie in radial planes which is believed to give better service. In Europe where these tires are very popular, many of them are made with steel wire in the carcass or in the belt, in addition to the conventional wire bead. If this becomes popular in the United States, the problem of selling scrap or disposing of wastes will become even more difficult.

In Figure 83, a schematic outline of the tire manufacturing process is given to show the possible origin of the various types of scrap which will end up as solid waste if it cannot be reused internally or sold for some value.

Inner tubes are somewhat different from tire casings in the waste problem generated. They contain very little textile material or metal and are nearly straight rubber compound. The waste is almost entirely paper input from raw materials and in-process trimmings. Inner tube scrap and rejects find a ready market with scrap dealers because they are clean compounds and are made exclusively from butyl or natural rubber which is very desirable to reclaimers. Scrap valves find a ready market on the metal scrap market. It was not possible to obtain waste data on inner tube production alone so the information is included in that for tires and tire products. Very few inner tubes are now used in passenger tires and then only in special cases such as wire wheel mountings. In 1968, at least 80 percent

FIGURE 83: SCHEMATIC DIAGRAM OF TIRE MANUFACTURING PROCESS SHOWING SOLID WASTE GENERATION

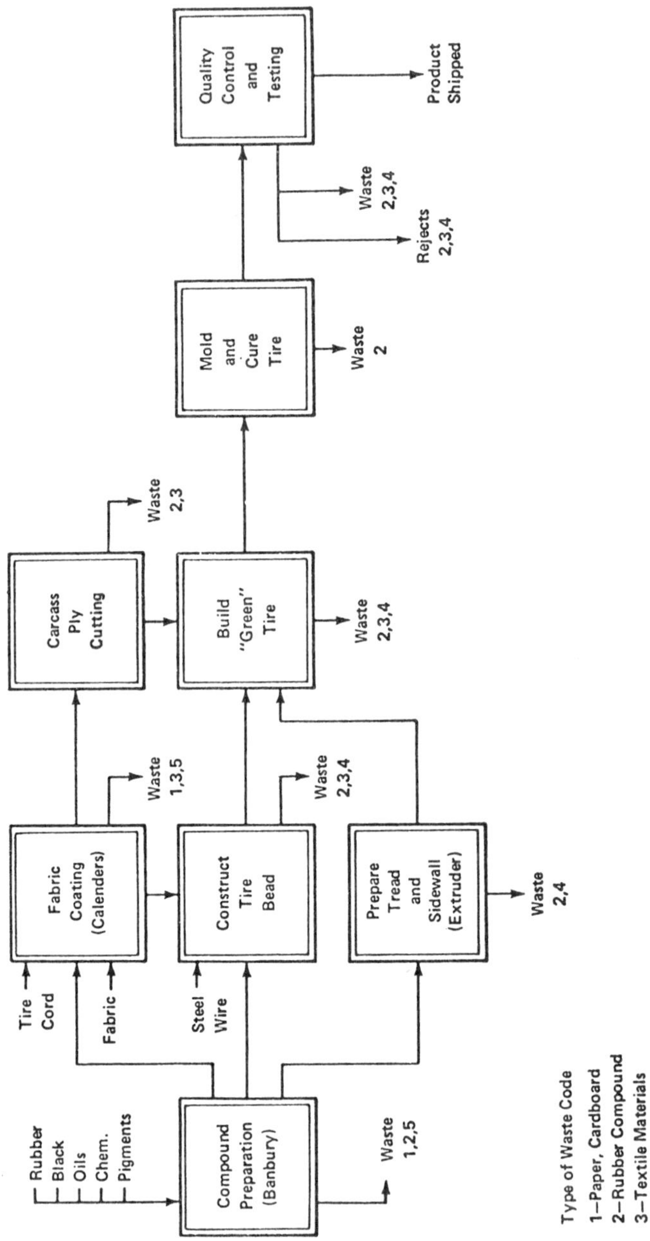

Source: EPA (21)

of truck, bus and heavier tires were operated with inner tubes because of the severe service. Heavy duty tubeless tires are made but the current consensus in the industry is that they will not ever be a major factor.

Tread rubber and repair materials generate little if any waste as they are uncured materials and can usually be reworked. It is estimated that some 20 million pounds of raw material packaging paper must be disposed of and that is included in the tire and tire products total. Tire flaps and sundries generate practically no waste and in fact are a primary means of using up in-process scrap from other tire operations.

Footwear Manufacturing

In 1968 the canvas and waterproof footwear industry produced 500 million pounds of finished product, or 190 million pairs of canvas footwear and 30 million pairs of waterproof footwear. The amount of solid waste generated is shown below. It was not possible to make a distinction between that of the canvas and of the waterproof type as they are made in the same plants.

Type of Solid Waste	Pounds of Waste per Million Pounds of Product Shipped
Paper, cardboard and wood	17,000
Rubber compound	90,000
Textile material	66,000
Metal	1,000
Other	35,000
Total Waste	209,000

The paper waste is comparable to that for tires but little of the raw materials can be received in bulk because of the danger of soiling the product with loose dust. Metal waste is small because few small metal parts are used, mostly of the zipper type. The rubber compound and textile material scrap is high because the industry produces small units by hand, assembled in a multitude of sizes and styles.

Both before and after combining with rubber, a great variety of small pieces must be cut from sheet goods leaving unavoidable waste trimmings. In the compression molding process an excess of rubber compound is used which appears as molding waste. Sponge innersoles contribute considerable scrap because of the odd shapes and sizes to which they must be cut. The trend to injection molding in place of compression molding for attaching soles to uppers will reduce the molding waste as the rubber compound quantity can be more accurately fixed.

Some of the rubber and textile waste is ground up and used in hidden parts of footwear. The remainder has no value on the market and must be discarded. The waterproof footwear industry has declined sharply and the canvas footwear industry has begun to decline slightly because of the competition of imports. Because of the hand work involved, lower labor costs off-shore make lower prices possible.

The quality is said to be lower but perfectly adequate for the American consumer who is more style conscious than ten years ago and less concerned with durability. Imports account for 10 to 15 percent of the canvas footwear domestic market and 20 to 25 percent of the waterproof market, or a total of about 72 million pounds of imported product.

In Figure 84 a schematic outline of rubber footwear manufacture with the sources of the various kinds of waste is given. Most production rejects are repaired and sold as seconds and thirds. Rejects that must be scrapped are reported to be less than 0.1 percent of production including those destroyed in testing.

FIGURE 84: SCHEMATIC DIAGRAM OF CANVAS AND WATERPROOF FOOTWEAR PROCESSES

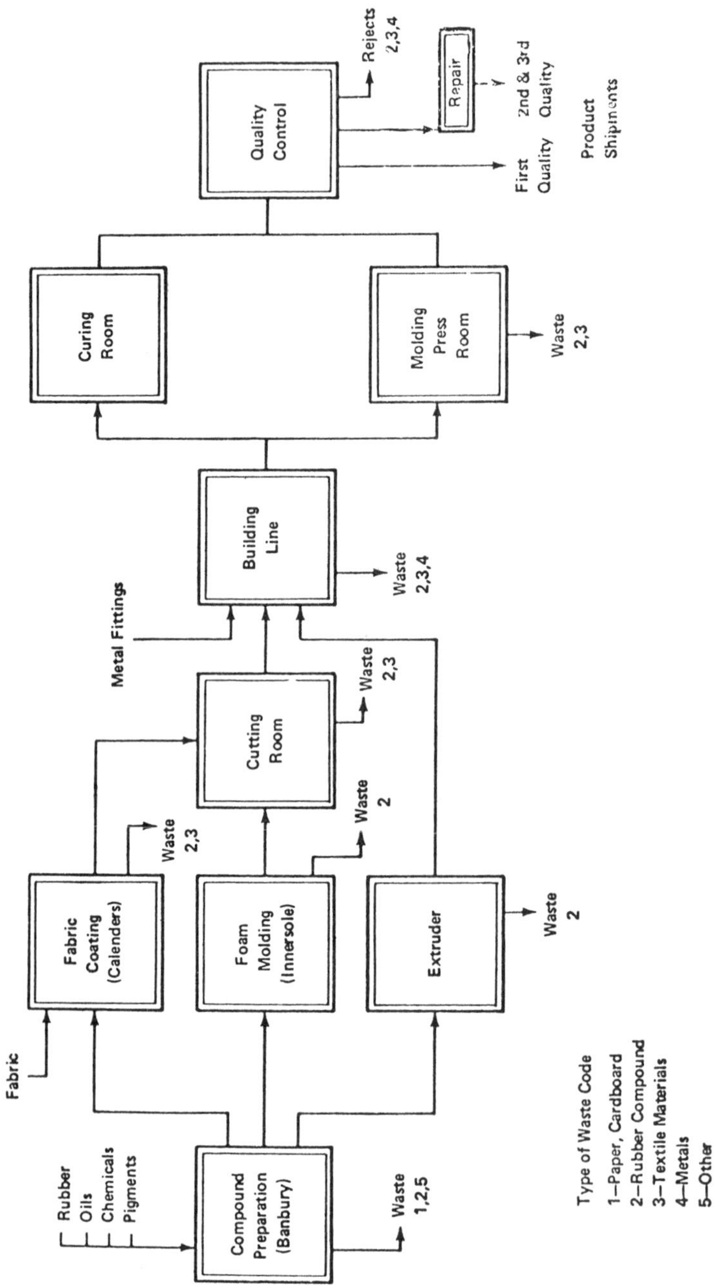

Source: EPA (21)

Belting Manufacturing

The belts and belting portion of the fabricated rubber products industry is comprised in the SIC classification 30691. This includes flat belting for conveying materials and for power transmission, V-belts for automotive, appliance and industrial power use, and a wide variety of belts for toys, business machines, instruments and other devices. They vary in size from complete belts weighing as little as two ounces up to mining conveyor belts weighing up to 12 pounds per linear foot. Belts are generally built of heavy textile fabrics or cord combined with cured rubber compounds but they may include metal wire or cables. This study covered what is estimated to be 85 percent of the production of the belting industry. The solid waste generated per million pounds of belt produced is as follows:

Type of Waste	Pounds per Million Pounds of Products
Paper, cardboard and wood	27,300
Rubber compound	41,000
Textile material	25,200
Metal	2,300
Other	30,200
Total Waste	126,000

Raw materials are almost all received in paper bags or cardboard containers. Rubber compound and textile waste is higher than for tires because less standardization and automation is possible, but is much less than for footwear because design is simpler and there is less variety. This waste is largely cured and uncured trimmings and very little of it can be reused in the same plant. The metal scrap is ends and damaged material from those types of belts which use metal reinforcement. The other waste is spilled pigments and earth materials which get mingled in general floor sweepings.

Belts which have minor defects are repaired in the plant and sold as first quality product. Flat belts are made in semicontinuous lengths and major defects can be cut out and the remainder sold as short lengths of first quality material. Splicing of lengths is also possible as they are usually spliced on the job in any case. A block diagram for belt processing is given in Figure 85. The belt industry is showing a steady growth supported as it is by the steady growth of the automotive and appliance industry and by the strong trend toward bulk movement of minerals, grain, chemicals, industrial products, and people.

Hose Manufacturing

The rubber hose segment of the fabricated rubber products industry is comprised in SIC 30692 and is made up of all types of household, retail service, appliance, industrial and marine hose and tubings which are made in part of rubber, with the exception of small medical tubing. The types range from very small laboratory and instrument tubings to very large and massively reinforced hydraulic mining and dredging hose. Some types are solid rubber compound, some such as firehose are nearly all fabric with a minimum of rubber, and some are composites heavily armored or reinforced with metal wire. The metal couplings are attached in the factory except for lighter hose and tubings. The one thing in common is that they are intended to convey liquids and gases and are at least semiflexible.

Most of the major producers make a wide varierty of hose types in a single plant and they were unable to provide data by type. The average solid waste generated per million pounds of hose produced is as follows:

Type of Waste	Millions of Pounds per Million Pounds of Product
Paper, cardboard and wood	33,300
Rubber compound	54,700

(continued)

Type of Waste	Millions of Pounds per Million Pounds of Product
Textile material	18,300
Metal	10,200
Other	49,300
Total Waste	165,800

The paper scrap is largely from raw material packages but contains as well, nonreturnable wire reel cores and textile yarn cores. Most of the reinforcing jackets are braided from textile yarns or wire and little or no woven or knit fabric is used. Most of the hose plants are old and not well adapted to automation, a large variety of sizes and styles are made in most plants, and the hose is built to rigid specifications. These factors all tend to make in-process spoilage a serious contributor to rubber compound and textile waste. Since the textile fabric is braided in place, there is little trimming and cutting waste as there is in the other parts of the rubber product industry.

There is little possibility of salvaging the steel wire or steel jacket material but the nonferrous materials in reject fittings are either renovated and reused as such or sold as high value merchant scrap. Because of rigid performance specifications and quality control standards rejects are high. In the past it was customary to cut rejects into short ends and sell them as substandard lengths for nonspecification applications.

Because of misuse of these products and severe manufacturer's liability rulings, it is no longer possible to dispose of most rejects in this fashion and they must be spoiled and become solid waste. The finished product rejects that must be scrapped run 8 to 12 percent of total production and their weight is included in the tabulation above.

Much hose is cured in a temporary lead sheath to permit the use of internal pressure. This lead is stripped in the process and reused over and over until it is too contaminated to work well. The contaminated lead is sold at scrap metal prices for refining and reuse. Some lead oxides and dross probably get into other waste and is reported as such. A schematic diagram of hose manufacture is given in Figure 86. The industry is thriving, thanks to growth in automotive and appliance demands, especially automotive air conditioning. The growth in marine shipments of petroleum and chemicals has required large amounts of large hose for product transfer.

Sponge and Foam Rubber Products Manufacture

Sponge and foam rubber products are included in SIC category 30693. This includes all the foam products which are made from rubber latex, which is a suspension of natural or synthetic rubber in water, such as cushioning, mattresses, pillows, and carpet underlays. The sponge products which are also included are made of solid rubber which has been expanded by chemical blowing agents and includes various insulation items, flotation items, seals for windows and doors, and carpet padding. The products are essentially rubber compounds as they rarely contain textile materials or metal inserts. Although the range of products is wide, the operating procedures are very similar for all plants. The solid wastes generated per million pounds of product are as follows:

Type of Waste	Pounds of Waste per Million Pounds of Product
Paper, cardboard and wood	16,000
Rubber compound	90,000
Textile material	200
Metal	Negligible
Other	27,000
Total Waste	133,200

Since a large part of the raw materials used are received as either bulk liquids (latices) or bulk powders (mineral fillers) the paper waste from raw material receiving is low. The other

FIGURE 85: SCHEMATIC DIAGRAM OF RUBBER BELTING PROCESSES

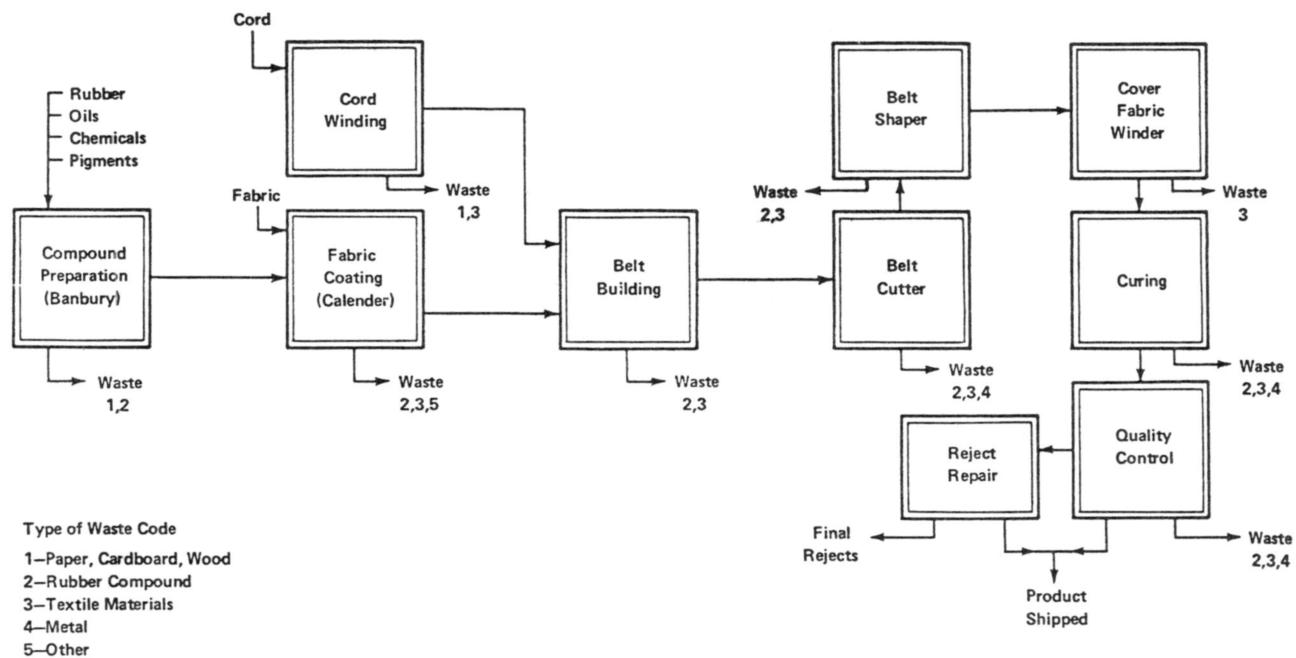

Source: EPA (21)

FIGURE 86: SCHEMATIC DIAGRAM OF RUBBER HOSE PROCESSES

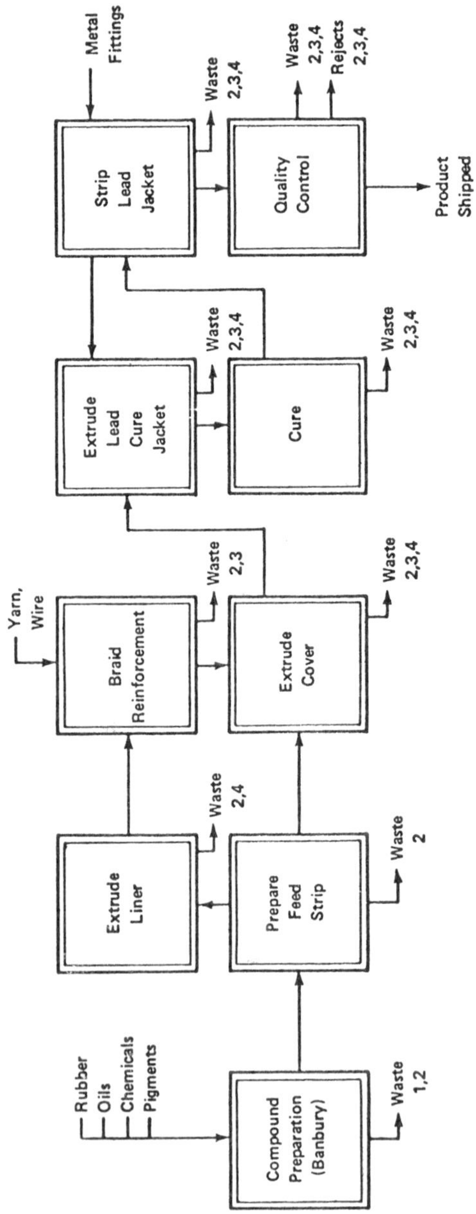

Type of Waste Code

1—Paper, cardboard, wood
2—Rubber compound
3—Textile material
4—Metal
5—Other

Source: EPA (21)

category includes liquid spillage which promptly dries to useless solid rubber. This is included in "other" with floor sweepings and other incidental waste. The rubber compound waste, which amounts to two-thirds of the total, is imposed by the process and the variety of products made. Foam is molded either in slabs for further division or in complete units such as mattresses, pillows, or chair cushions. In either case the open molds must be overfilled and the excess trimmed away.

The slab stock is used as blanks from which a multitude of shapes and sizes are cut on order. This custom sizing makes a large amount of three-dimensional trimmings which do not find a ready market. Because of the low density of foam and sponge, this material is very bulky and expensive to store; if finely ground it can be a serious fire hazard. The ground material cannot be reworked successfully and finds only a limited market as reprocessing scrap. Attempts have been made to bond this foam scrap with more latex to make secondary padding for furniture but it has not been commercially feasible. There has been a fairly modest growth of sponge and foam products through the years. Growth would have been much greater except for the inroads of urethane foams which have many of the same values but are not included in the rubber products industry.

Mechanical Rubber Goods Manufacture

The mechanical goods segment of the fabricated rubber products industry includes the following SIC categories:

30695	Mechanical rubber goods
30696	Rubber heels and soles
30697	Drug and medical sundries
30698	Other rubber goods

It excludes those groups already described. Mechanical rubber goods includes tens of thousands of molded and extruded items used by industry and as components of consumer hard goods. It includes such things as seals, rolls, inflatable goods, battery cases, fuel cells for aircraft, dock bumpers, and many others. Rubber heels and soles includes these items and the slab stock from which they may be cut for use by the shoe industry and the shoe-repair services. It is not a part of the rubber footwear industry. Drug and medical sundries are such things as hot water bottles, douches, medical tubings, and prosthetics. Other rubber goods is a large conglomerate of items which defy classification such as coated fabrics, rubber thread, balloons, and custom compounded rubber for others. The solid waste generated by this segment of the industry per million pounds of product is as follows:

Type of Waste	Pounds per Million Pounds of Product
Paper, cardboard and wood	43,700
Rubber compound	93,800
Textile material	28,500
Metal	14,200
Other	68,000
Total Waste	248,200

It is impossible to give a representative process description because of the great number of ways of making this conglomerate of items. They usually involve short runs of hand-built items of considerable complexity. Most of the items are basically rubber compound with fabric and metal inserts and reinforcement. The ratio of waste is the highest for any segment of the rubber products industry for the above reasons and all categories of waste are high. Only in the case of soles and heels can trimmings and rejects be ground and re-used in the operation and regrind is said to be 50 percent of the rubber compound waste. The large number of small plants makes it especially difficult to collect and classify the solid wastes generated other than by casual methods. The growth of this segment of the industry is not spectacular as it will have only doubled in quantity in 15 years.

Wire and Cable Manufacture

The wire and cable industry, which is traditionally considered part of the fabricated rubber products, is described as SIC 3357, Insulating of Nonferrous Wire. At one time all such power, communication, and other electrical service wires and cable were insulated and jacketed with rubber or rubber-like materials by conventional rubber processes. The rubber component was never very large in value or volume compared to the nonferrous metal used in the industry, but at one time it made up a sizable part of the rubber consumed. In the period covered by this report, the rubber usage is low, varying from three percent to one percent of total rubber consumption, because of its large replacement by plastics in many kinds of wire. The industry does pose a rather unusual disposal problem.

Some 175 million pounds of rubber compounds are mixed in these operations much as in tire plants and applied to the wire and cable in various ways, but usually by simple extrusion. The finished product is of high value and made to exacting standards on a continuous basis. Rejects develop locally in testing and these lengths are cut out. Rejects are stripped of rubber compound, which is discarded and the core rerun. If the core is damaged the rubber compound is burned off and the nonferrous core sold in the scrap metal market. The solid wastes generated in 1968 are estimated below as pounds per million pounds of compound.

Type of Waste	Pounds per Million Pounds of Rubber Compound
Paper, cardboard and wood	11,000
Rubber compound	20,000
Textile material	1,000
Other	2,800
Total Waste	34,800

These numbers refer only to the rubber compound used and not to the total weight of the industry product. No metal waste is given as this is a concern of the wire industry proper. The paper waste is that from the rubber compound raw materials. The rubber compound is that which is stripped or burned off of rejects. The textile waste is the small amount stripped or burned with the rubber component. Future consumption of rubber compound by the wire and cable industry should remain where it was in 1968, or decline slowly.

RECOVERY OF PLASTICS PRODUCTS FROM SOLID WASTE

Plastics can be recycled. They can be reground and reformed into new finished goods and some one billion pounds were used in this fashion by the plastics industry during 1973. Recycling is a regular process in most fabricating plants. However, techniques for recovering plastics from post-consumer waste have not been developed.

There are no real data available to indicate the quantities of plastics recycled. Investigations, conducted predominantly through local interviews and correspondence, (3), revealed that waste production in plastics fabrication is quite high, ranging from 5 to 15% for all thermoplastics and from 10 to 30% for all thermosets. Thermoset wastes, of course, cannot be remelted and go to disposal. Fabrication waste reuse related to thermoplastic production varies from operation to operation and from company to company. No pattern could be discerned. Once plastics leave fabrication points, they are not recovered and there is no recovery from obsolete products.

Fabrication wastes, unless they are immediately reused in the plant where they occur or are disposed of, are acquired by plastics scrap processors who specialize in regrinding scrap, color blending, and remelting. These processors frequently operate on a contract basis, returning wastes to the organization that provided them, reformulated into a product that can be reused. These processors also supply manufacturers that use scrap plastics as an input material. Midwest Research Institute (3) was able to identify 13 such companies in the United States.

Plastic scrap, as prepared by processors, is consumed in the toy industry, in the manufacture of cheap housewares, in the manufacture of plastic pipe, in the production of artificial flowers, and in similar applications where [1] plastics properties and performance are not paramount; [2] relatively noncritical processes are used (compression molding or heavy extrusion); and [3] the cost of plastic resin is a high proportion of total product cost.

Only one instance is known where a plastics product is recycled after doing service in the home. The project involved a San Diego dairy that used polyethylene milk bottles. Following Earth Day activities in the San Diego area, the dairy worked out an arrangement, with the help of Dow Chemical, producer of polyethylene resins, whereby bottles returned through the milkman or placed into special supermarket bins were ground by the dairy and the material was sold to a manufacturer of drainage tile. Approximately 8% of the bottles were returned; consumers were not paid for the plastic.

Plastic fabrication wastes are worth $10 to $20 per ton for clear polyethylene, $40 to $60 per ton for clear polystyrene, $90 to $110 per ton for clear vinyl and $120 to $160 for clear cellulose acetate. These are prices paid by processors. Virgin plastics sell for around $225 to $400 per ton for polyethylene, $310 to $350 for polystyrene, $240 to $400 per ton for vinyl chloride resins, and $1,240 per ton for cellulose acetate.

Plastics presently constitute approximately 3.8% of municipal solid waste, but plastic consumption is growing rapidly. Essentially no recovery of plastics as a material from mixed waste now takes place according to the EPA (19).

Of all the materials in mixed waste, plastics are probably the moxt difficult to extract. Recycling of certain plastic wastes from fabrication plants is practiced. In industrial plants it is possible to keep different types of plastics, (e.g., polyethylene, polystyrene, and polyvinyl chloride) separated at the source. Once plastics have entered use, and especially after they have been discarded into the waste stream, they are extremely difficult to separate.

Supply of waste plastics is simply not possible with existing technology because plastics cannot be separated from paper and other materials in mixed waste with similar physical characteristics. Experiments have been conducted with new separation techniques but are a long way from full-scale application.

Among the major factors that combine to minimize interest in plastics salvage and to limit severely the extent of recycling are the following.

[1] Techniques for separating plastics from mixed wastes are not available.

[2] Identification of plastics is difficult. Segregation of waste plastics by resin type appears impossible. There are currently over 700 different grades of polyethylene alone.

[3] Degradation of resin properties and performance occurs during the initial fabrication, through aging, and in any reclamation process.

[4] The technology for purification and upgrading of contaminated, degraded, and colored plastics simply does not exist.

[5] Economics of collecting and shipping plastics to points of reuse are generally unfavorable.

[6] Effective market mechanisms for trade in reclaimed plastic materials have not developed.

[7] Standard grades and specifications for reclaimed or salvaged plastics are not established or not widely accepted.

[8] Plastics users distrust the uniformity of reclaimed plastics and are generally unwilling to risk adding another variable to the "inexact science of molding."

A fundamental obstacle to plastics recovery and reuse springs directly from the synthetic origin of the resins. Because resins are created in a relatively pure form, it has not been necessary to develop the technology for purifying and refining plastics.

Unlike the metals production process, which begins with an impure ore which is progressively concentrated, smelted, refined, and freed from impurities to create the desired alloy, plastics processing begins with high purity virgin polymer to which various nonresinous additives, colorants, and reinforcements are added. All plastics processing is downhill from pure and uniform resin to blended, colored, shaped, and partially degraded plastic products.

The recovery of metal scrap is made possible by decades of metallurgical technology designed to upgrade ores and concentrates. Likewise, paper stock recovery is supported by the traditional steps of fiber separation, refining, washing, and bleaching cycle which also converts cut timber into papermaking furnish although it is accomplished in different types of equipment than virgin pulp making. Production processes based upon purification, refining, and upgrading of crude materials are capable of accepting scrap as though it were partially processed ore or charge stock. No analogous technology for plastics processing has yet developed. Thus plastic scrap is never purified; rather, the regrind is colored darker and used for less critical products. Practical means of removing unwanted contaminants are largely nonexistent.

The general rule governing the reuse of plastic scrap seems to be: "If plastic leaves the machine on which it was formed, it is no longer usable scrap; it's waste!" The dominant pattern of plastics recycling, for these reasons, is one in which the wastes of a process using high quality resin are either blended back into the process in limited quantities or are resold for use in a less exacting application, sometimes passing through a secondary materials operation where color blending, repelletizing, and similar scrap preparation steps are taken.

The problems of segregating plastics from the total municipal solid waste mix are certainly not unique to plastics. This is a formidable problem even for those materials that are potentially more readily and economically separated. However, as the result of several projects sponsored by the Solid Wastes Office of the U.S. Environmental Protection Agency, considerable progress is being made to develop practical separation systems for several materials, including plastics.

Since plastics constitute such a small proportion of the total solid waste stream, sophisticated systems to separate different types of plastics from each other may not become ecologically desirable or economically feasible until much larger volumes are achieved.

Except possibly on a limited basis, therefore, plastic fractions obtainable from the solid waste stream do not yet appear to be truly subject to recycling. The reuse of some types of mixed plastics, however, is a strong possibility technologically, and work in this area is underway. For example, the utilization of a flux or binder material such as chlorinated polyethylene for combining different types of plastics is being studied and appears to be quite promising (2).

SEPARATION OF PLASTICS FROM MIXED WASTE

The major objective of Bureau of Mines research as described by Holman, Stephenson and Adams was to show that heterogeneous thermoplastic concentrates from urban and industrial refuse could be separated into reusable generic plastic types. The optional methods for separating plastics into major types are essentially those based on shape, density, selective dissolution, and electrical properties. A separation scheme based on a sink-float-elutriation separation process was developed and applied to the separation of urban and industrial plastic wastes. Methods for utilizing these reclaimed secondary thermoplastics, such as hot pressing, injection molding, extrusion, and thermal decomposition, were also investigated.

Air Classification

A large portion of collected waste plastic such as film, fiber, and foam can be separated from the molded plastic and container fragments by air classification. Numerous experimental and commercial types of air classifiers are available.

A process developed by *J.F. Laundrie; U.S. Patent 3,814,240; June 4, 1974; assigned to the U.S. Secretary of Agriculture* transports, in suspension, a mixture of wastepaper and thermoplastic film in a hot gas stream causing the thermoplastic film to contract upon itself and form particles, thereby decreasing its specific surface to such an extent as to allow removal of the contracted plastic particles from the mixture by standard mechanical methods, such as air classification or screening. The proces effectively removes all thermoplastic film from the wastepaper thus allowing for more efficient paper utilization through higher recycling rates.

An inexpensive model, powered with a household vacuum cleaner, was designed and built in the Rolla laboratory. The design is shown schematically in Figure 87. This vertical air classifier is capable of classifying 50 to 75 pounds of mixed plastics per hour. The air velocity in the 3-inch diameter vertical column is approximately 500 to 750 feet per minute during operation.

FIGURE 87: SCHEMATIC DIAGRAM OF AIR CLASSIFIER

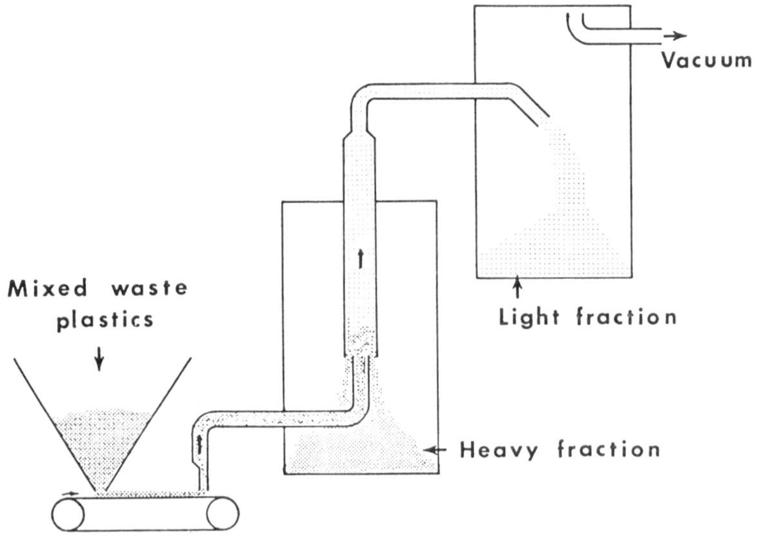

Source: J.L. Holman, J.B. Stephenson, and M.J. Adams (20)

Liquid Media Separation

The densities of virgin thermoplastics are nearly the same as that of water, and vary in a manner that favors gravity methods in aqueous solutions for separating them by types. Measurements of the densities of various pelletized virgin plastics gave the data in Table 59.

TABLE 59: OBSERVED DENSITIES OF VIRGIN PELLETIZED PLASTICS

Type	Density, g/cu cm
PP	0.90
LDPE	0.92
HDPE	0.94-0.96
PS	1.05-1.06
PVC	1.22-1.38

Source: J.L. Holman, J.B. Stephenson, and M.J. Adams (20)

As can be seen from these data, the polyolefin family (LDPE, HDPE, PP) of materials has a density range from 0.90 to 0.96 g/cu cm, PS compounds 1.05 to 1.06 g/cu cm, and PVC compounds 1.22 to 1.38 g/cu cm. Although additives may alter these densities, laboratory results on plastic waste have confirmed that most scrap plastics are in the density ranges given in the table. Experimental separations of the five major types of plastic from wastes were conducted in unagitated vessels with distilled water, a calcium chloride solution, and two alcohol-water combinations. These media gave virtually complete separations of hand sorted waste plastic samples. Based on this sink-float method, a theoretical scheme of separation for the five most plentiful components of waste thermoplastics in packaging wastes was devised and is given in Figure 88. This scheme relies on the use of fluids differing in density and would entail practical requirements for rinsing, waste disposal, and control of liquid density.

FIGURE 88: SINK-FLOAT SEPARATION SCHEME FOR PLASTICS

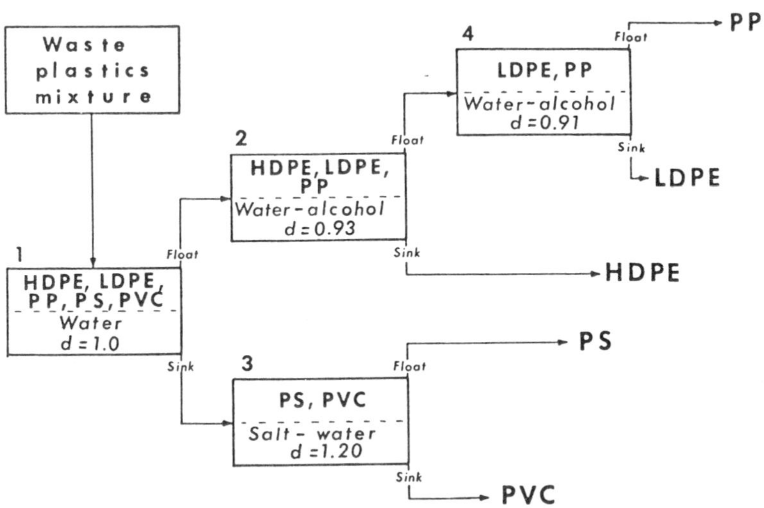

Source: J.L. Holman, J.B. Stephenson, and M.J. Adams (20)

It was deemed desirable that a separating system be devised which would use only water as the parting medium. A small hydraulic separator was designed and built for separating chopped waste plastics into three fractions: polyolefins, polystyrene, and polyvinyl chloride. The schematic diagram of this separator, a combination sink-float and elutriation unit, is shown in Figure 89. Mixtures of chopped plastics were fed into the sink-float separator where the polyolefins were floated off and the other components of the mixture sank. The sink portion was transported via airlift to the elutriation column where the PS was carried out the overflow and the PVC sink fraction was air lifted into the underflow receptacle. The segregated fractions were all caught on screens while the water was recycled to the main reservoir.

A laboratory model has been used successfully to separate mixtures of virgin pelletized plastics and chopped waste plastics at rates from 5 to 50 pounds per hour. An important consideration in separating plastics by hydraulic methods was the addition of 0.3 cc surfactant/gal of water. The performance of the laboratory model hydraulic separator has been successful enough to recommend it as a prototype design for larger units capable of processing the plastic waste output of municipal waste treatment plants.

FIGURE 89: SCHEMATIC FLOWSHEET OF WASTE PLASTIC HYDRAULIC SEPARATOR

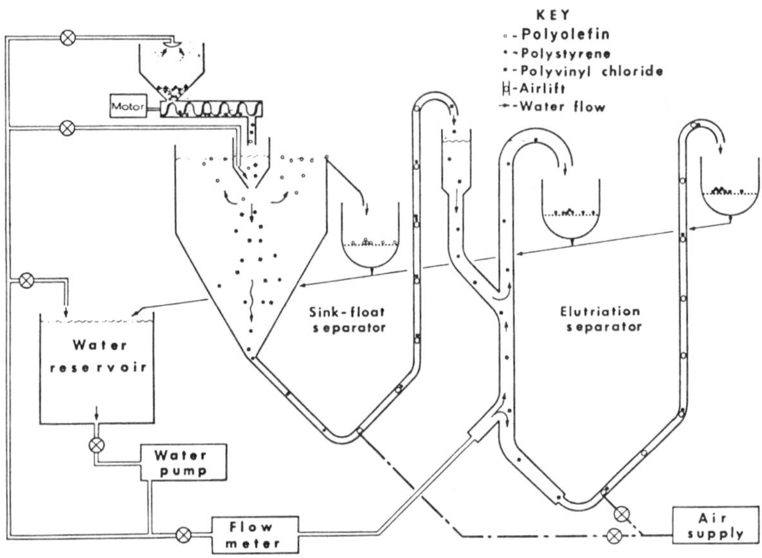

Source: J.L. Holman, J.B. Stephenson, and M.J. Adams (20)

A method which has been used successfully to reduce the residual metal from waste wire insulation tailings is the mineral jig. In this unit the mixed waste material is fed onto a pulsating liquid-solid bed. As the liquid is pulsed, the gravel bed is forced apart repeatedly allowing the heavy material (metal) to sink through the bed and eventually through the slotted screen at the bottom into a collecting chamber. At the same time, water fed into the unit carries off the lighter material (plastic) in the overflow. A schematic diagram of a jig separator is shown in Figure 90.

FIGURE 90: SCHEMATIC DIAGRAM OF JIG SEPARATOR

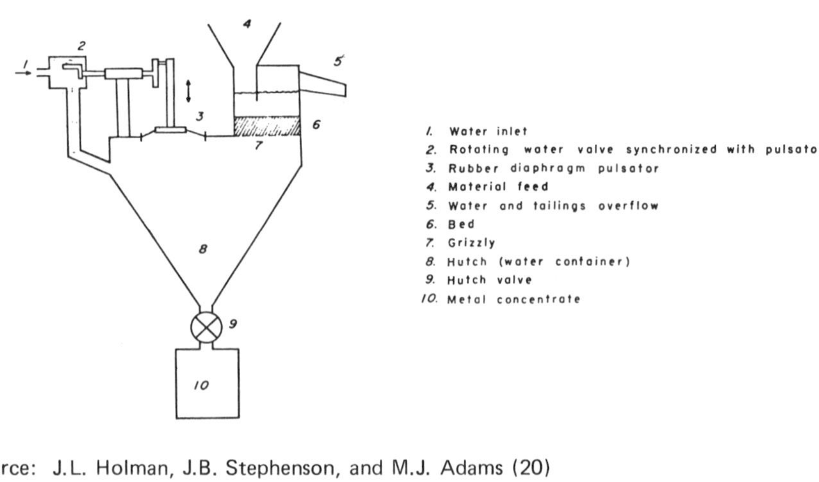

Source: J.L. Holman, J.B. Stephenson, and M.J. Adams (20)

Screening and Electrostatic Separation

In some instances vibrating screens can be used to concentrate or remove a specific component in mixed waste. Such is the case with wire insulation tailings. The residual metal wire, in most cases being of smaller size than the insulation, can be concentrated by screening. This method can be used to upgrade the metal content prior to metal recovery by processes such as jigging or electrostatic separation.

In an electrostatic drum type separator, the material is fed onto an electrically grounded drum which is rotating in a highly charged electric field. The plastic is pinned to the drum and removed by a fixed wiper blade and the wire is thrown into another compartment. Adjustable compartments are available to receive the products, and the material in the middle compartment may be rerun for additional separation. Figure 91 shows a schematic diagram of the operation of the electrostatic separator. The electrostatic separator has been used with some success to separate plastic film and styrofoam.

FIGURE 91: SCHEMATIC DIAGRAM OF ELECTROSTATIC SEPARATOR

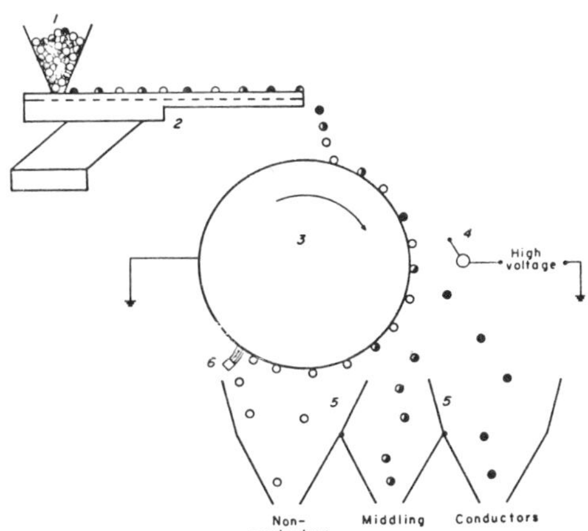

1. Hopper containing particles
2. Vibratory feeder
3. Grounded rotor
4. Adjustable cylinder and wire electrode
5. Adjustable knife edge receptacle divider
6. Rotor wiper

Source: J.L. Holman, J.B. Stephenson, and M.J. Adams (20)

Solvent Extraction

R.B. Goss; U.S. Patent 3,595,741; July 27, 1971; assigned to Riverside Paper Corporation has developed a solvent extraction method for reclaiming stock from waste materials containing natural or synthetic plastic or resin material. The miscella is separated from the stock and the solvent is recovered by distillation during which a hydrocarbon carrier is added to the miscella to form a favorable liquid waste residue which is readily recoverable from the distillation apparatus.

Proposed Combination Processes

Experimental work by the Bureau of Mines at Rolla, Missouri on collected waste plastic has produced some guidelines for systematic methods of processing. The first step, as already mentioned, was chopping the material to uniform size. This was done in a rotary knife-bladed granulator, equipped with interchangeable screens with hole sizes of one-eighth to one-half inch. The chopped material was then air classified to remove the plastic film, foam, dust, and unattached paper. In some cases this process might be accomplished during granulation. During air classification, about one-fourth of the material was removed in the air-light fraction. The air classifier was shown in Figure 87.

The next process was that of cleaning the air-heavy fraction to remove contaminants such as dirt, labels, adhesives, and residual contents. A number of washing solutions and agitation methods were tried. The best results were obtained by using a 1-normal solution of sodium hydroxide containing 150 grams of sand per liter of solution in an attrition scrubber. Approximately 98% of the particles were clean after a 1-hour period of this washing treatment. Tap water, containing the same amount of sand, cleaned 96% of the particles and would be more practical on a large scale in view of the cost of adding and purifying the reagent for reuse.

The final operation was that of separating the air-heavy cleaned plastics into major types. This was accomplished using the hydraulic separator which was described earlier and shown in Figure 89. Three fractions were produced from this separator: [1] the float fraction which contained polyolefins (PE and PP); [2] the elutriator overflow (PS); and [3] the elutriator underflow (PVC). A sink-float analysis was performed on fractions obtained at various feed rates. These data are presented in Table 60. In most cases, over 96% of the material reported in the correct density level.

TABLE 60: QUALITY OF SEPARATION OF PLASTICS AT VARIOUS FEED RATES BY HYDRAULIC SEPARATOR

Feed material	Feed rate lb/hr	Estimated impurity in fractions, wt-pct[1]		
		Polyolefin	PS	PVC
Pellets of virgin plastic.	17.6	0.1	0.1	0.1
	24.0	.1	.1	.1
	36.6	.1	.1	.1
	50.0	.1	.8	1.3
Collected waste chopped to minus 1/4 inch.	5.1	.1	2.2	.3
	8.6	.1	3.9	.4
	19.7	.1	3.9	.4
	30.0	.1	3.7	.1

[1] Impurities were extracted from separated fractions by sink-floating in solutions having densities of 1.00 and 1.20 g/cu cm.

Source: J.L. Holman, J.B. Stephenson, and M.J. Adams (20)

A processing system based on the above unit operations has been proposed and is shown in Figure 92. This system would conceivably operate in conjunction with a raw urban refuse processor which would provide a plastic concentrate fraction for the feed material. The proposed system would further upgrade the plastics fraction and concentrate three major types of plastics for selective reuse by one of several methods. Unit operations of the type used for minerals processing would be utilized in a system of this type.

A plastic rich sample from the Black Clawson FIBRECLAIM Urban Refuse Recovery System, at Franklin, Ohio, was obtained for upgrading and separation study. According to Black Clawson, the sample had been upgraded from 2.5% plastic by repulping and screening. This sample represented 13% of the refuse received at the Black Clawson plant. Oven drying and petrographic examination showed that the material contained 55% moisture, 14% plastic, 0.6% metal and 30.4% mixed fiber, wood, glass, etc. The dried sample was chopped to minus one-half inch and air classified.

FIGURE 92: PROPOSED FLOW DIAGRAM OF PROCESSING SYSTEM FOR RECLAIMING WASTE PLASTICS

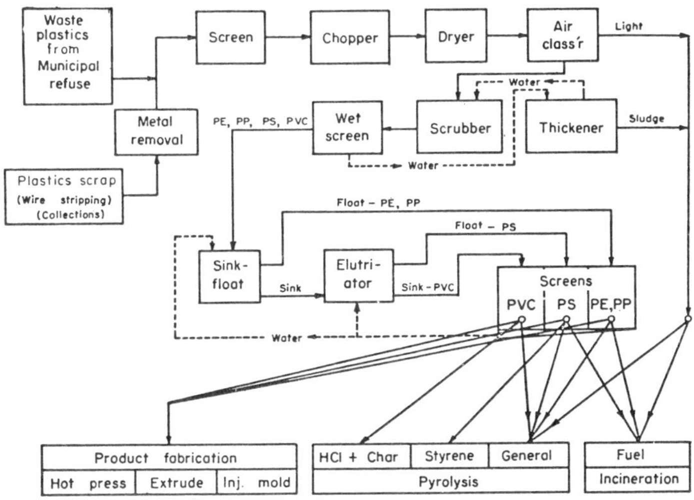

Source: J.L. Holman, J.B. Stephenson, and M.J. Adams (20)

The 25 weight percent air-heavy fraction contained 45% of the total plastic in the dried sample. During air classification, approximately 55% of the total plastic reported in the air-light fraction. It contained primarily fiber and film. Upgrading the air-heavy fraction to 95% plastic was accomplished by chopping and screening the material to 3- to 10+ mesh size and then concentrating by electrostatic separation.

In the first electrostatic separation, 8 to 10% moisture was necessary to obtain the best separation of plastic from the balance of material. The thrown fraction, containing the metal and fiber rejected in the first electrostatic separation, was dried and then separated further electrostatically. The metal concentrate from the second pass contained 75% metal, mostly aluminum foil, which amounted to a 60% recovery of the metal from the original sample. This procedure recovered 99% of the plastic contained in the air-heavy fraction. The total plastic recovery from the original sample was approximately 45% owing to the loss of plastic film to the air-light fraction.

The recovered plastic was then attrition scrubbed in hot tap water (50° to 60°C), containing 200 grams sand per liter of water, for a 1-hour period and then was rinsed with tap water. After the scrubbing treatment it was apparent that the plastic was not entirely clean, although most of the loose dirt and attached paper had been removed. The final separation of the plastics into major types was achieved by hydraulic separation with the unit previously described and shown in Figure 89.

Sink-float analyses of the hydraulically separated fractions are shown in Table 61. It can be seen from the table that the polyolefin and PS fractions are relatively uncontaminated compared with the PVC fraction. It is understandable that the lowest purity fraction would be PVC, because most of the thermosets and composites would fall in the PVC density range. The overall process, for the recovery of plastics, metal, and fiber is shown in the flow diagram, Figure 93.

TABLE 61: ANALYSES OF FRACTIONS

Hydraulic Separation		Sink-Float Analysis of Fraction, Weight Percent, by Density, g/cu cm			
Fraction	Weight Percent	<1.0	1.0 - 1.2	1.2 - 1.5	>1.5
Polyolefin	48	97	3	- - - -	- - - -
PS	33	0.5	95	4.5	- - - -
PVC	19	- - - -	28	40	32

Source: J.L. Holman, J.B. Stephenson, and M.J. Adams (20)

FIGURE 93: FLOW DIAGRAM OF RECOVERY PROCESS

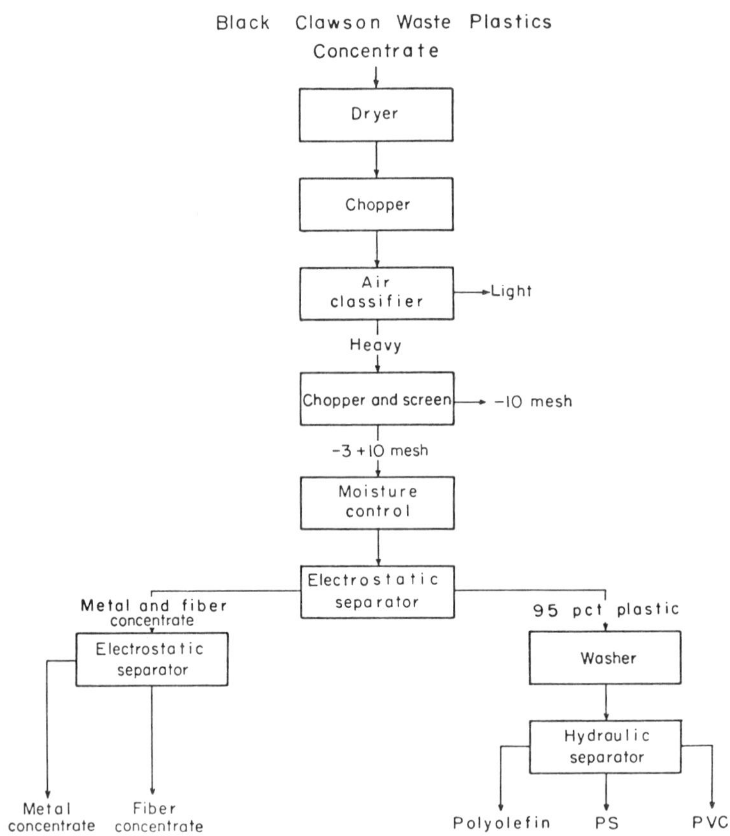

Source: J.L. Holman, J.B. Stephenson, and M.J. Adams (20)

A synthetic waste plastics concentrate sample was received from the Bureau's College Park Metallurgy Research Center, College Park, Maryland. One-fourth of the sample was characterized by hand sorting. Table 62 gives the sample composition as determined from the characterization study. The as-received sample was -3 inch pieces with a bulk density of 0.09 g/cu cm. The first step in separating the mixture into major types was chopping or shredding to -5/16 inch size in order to be compatible with the processing equipment. This operation increased the bulk density to 0.18 g/cu cm.

TABLE 62: ANALYSIS OF SYNTHETIC WASTE PLASTICS CONCENTRATE DETERMINED BY HAND SORTING

Material	Wt-pct
Polyolefin:	
Film	61.6
Other	18.9
Polystyrene (PS):	
Styrofoam	7.1
Other	1.4
Vinyl (PVC):	3.5
Paper	0.6
Miscellaneous (other thermoplastics, thermosets, metal, etc.)	6.9
Total	100.0

Source: J.L. Holman, J.B. Stephenson, and M.J. Adams (20)

The shredded material was subjected to air classification in a suitable unit. The air velocity was adjusted to remove most of the film and foam in the air-light fraction. The material was classified as 69 weight percent air-light and 31 weight percent air-heavy. Tests were conducted on the air-light fraction to separate the film and foam fractions by electrostatic separation. Samples containing 5 to 20% moisture in 5% increments were tested. The best separation was obtained at the 5% moisture level where the recoveries and grades for plastic film were 99 and 97%, respectively, and for foamed plastic 70 and 90%. These results are illustrated in graphic form in Figure 94.

FIGURE 94: EFFECTS OF MOISTURE ON ELECTROSTATIC SEPARATION OF PLASTIC

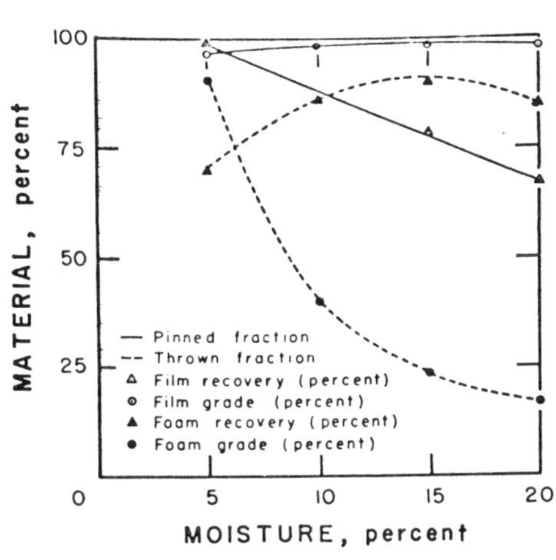

Source: J.L. Holman, J.B. Stephenson, and M.J. Adams (20)

The air-heavy material resulting from air classification was further separated into primary plastic types by hydraulic separation. In normal practice, using plastic derived from urban refuse, a cleaning step would be required prior to hydraulic separation. The results of hydraulic separation are shown in Table 63. A sink-float analysis on each fraction was performed by static sink-floating each in water and calcium chloride solutions having densities of 1.20 g/cu cm and 1.50 g/cu cm, respectively. The over-all process is shown in Figure 95.

TABLE 63: ANALYSES OF FRACTIONS OBTAINED FROM HYDRAULIC SEPARATION OF AIR-HEAVY SYNTHETIC WASTE PLASTIC

Hydraulic separation		Sink-float analysis of fraction, wt-pct, by density, g/cu cm			
Fraction	Wt-pct	<1.0	1.0-1.2	1.2-1.5	>1.5
Polyolefin.	72.1	[1]99.9	0.1	-	-
PS.........	6.0	2.5	70.6	26.9	-
PVC........	21.9	-	18.3	79.8	1.9

[1]0.5 wt-pct styrofoam included in this amount was caught in the feed cone as the water flow was not enough to force it into the main sink-float tank.

Source: J.L. Holman, J.B. Stephenson, and M.J. Adams (20)

FIGURE 95: FLOW DIAGRAM OF PROCESS TO SEPARATE WASTE PLASTICS CONTAINING HIGH PERCENTAGES OF FILM AND FOAM

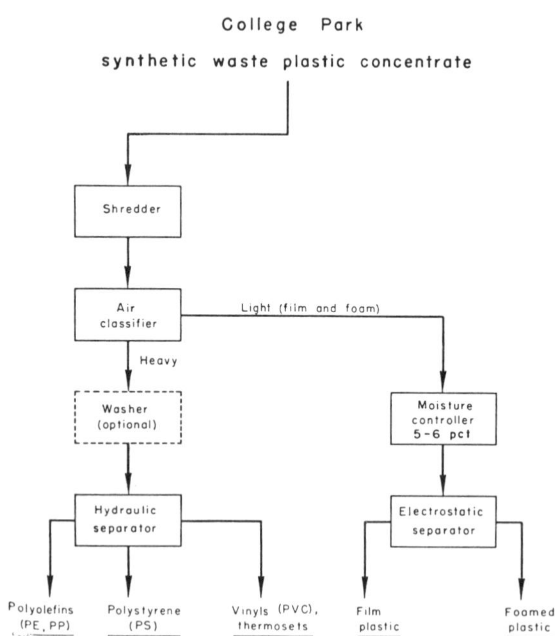

Source: J.L. Holman, J.B. Stephenson, and M.J. Adams (20)

REFABRICATION

One method for reusing plastic waste, after separation into major types by the preceding operations, is the fabrication directly into new products by mechanical or other means. Mechanical methods for refabrication include compression molding, injection molding, and extruding. In the laboratory, wafers 2 inches in diameter were easily made from chopped waste plastics in a laboratory press at pressures of 3,000 to 4,000 psig and temperatures of 150° to 170°C. Although the materials fused into impervious solids, the flow did not homogenize the plastics and the particles retained their bright colors. The method offers many ways of varying the ingredients of the mixture with virgin plastics, plasticizers, inert fillers and binder materials. Fabrication by hot pressing of traffic markers, parking barriers, paving blocks and tiles, and park benches and fixtures has been suggested as a useful way to utilize the waste plastics from urban refuse.

Injection molding experiments conducted with mixtures of waste PE, from bottles collected in the Rolla campaign, and virgin-injection molding grade PE were successful in producing fan fairings. However, fairings made from 100% scrap contained one small area where the mold did not fill. Melt index measurements made on the material indicated a melt index of less than 1, which is low for injection-molding grade but normal for blow-molding grade plastic from which bottles are fabricated. The addition of a small amount of high melt index material mixed with the waste to raise the overall melt index, causing improved flow, would undoubtedly have improved the filling of the mold.

The results of physical and mechanical testing of specimens cut from the fairings of virgin, mixed virgin and waste, and all waste material yielded the information shown in Table 64. Although more uniform material is required for injection molding than for hot pressing, it is a suitable high volume method for fabricating products from waste plastics.

TABLE 64: COMPARISON OF PROPERTIES OF INJECTION MOLDINGS FROM VIRGIN AND WASTE PE

Material	Hardness, shore units	Tensile strength, psi	Elongation, pct	Melt index, g/10 min	Density, g/cu cm
Virgin.............	39.6	4,120	500	0.77	0.960
Mixed..............	38.8	4,090	525	.74	.955
Waste..............	38.4	3,940	380	.37	.955

Source: J.L. Holman, J.B. Stephenson, and M.J. Adams (20)

Laboratory experiments using a 1-inch extruder revealed that uniformity of melt index and a low level of contamination are more critical than for the other fabrication methods already mentioned. Extruded sheet 3½" wide by 1/16" thick were made from three pelletized virgin PE materials and six granulated PE wastes and mixtures. Five tensile specimens were cut and tested from each of these extrusions and the average results are given in Table 65.

It is evident from the table that the tensile values for HDPE bottle waste are almost as high as those from virgin material. This is consistent with the fact that the waste bottle fraction represents a more uniform grade material than the separated waste PE which contains a mixture of high and low density PE having large differences in melt properties.

Mixed wire insulation wastes were fabricated into tricycle tires by intrusion molding (low-pressure injection molding) after the material had been air classified to remove fiber and dust and then jigged to lower the metal content to below 0.5%. These trial moldings were fabricated without additives. In practice, black coloring and plasticizer additives might be desired.

TABLE 65: TENSILE DATA FROM EXTRUDED PE SHEET

Material	Tensile strength, psi	Elongation, pct
Waste HDPE from bottles.........	3500±200	850±200
Separated waste PE.............	1720±150	15±10
Mixtures of PE:		
75 percent waste 25 percent virgin HD	1970±150	50±30
50 percent waste 50 percent virgin HD	2180±200	275±200
25 percent waste 75 percent virgin HD	2455±400	1000±120
Virgin HDPE.....................	3840±320	1000±120
Virgin medium density PE........	2560±200	600±60
Virgin low-density PE..........	1390±90	65±11

Source: J.L. Holman, J.B. Stephenson, and M.J. Adams (20)

A process developed by *E. Beck and H. Schulz; U.S. Patent 3,510,067; May 5, 1970; assigned to Fellner & Ziegler GmbH, Germany* is one in which thermoplastic foil waste is converted to reusable granular material by mechanically beating the waste to thereby comminute it and then continuing the beating operation to produce frictional heat which causes densification and agglomeration of the particles. The resulting granules are cooled while the beating motion is continued, thus permitting the granular material to be issued without sticking. A correspondingly controllable single beater unit equipped for timed supply of coolant is employed to perform the process in a continuous sequence of operations.

A process developed by *C.P. Ronden; U.S. Patent 3,535,408; October 20, 1970; assigned to Cupples Container Company* is one in which scrap foamed vinylaromatic polymeric material, and particularly scrap foamed polystyrene, is recovered by forming the scrap into dense granules, impregnating the granules with a normally liquid low boiling hydrocarbon blowing agent, blending the blowing agent-containing granules with fresh foamable polymeric particles, and subjecting the resulting mixture to a processing step, such as sheet extrusion, involving heat plastification of the polymeric material.

A process developed by *P.C. Neil; U.S. Patent 3,733,160; May 15, 1973; assigned to Standard Oil Company* is one for handling scrap from an extruder followed by a thermoformer. The system includes mixing apparatus, extruder, thermoformer, trimmer and grinder. Ground plastic from the mixing apparatus is fed directly to the extruder which provides a thin, continuous film of plastic.

In the direct feed mode this film moves from the extruder directly into the thermoformer and then to the trimmer. Articles are formed in the film by the thermoformer and are cut from the film by the trimmer to leave a skeletal film which is fed directly into the grinder. In the bypass feed mode, film from the extruder or from the thermoformer is fed directly into the grinder. A mixing apparatus is disclosed for mixing plastics of different bulk densities, such as ground film from the thermoformer, extruder and trimmer. Also, a cutting apparatus is disclosed which reduces noise and overloading problems.

A process developed by *J.D. Robertson; U.S. Patent 3,797,702; March 19, 1974* involves a scrap recovery and feed system wherein scrap plastic is recovered from a plastic manufacturing process, such as in the manufacture of sheet, film, formed or molded objects. The scrap plastic is chopped up and fed back to the processing machine along with the base material, such as a virgin plastic material and various additives. The scrap is fed with an auger from an upper hopper toward the outlet opening of a lower hopper containing the base material. The materials are mixed just prior to entry in the extruder or other processing machine.

CHEMICAL RECOVERY BY PYROLYSIS

Pyrolysis, use of flameless high temperatures to break down organic compounds, is gaining as a means of solid waste disposal including plastics. Since pyrolysis does not involve incineration, it minimizes potential air pollution. In addition, pyrolysis offers possible recovery of chemicals or synthetic gas for use in the production of methane or higher hydrocarbons.

Pyrolysis, also known as destructive distillation, is a process where organic material is heated in a retort in the absence of oxygen or in a controlled oxygen environment. Investigations of pyrolysis by the Bureau of Mines and other organizations have led to the conclusion that this method of disposal of plastic wastes may have an economic advantage by generating more by-product fuel gas than is needed to supply heat for the reaction. That other products from this reaction, such as organic liquids and char also have value as fuel, is expected since most plastics are derivatives of petroleum.

Pyrolysis then is a recycling method for plastics which will not only reduce the volume of waste, but from which valuable chemicals and fuel grade products can be recovered. Polystyrene reportedly is one of the few polymers that yields a relatively high percentage of monomer when thermally decomposed.

Polyvinyl Chloride

In earlier work by the Bureau of Mines, the production of hydrochloric acid and a char were advocated. It was also illustrated, on the basis of freight charges in the southwestern United States, that it costs twice as much to ship HCl in tank cars in the form of concentrated hydrochloric acid (20° Baumé, 31.45% HCl) as it would to ship an equivalent amount in boxcars in combined form as polyvinyl chloride. Research to determine the feasibility and optimum parameters for the recovery of HCl from waste PVC was conducted. About half the weight (50 to 60%) of virgin PVC resin is HCl, but the addition of plasticizers and fillers dilute this amount. For example, two commercial blow-molding grade PVC compounds contained 46 and 54% HCl. The PVC waste from the Rolla collection contained about 30 weight percent HCl.

Thermal decomposition experiments of PVC conducted in a 1-liter reaction kettle using helium carrier gas showed that HCl evolution begins at 220°C and proceeds with increasing rapidity as the temperature is raised. A schematic diagram of the decomposition apparatus is shown in Figure 96.

FIGURE 96: FLOW DIAGRAM OF APPARATUS FOR DISTILLATION OF HCl FROM PVC WASTES

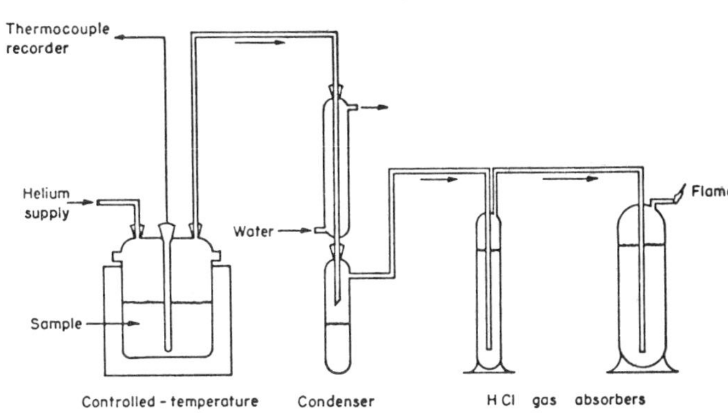

Source: J.L. Holman, J.B. Stephenson, and M.J. Adams (20)

Results of 1-hour thermal treatments of two virgin samples of PVC molding compounds and the Rolla waste PVC concentrate are shown graphically in Figure 97. Dissociation was virtually complete after 1 hour at 300°C, liberating 99% of the HCl in the virgin samples and 92% in the waste. The other products of the dissociation, the char, the liquid condensate, and the off-gas all have fuel value.

FIGURE 97: LIBERATION OF HCl FROM PVC VERSUS TEMPERATURE

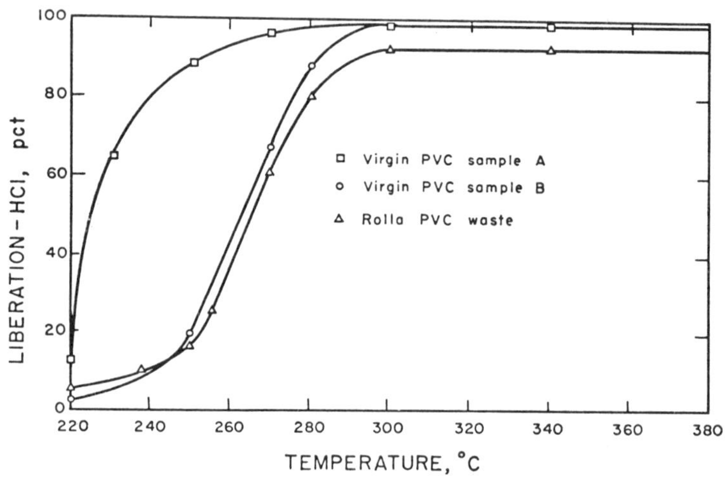

Source: J.L. Holman, J.B. Stephenson, and M.J. Adams (20)

Isothermal bomb calorimeter measurements were made on these products and on the major types of thermoplastics the results of which are shown in Table 66. The char from the waste PVC was 44% of the original sample and contained 6.3% water, 15% ash, 51% volatile matter and 27% fixed carbon. From 10 to 15% of the PVC from the Rolla collection dissociated as liquid condensate, which contained benzene, toluene, xylenes, and a heavy oil, and when burned left 1% ash. The off-gas, 10 to 19% of the waste PVC, was shown by mass spectral analyses to contain water, methane, ethane, propane, and butane.

TABLE 66: HEATS OF COMBUSTION OF PLASTICS AND BY-PRODUCTS FROM ISOTHERMAL BOMB CALORIMETER MEASUREMENTS

Sample	Btu/lb
Waste PVC reactor.	
Residue	12,300
Liquid condensate	14,000
Offgas	[1]20,000
PVC waste	9,900
HDPE	18,700
PS	16,900

[1]Estimated

Source: J.L. Holman, J.B. Stephenson, and M.J. Adams (20)

The optimum decomposition temperature for waste PVC was determined to be 350°C. Most of the unrecovered HCl was found in the char, having reacted with the calcium carbonate filler material to form calcium chloride. Compounding PVC with basic fillers, such as calcium carbonate, potassium carbonate, magnesium oxide, and others, reduces the outflow of HCl during pyrolysis. Water leaching of the char provided a 98% removal of HCl from the PVC waste.

Thermal decomposition of waste wire insulation tailings at 300°C for 1 hour resulted in 72 to 93% liberation of HCl. The percentages of char, liquid condensate, and off-gas were very similar to those from the waste PVC tests. It was noted that liberated HCl reacted with the residual aluminum and copper metal to form metal chlorides which remained in the char fraction. Water leaching of the char lowered the residual chlorine in the char to 1%, thus providing comparable HCl recovery to that from the PVC waste.

Tests at 400°C showed very little increase in HCl liberation, but the liquid condensate increased from 6 to 10%. At the higher temperature a smaller amount of char and an increased amount of off-gas were obtained. Ash from the char contained minor amounts of Al, Fe, Cu, Mg, Pb, Sn, and Ti and preponderant amounts of Ca and Si.

It is evident that waste PVC merits consideration in chemical processing as a resource for HCl and fuel. Waste PVC could offer advantages, for the consumer in outlying regions, owing to its flexibility in supply, storage, and use; to the low hazard in its shipping and storage; and to low shipment costs.

Polyvinylidene Chloride

A process developed by *R.D. Böhme and R.A. Wessling; U.S. Patent 3,852,223; December 3, 1974; assigned to The Dow Chemical Company* is a method for disposing of halogen-containing plastic comprising preheating such plastic at dehydrohalogenating temperatures in the presence of a catalytic amount of a Lewis acid and recovering the halogen-containing by-products.

Polytetrafluoroethylene

A process developed by *B.C. Arkles and R.N. Bonnett; U.S. Patent 3,832,411; August 27, 1974; assigned to Liquid Nitrogen Processing Corporation* is one in which high yield monomeric tetrafluoroethylene may be recovered from polytetrafluoroethylene and polytetrafluoroethylene composites by pyrolyzing the polymer in the presence of high temperature steam. The steam acts as a condensable carrier gas for the low molecular weight pyrolysis products and should be present in an amount so that the mol ratio of steam to pyrolysis products is at least one to one. The steam also acts as a means for heating the polymer to the necessary decomposition temperatures.

Polyurethanes

A process developed by *N. Matsudaira, S. Muto, Y. Kubota, T. Yoshimoto and S. Sato; U.S. Patent 3,404,103; October 1, 1968; assigned to Bridgestone Tire Company Limited, Japan* is one in which polyurethane made from a polyether and an isocyanate is thermally decomposed in the presence of an amine and an alkali metal oxide or hydroxide or an alkaline earth metal oxide or hydroxide. The products obtained are a polyether and an amine corresponding to the isocyanate used to prepare the polyurethane wherein the –NCO groups of the isocyanate are substituted with amino groups.

A process developed by *F.F. Frulla, A. Odinak and A.A.R. Sayigh; U.S. Patent 3,708,440; January 2, 1973; assigned to The Upjohn Company* is a method for converting scrap polyisocyanurate foam into a polyol which can be used, without the need for any further treatment, as the polyol component in the preparation of polyurethane and related foams. The process comprises heating the scrap at about 175° to 250°C in the presence of a mixture of an aliphatic diol having from 2 to 6 carbon atoms, inclusive, and a boiling point above about 180°C and a dialkanolamine having from 4 to 8 carbons, inclusive, the dialkanolamine constituting from about 2 to 20% by weight of the mixture.

Polyolefins

A process developed by *Y. Kitaoka, K. Murata, K. Hama, M. Hashimoto and K. Fujiyoshi; U.S. Patent 3,832,151; August 27, 1974; assigned to Mitsui Shipbuilding and Engineering Co., Ltd. and Mitsui Petrochemical Industries Limited, Japan* is one in which polyolefin wastes are thermally decomposed, the product is cooled and separated into gaseous product and liquid product, and carbon sludge in the reactor content is dispersed to avoid the deposition of such carbon sludge and a part of the reactor content is discharged to reduce the carbon sludge in the reactor.

A process developed by *H.G. Schutze and J.A. Dinwiddie; U.S. Patent 3,736,271; May 29, 1973; assigned to Esso Research and Engineering Company* is one in which waste polymers such as scrap polypropylene which heretofore have been buried, burned, or otherwise disposed of with adverse effects on the ecology are converted to useful products by converting the polymer into a liquid state either by heating the polymer to a temperature above its melting point or dissolving the polymer in a suitable solvent and then initiating a free radical decomposition reaction by the addition of a suitable initiator. Inorganic materials in the polymer are removed while the polymer is in the liquid state by centrifugation, settling, filtration or solvent washing.

Polystyrene

A process developed by *M.E. Banks, W.D. Lusk and R.S. Ottinger; U.S. Patent 3,829,558; August 13, 1974; assigned to the U.S. Secretary of the Department of Health, Education and Welfare* involves disposing of waste polystyrene plastic without polluting the environment by passing the plastic to a reactor, heating the plastic in the presence of a gas to at least the decomposition temperature of the plastic, and recovering decomposition products therefrom. The preferred embodiment utilizes a heated inert carrier gas as the source of heat. Figure 98 shows a suitable apparatus.

FIGURE 98: APPARATUS FOR PYROLYSIS OF WASTE POLYSTYRENE

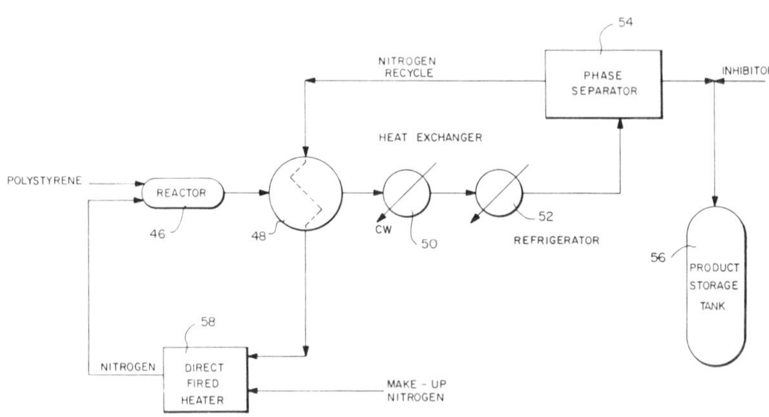

Source: M.E. Banks, W.D. Lusk and R.S. Ottinger; U.S. Patent 3,829,558; August 13, 1974

The polystyrene is fed into stainless steel reactor **46**. Nitrogen is used as the inert carrier gas and is fed into the reactor at about 1230°C so as to provide a degradation temperature of about 600°C. The effluent from the reactor passes to heat exchanger **48** where the gaseous reaction products are cooled to about 250°C by heat exchange with the nitrogen recycle from the refrigeration step. From the heat exchanger the gaseous reaction products are sent to secondary heat exchanger **50** where they are further cooled with water to about 50°C. Refrigeration is used to cool the stream to about 20°C in refrigerator **52**, removing all but 0.6% of the styrene monomer from the gaseous nitrogen.

The stream then goes to phase separator **54** and an inhibitor is added to the liquid monomer before it is pumped to the product storage tank **56**. The nitrogen leaving the refrigerator via the phase separator is recycled to the direct fired heater **58**, cooling the effluent reactor gases on the way. In the direct fired heater step, makeup nitrogen is added to the recycled nitrogen stream, now at 460°C, and heated up to 1230°C using oil/gas combination burners.

Mixed Plastics

Figure 99 from the patent cited just above is a process for pyrolyzing mixed waste plastics. When degrading mixed waste plastics, the plastics are charged to a semicontinuous stainless steel reactor **60**. Nitrogen, used as an inert heat carrier, is preheated to 1230°C in a direct fired heater **62**, using oil/gas combination burners. Degradation of the waste plastics mixture is accomplished at a constant temperature in the reactor of 600°C.

FIGURE 99: APPARATUS FOR PYROLYSIS OF WASTE MIXED PLASTICS

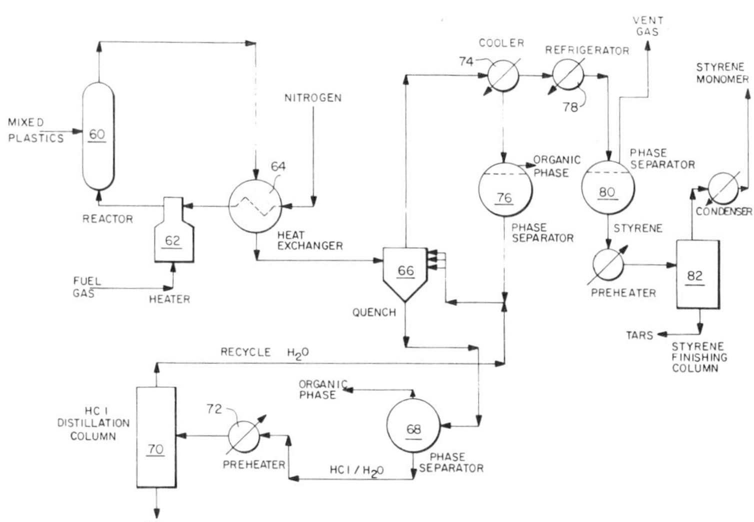

Source: M.E. Banks, W.D. Lusk and R.S. Ottinger; U.S. Patent 3,829,558; August 13, 1974

After leaving the reactor, the gaseous decomposition products are cooled by heat exchange with incoming nitrogen being sent to the direct fired heater for preheating in a heat exchanger **64**. The stream temperature is then lowered to 90°C with a water spray quench at **66** which removes the hydrogen chloride and water from the stream. The condensed hydrogen chloride/water stream is sent through a phase separator **68** for recovery of heavy hydrocarbons and then on to a hydrogen chloride distillation column **70** through preheater **72**.

The gaseous stream leaving the quench is cooled to 50°C with a cold water heat exchanger **74**, sent to a phase separator **76** to remove organics other than styrene, and then refrigerated to 20°C by refrigerator **78**, which removes all but 0.6% of the styrene monomer from the gas stream. After passing through phase separator **80**, the condensed styrene stream is further purified in a vacuum finishing column **82**, treated with an inhibitor, and sent to storage.

When using steam instead of nitrogen as the inert carrier gas, low pressure steam is sent through the direct fired heater and heated to 1230°C prior to being sent to the reactor. Basically, the same method steps are used for the recovery of the degradation products.

In another aspect of the process, the mixed plastic wastes are thermally degraded in the presence of an oxidizing atmosphere, the preferred such atmosphere being air. The reactor, generally indicated by the numeral **84**, has a waste plastics inlet **86** and an air inlet **88**. The oxidation products are drawn off at **90**. The reactor system is adiabatic with air entering at ambient temperature. The product distribution, reactor exit temperature, and air input were calculated for systems of polyvinyl chloride, polyethylene, polystyrene, and a mixture of equal parts by weight of three plastics. The results of the determination are shown as follows. The result of the combustion according to this aspect of the process is to completely combust the organic material with the subsequent harnessing of the heat energy produced.

Waste Plastics/Air Adiabatic Reactor*

Products (Grams)	Polyvinyl Chloride	Poly-ethylene	Poly-styrene	Mixture (Equal parts by weight of the three)
CO_2	1.41	3.06	3.19	2.69
CO	traces	0.06	0.08	<0.01
H_2O	0.30	1.27	0.66	0.77
N_2	12.00	13.19	11.74	12.85
NO	traces	0.06	0.06	0.05
O_2	2.12	0.34	0.34	1.01
HCl	0.55	—	—	0.18
Cl_2	0.04	—	—	traces
Others	traces	0.02	0.02	0.01
Temperature out of Reactor	1200°K	2130°K	2200°K	1870°K
Weight of Air Input	15.40	16.99	15.10	16.55

*All data Based on 1 Gram of Plastic

RECOVERY BY SOLVENT EXTRACTION

A process developed by *E.A. Hafner; U.S. Patent 3,836,486; September 17, 1974; assigned to Hafner Industries, Inc.* is one in which reusable vinyl chloride polymer and copolymer and other components such as plasticizers are reclaimed from scrap plastics by contacting the scrap plastics with a solvent for the vinyl chloride polymer, contacting the resulting liquid phase with a substantially nonaqueous nonsolvent for the polymeric portion, and then separating the resulting insoluble phase containing the vinyl chloride polymer.

Following the removal of this polymer portion, the filtrate is fractionally distilled to recover the solvents, nonsolvents, plasticizers and other ingredients. Typical solvents are methyl ethyl ketone, methyl isobutyl ketone, cyclohexanone, dimethylformamide, tetrahydrofuran and ethylene dichloride. Typical nonsolvents are methanol, ethanol, isopropanol and n-butanol. The recovered vinyl chloride polymer compositions may contain pigments and fillers and exhibit improved performance on recycling in the fabrication of plastic articles.

MANUFACTURE OF COMPOSITE MATERIALS FROM WASTE PLASTICS

A process developed by *G. Haveman; U.S. Patent 3,516,841; June 23, 1970; assigned to The Dow Chemical Company* involves converting an aluminum-plastic laminate structure into an aluminum oxide-filled plastic raw material which can be used for fabricating other products.

A process developed by *P. Lamort and J.P. Lamort; U.S. Patent 3,806,562; April 23, 1974* is a method for production of composite products from waste plastic materials utilizing plastics having separate and distinct softening temperatures. The waste material has various thicknesses and densities and is compacted into a flat form having superposed layers packed one against another but not agglomerated. The rolled mixture is cut into strips, the strips are cut to form rods, and then the layers are separated.

The resulting rods are separated by density into light and heavy rods and then each group of light and heavy rods is separated by thickness. The various groups of rods may be recombined in proportions to form various molding mixes. Fillers such as sand, cement, plaster, stone, brick, and gravel may be added to the mix. The mix may be formed by rollers into a sheet or may be spread onto a surface and rolled to provide a compacted surface covering.

A process developed by *H.J. Brown; U.S. Patent 3,852,046; December 3, 1974* is one in which waste plastics comprising a mixture of two or more thermoplastic materials are treated by grinding the products to form a particulate thermoplastic mixture, washing the particulate mixture to remove nonplastic materials, placing the thermoplastic mixture in a mold or shaped vessel, and heating the mixture to a temperature above the softening point of the lower melting thermoplastics and below the temperature of the softening point of the highest melting thermoplastic of the mixture.

In another embodiment, particulate thermoplastic is used as a filler for asphalt compositions. In a further aspect, waste thermoplastics are mixed with combustible cellulosic materials and formed into logs and the like which thermoplastic mixture has been treated to remove halogen-containing thermoplastic such as polyvinyl chloride.

A process developed by *S. Ooba and S. Hirayama; U.S. Patent 3,857,799; December 31, 1974; assigned to Fuji Photo Film Co., Ltd, Japan* is a method for regenerating waste plastic material. The process involves mixing a waste plastic material selected from the group consisting of thermoplastic and thermosetting resins with [1] a modified polyester and [2] at least one compound selected from the group consisting of a vinyl acetate homopolymer, an ethylene-vinyl acetate copolymer and a tacky polyolefin, the polyolefin selected from the group consisting of a polyolefin having a low polymerization degree and atactic polypropylene, the amount of components [1] and [2], in combination, being effective to adhere the waste plastic material in a moldable form.

The modified polyester is prepared by heating polyethylene terephthalate in the presence of a glycol to substitute the glycol for 30 to 100% of the repeating ethylene glycol units of the polyethylene terephthalate, the glycol being selected from the group consisting of polymethylene glycols, polyethylene glycols, alkylene glycols and glycols represented by the formula $HOCH_2-X-CH_2OH$ in which X is an organic group having at least one aromatic ring with the proviso that the glycol is not ethylene glycol. The resulting mixture is then molded into the desired shape.

RECOVERY OF RUBBER PRODUCTS FROM SOLID WASTE

Processes for reclaiming rubber and the water pollution aspects of those processes have been discussed earlier in the volume because the waste disposal problems are so closely related to those in the molded, extruded and fabricated rubber products industry. Additional attention will be given here, however, to the statistical aspects of the reclaiming of rubber tires in view of their widespread occurrence as scrap or solid waste.

In 1971, 266 million tires for passenger cars, trucks, and motorcycles were shipped by domestic tire manufacturers or imported for consumption in the United States (Table 67). In that year, tires taken out of service totaled 240 to 250 million (19).

TABLE 67: TIRE SHIPMENTS, 1971

Category of tire	Number of tires (millions)
New replacements (passenger cars, motorcycles, and trucks)	153.4
Retread replacements:	
Passenger cars	36.0
Trucks	10.2
Original equipment (passenger cars, motorcycles, and trucks)	56.0
Exports (passenger cars, motorcycles, and trucks)	2.0
Imports:	
Passenger cars and motorcycles, original equipment	9.1
Truck and bus tires, original equipment	1.5
Total	268.2

Source: EPA (19)

Of the tires taken out of service, approximately 46 million were retreaded, 7 million were consumed by the rubber reclaiming industry, and roughly 2 million were consumed by tire splitters. The remaining 185 to 200 million tires, weighing approximately 2.4 million tons, were disposed of by retailers, retreaders, or consumers or were left on discarded vehicles. See Table 68 for tire disposal statistics for 1969. The most common disposal method is land disposal, but many tires become litter or are left to accumulate at various locations.

TABLE 68: TIRE DISPOSAL, 1969

Point from which final disposal is made	Number of tires (millions)
Retailer	78.3
Retreader	64.3
Consumer	4.0
Discarded vehicle	37.3
Total	183.9

Source: International Research and Technology Corporation (22)

There are three means by which rubber tires are recycled: retreading, conversion by rubber reclaimers into new products, and physical conversion by tire splitters into new products. The reader is also referred to the earlier sections on fabricated rubber products (page 182) where some attention is also given to retreading and the water pollution potential of retreading processes.

Retreading: As percents of new tire shipments, retreaded passenger tires have dropped from 25% in 1963 to 17% in 1968. Truck tire retreads dropped from 32 to 28% in the same period.

Tire performance and consumer preference are two critical factors inhibiting the expanded usage of retreads. Poorly constructed retreads do not perform as well as new tires, especially at high speeds. This is primarily because the bond between the carcass and tread can fail, causing the tread to come loose from the tire. This bond could be improved if the precise chemistry and carcass dimensions of every retreaded tire were known. However, tires of various manufactures have slightly different chemistry or size, making the development of a perfect bonding agent and exact tread fitting difficult.

Consumers have a preference for new tires because retreads are felt to be inferior. Manufacturers sell an increasing variety of low-priced new tires, which in essence compete with retreads.

A factor that would ultimately restrict retreading is the technical suitability of old carcasses for retreading. Only about 35% of the discarded passenger tires are suitable for retreading. This is largely because consumers allow tires to become too worn before replacing them and because driving on underinflated tires weakens the tire carcass. Better tread bonding, leading to consistent tire performance, and consumer education and cooperation would be required to increase tire retreading.

Rubber Reclaiming and Tire Splitting: Rubber reclaiming and tire splitting are limited by the market potential for the products they produce. Reclaimed rubber is not technically suitable for making new tires. The total production of other lower grade rubber products produced by reclaimers and tire splitters, e.g., doormats, hoses, and belts, is extremely small relative to tire discards. In 1969 the entire output of these industries was only about 290,000 tons.

DISPOSAL OF PLASTICS PRODUCTS

The disposer can be a public agency, a private contractor, an individual householder, or an industrial firm. According to a national survey, 12% of household and commercial wastes are self-collected and transported to disposal sites, 32% are handled by private contractors, and 56% by municipal collection agencies. Considering only industrial wastes, approximately 55% are disposed of by the industrial concern on his own site, and the remaining 45% are handled by private contractors who dispose of them either in private or municipal sites.

A total of approximately one billion pounds of nuisance plastic (NP) was generated by manufacturing operations in 1970 (Table 69). Fabrication and converting operations are responsible for most of the industrial NP, and together these account for almost 60% of the total. PVC predominates in the industrial wastes (Table 70); it accounted for approximately 34% of the total in 1970.

As mentioned previously, large quantities of PVC wastes are generated, because PVC is more difficult to process than the other major thermoplastics, and it is used in numerous formulations. The second major industrial waste is low-density polyethylene; it accounted for about 30 weight percent. It is believed these wastes from LDPE and PVC will continue to predominate in the industrial plastic wastes generated during this decade (Table 70). Also note from this table that NP from industrial sources will triple during this ten year period and will amount to approximately 3 billion pounds by 1980.

According to government statistics, over 260 million tons of refuse (probably closer to 400 million tons) are generated from urban and industrial sources today; of this total, only 190 million tons are collected by private and municipal collection agencies. It was estimated that 6.5 billion pounds of nuisance plastics (based on the big five thermoplastics) were generated in 1970 from all sources. If all plastic wastes (thermoplastics and thermosets) are included, then approximately 9.7 billion pounds or 4.8 million tons were generated in 1970.

Thus, plastic wastes from all sectors were 1.4 to 1.2 weight percent of urban and industrial solid wastes. By 1980, the volume of all nuisance plastics should rise to about 14 million tons, and urban and industrial solid wastes to about 650 million tons. At this time, therefore, the calculated value for the concentration of nuisance plastics in the urban and industrial solid wastes will be about 2.2 weight percent.

An examination of the sources of all nuisance plastics derived from the big five thermoplastics in the disposal area and the volumes generated by these sources shows that the major NP source is that from packaging (Table 71); it corresponds to about 60% of all

NP in the disposal area. More surprisingly, however, is manufacturing wastes. Thus, these two types of NP account for 75% of all plastics in the disposal area. Notice that housewares, which is the third category, only account for about 6% of the total NP. Moreover, these other types of consumer NP would be difficult to recycle, because they are usually part of a composite.

TABLE 69: SOURCES OF INDUSTRIAL NP (1970)

Operation	Million lb.†	Wt. %
Polymerization	200	20.0
Compounding/Reprocessing	110	11.0
Fabrication	320	32.0
Converting	260	26.0
Other‡	110	11.0
TOTAL	1000	100%

†Numbers have been rounded to the nearest 10 million lb.

‡Wastes generated in distributing products from manufacturer/packager to consumer.

Source: Arthur D. Little, Inc. (16)

TABLE 70: INDUSTRIAL WASTES (NP) ACCORDING TO PLASTIC TYPE

Type of plastic	1970 Million lb.	1970 wt.%	1975 Million lb.	1975 wt.%	1980 Million lb.	1980 wt.%
HDPE	68	6.8	142	7.8	260	8.5
LDPE	316	31.7	582	31.9	977	32.0
Polypropylene	84	8.4	200	10.9	402	13.2
Styrene polymers	193	19.4	340	18.6	547	17.9
PVC	336	33.7	563	30.8	867	28.4
ALL TYPES	997	100.0	1827	100.0	3053	100.0

Source: Arthur D. Little, Inc. (16)

TABLE 71: NUISANCE PLASTICS IN THE DISPOSAL AREA

Type of product	1970 Million lb.	1970 wt. %	1975 Million lb.	1975 wt. %	1980 Million lb.	1980 wt. %
Packaging	3925	60.1	6445	56.2	10,170	54.0
Footwear	90	1.4	140	1.2	190	1.0
Records	95	1.4	140	1.2	205	1.1
C and A Film†	130	2.0	195	1.7	285	1.5
Industrial Wastes	1000	15.3	1830	15.9	3,050	16.2
Toys	310	4.7	555	4.8	945	5.0
Transportation	90	1.4	250	2.2	470	2.5
Appliances	100	1.5	230	2.0	440	2.3
Furniture	60	0.9	170	1.5	355	1.9
Wire and Cable	40	0.6	95	0.8	480	2.5
Novelties, Disposables	100	1.5	200	1.7	400	2.1
Others §	120	1.8	230	2.0	430	2.3
Housewares	425	6.5	885	7.7	1,270	6.7
Construction	50	0.8	100	0.9	150	0.8
TOTAL	6535		11,465		18,840	

† Construction and Agriculture

§ Includes business machines, instruments, luggage, sporting goods, apparel.

Source: Arthur D. Little, Inc. (16)

The data in Table 71 indicate that the percentage of plastic packaging wastes in the disposal area will drop slightly during the present decade. This decrease will occur because more and more plastic items with long service lives will enter the solid waste stream. For example, in 1980 wastes from such durable goods as wire, cable, and furniture will account for a much greater portion of the NP (2.5 and 1.9%, respectively). Even if plastic beverage bottles generate 2 billion pounds of NP by 1980, plastic packaging wastes as a percentage of all NP in the disposal area will still not increase during this decade.

Of all nuisance plastics in the disposal area, polyolefin wastes will continue to dominate throughout the present decade (Table 72). It was estimated these wastes will account for approximately 65 weight percent of the total NP. Note from the table that PVC wastes from all sources represent about 17% of the total, while PVC wastes from packaging are only 6%. It is believed that the concentration of PVC wastes from all sources will either remain constant or perhaps decrease slightly over the current ten year period. Accordingly, the percentage of PVC wastes from industrial and consumer sources is, on the average, only 0.24% of all urban and industrial solid wastes.

INTERMEDIATE DISPOSAL

Whether the NP is generated by a consumer or a manufacturing firm, it usually goes into an intermediate disposal area or storage facility. In the case of the householder, this is normally a garbage can, and in the case of the manufacturer it may be a large trash bin that is emptied mechanically.

Plastic wastes in storage are troublesome primarily because they have low specific gravities and, more significantly, they have low bulk density and occupy large volumes of space. This is particularly true of empty plastic bottles. The average density of completely compressed plastics is approximately 70 pounds per cu ft, whereas the density of uncrushed plastic wastes can be 1.6 pounds per cu ft. A study by the Bureau of Mines indicated that plastic wastes from packaging have a density of 3 pounds per cu ft. Compare this with the average density of solid waste in the United States in the uncompressed form which varies from 7 to 18 pounds per cu ft. Thus, uncompressed plastic waste is ¼ to ⅐ as dense as most uncompressed solid waste.

Accordingly, assuming a 1.4 weight percent value for plastics in solid wastes in the intermediate disposal area, where there normally is little compression, then on a volume basis, this would amount to 5.6 to 9.8%. To keep this in perspective, note that other nonplastic packaging wastes such as paper cartons, boxes, tin cans, glass bottles, and aerosol containers also have low bulk density. Recently, appliance manufacturers have developed trash mashers that compact solid wastes to approximately ¼ of their original volume in 60 seconds. These units are specifically designed to handle these bulky items.

TABLE 72: PLASTICS IN THE DISPOSAL AREA FROM ALL SOURCES*

Type of plastic	1970		1975		1980	
	mm lb	wt.%	mm lb	wt.%	mm lb	wt.%
Polyolefin	4,218	66.9	7,276	66.1	11,884	66.0
Styrene polymers	993	15.7	1,830	16.6	3,182	17.7
PVC	1,096	17.4	1,908	17.3	2,937	16.3
All types	6,307	100.0	11,014	100.0	18,003	100.0

*Does not include business machines, instruments, luggage, sporting goods, novelties, disposables, and construction.

Source: Arthur D. Little, Inc. (16)

The costs of intermediate disposal to the consumer or the industrial manufacturer are not included in the $4.5 billion cost often cited for handling the 3.5 billion tons of solid wastes in the U.S. Furthermore, these costs are extremely variable and would be difficult to estimate. For example, what is the cost to the householder to separate his refuse? How should this be estimated? The facts are that in many cases the household consumer does separate his refuse when he stores it. Commonly he separates wet from dry refuse, and garbage-disposal units in many homes automatically remove much of this wet refuse. In some communities, for example, solid waste is separated into combustibles and noncombustibles and some householders separate old newspapers for collection. According to the former Bureau of Solid Waste Management, at least 30% of the householders in the U.S. practice some form of separation in the home.

COLLECTION

In collection, the low bulk density and the high area-to-weight ratio of plastic wastes are again troublesome. For example, thin plastic films are easily scattered in the collection process during the transfer of refuse from the garbage can and during discharge of the refuse at the final disposal site. The use of polyethylene refuse bags helps avoid this problem, and today more and more communities use these to reduce the labor costs of collection. If a compactor truck is used for collection, some compaction can occur, although this is seldom greater than a 50% reduction in volume. Based on a recent government survey of 47 communities, about 70% of the community-owned collection vehicles are compactor trucks, 23% open dump trucks, and 8% enclosed noncompactor trucks.

The collection process typically involves considerable manual handling, and it is generally a labor-intensive process. Therefore, collection costs often constitute between 50 and 75% of the total disposal cost and vary from five to twenty dollars per ton. Again, the low density of plastic wastes directly affects the cost of collection. Not only are trucks filled up sooner; but if a large number of garbage cans are required to hold the low bulk density waste, the refuse collector makes more than one trip between the home and the refuse truck and the labor costs increase.

Industrial firms have similar collection problems. Often their fees for refuse collection services, in contrast to those of the householder, are assessed on the basis of the volume of the container and the frequency of collection. Using volume as a basis for the service charges further reaffirms the significance of noncrushable plastic containers, which increase refuse bulk at a net cost to the waste generator. As a result, some manufacturers today practice on-site containerization of refuse, and some refuse containers are equipped with hydraulic compactors that not only assist in filling the container but provide a level of compaction comparable to that of a packer truck.

TREATMENT PROCESSES

When refuse has been collected and accumulated in sufficient quantity to justify some processing prior to ultimate disposal or further handling, a number of operations can be carried out to recover usable values or to improve the economics of subsequent disposal. These include salvage, compaction, shredding, pulverizing or milling, composting, and incineration. Generally these treatment processes represent some level of technology, and they increase capital costs. Also, they represent areas now receiving the greatest attention on the part of investors and municipal authorities.

Baling

With the decreasing availability of landfill sites near the areas of high refuse generation, longer haul distances to alternate disposal sites are being required; therefore, further densification of the refuse to reduce transfer costs has become more attractive. Most compaction today is carried out with comparatively conventional baling equipment. In some cases, baling has been preceded by shredding to reduce the refuse to a more easily deformable state.

Baling equipment can increase the density of refuse to as much as 65 pounds per cu ft, which is approximately 30% greater than that achieved in good sanitary landfill operations. Compaction costs range from 50 cents to as high as three dollars per ton. The differences in cost reflect partly economies of scale and partly the degree of treatment. In general, plastic films deform readily in compactors without shredding and, other than their low inherent specific gravity, they do not affect the ultimate bale density. Plastic bottles and toys undoubtedly offer greater resistance to compaction, but there is little information on problems they cause in baling operations.

Pulverizing, Shredding, and Milling

Any treatment that converts refuse into a more particulate state, whether it is done by pulverizing, shredding, or milling, is advantageous in any subsequent process. Pulverized refuse, for example, is most useful where it is intended for the restoration of land sites and landscape improvements.

Pulverization costs range from one to three dollars per ton depending on the particle size of the final product and the economies of scale. The effect of plastics in the pulverization operation depends not only on their properties such as hardness, impact resistance, flexibility, resilience and rebound characteristics, but also on the physical form in which they are introduced. For example, rotating-drum-type pulverizers have little disintegrating action on plastics in general, although brittle or thin wall articles such as records, rigid foam, packaging trays, etc., are cracked and broken by the churning and tumbling action.

Most plastics are ungrindable in conventional hammer mills, because many plastic articles such as those made from plasticized PVC or low density polyethylene bottles are resilient and have a high degree of elasticity. On the other hand, rigid polystyrene moldings or those made from phenolic resins and other hard plastics are readily shattered and broken up in a mill. Recently, certain types of hammermills have become available that can easily accomodate plastic wastes.

Melting

A process developed by *M. Yamada; U.S. Patent 3,788,567; January 29, 1974; assigned to Mitsubishi Petrochemical Co.* is a device for disposing of synthetic resin waste wherein the waste material of different kinds, such as polyvinyl chloride resin waste, and polyethylene resin waste mixed with sand, paper, metal pieces, etc. is processed by melting and binding the synthetic resin component in such waste material for subsequent use.

ENVIRONMENTALLY DISPOSABLE PLASTIC PRODUCTS

Degradable plastics are plastics which will predictably environmentally decompose to a powder or liquid form and eventually be encompassed in the natural environment. Now, just how do plastics degrade? Well, they can be degraded by three basic mechanisms: biodegradation, solubility and photodegradation. This technology is based on over twenty years research and experience in making plastics more durable, an area in which they were highly criticized in the past, mainly for poor weatherability.

Biodegradable Plastic Products

Several plastics do degrade; however, most plastics are not biodegradable. Yet, this very characteristic makes plastics very stable materials for landfill operations because they do not break down and pollute the water table with decaying matter or chemical residues. Nor do they continue to settle in landfills over the years.

The biodegradables, are those plastics which, because of their chemical structure, are susceptible to being assimilated by microorganisms such as fungi and bacteria through enzyme action. This mechanism requires heat, oxygen and moisture.

Most naturally occurring polymers such as cellulose and natural rubber are attacked by a variety of fungi and bacteria. However, minor modification of these polymers such as acetylation of cellulose or vulcanization of rubber produce biodegradation-resistant materials.

Laboratory and field investigations on synthetic polymer deterioration have shown that some plastics have been attacked by microorganisms. The time span, however, for such deterioration is lengthy and can cover more than a decade. Actually, the filler and additives incorporated in plastics have much to do with the rate of degradation. These chemicals are of lower molecular weight and can act as a source of food for the microorganisms. When they have been consumed, their loss will result in the loss of the material's elongation, flexibility, and there may be accompaning discoloration and staining.

The almost total resistance of potential waste plastics such as polyethylene, polypropylene, polyvinyl chloride, and polystyrene to biodegradation has led to the modification of their basic structure by specific functional groups to the monomer stage during polymerization. However, this copolymerization approach has been unsuccessful to date in producing completely degradable materials.

In addition to modifying plastics materials via additives and copolymerization, experimental work is being conducted to obtain selected mutants and microorganisms to better degrade plastics. For example, Enzymes Inc. has prepared mutant microorganisms by Cobalt 60, Strontium 90 and 500,000 volt x-ray radiation. These new mutant strains of soil microorganisms show increased capability for degrading the resistant types of synthetic polymers. The use of these specialized bacteria and fungi as inoculants in waste disposal areas may do much to help the problem of degrading waste polymers.

Recently a study was conducted under the sponsorship of the Environmental Protection Agency which covered the biodegradation of all thermoplastic plastics. It indicated that plastics themselves are fairly immune to enzyme attack. In fact, only the aliphatic polyesters and urethanes derived from aliphatic ester diols and low molecular weight, under 500, unbranched polyethylene derivatives can be assimilated.

The Union Carbide Corporation has formulated polycaprolactone resins which give predictable degradable rates. This material is not attacked by airborne spores. It must be in contact with a nutrient soil environment to be assimilated. Degradation is thickness sensitive and influenced by environment. For example it would not degrade in an arctic environment. This material is available in all forms and can be pigmented and fabricated by normal processing techniques. Table 73 shows a 95% weight loss of a 40,000 molecular weight polycaprolactone resin container after a 12 month soil burial test

TABLE 73: RESULTS OF CAPROLACTONE POLYESTER SOIL BURIAL—SMALL CONTAINERS

Months of Burial	Weight Loss Percent
0	0
2	12
4	29
6	48
12	95

Source: J.B. Titus (23)

A process developed by *G.C. Newland, R.M. Schulken, Jr. and R.C. Harris; U.S. Patent 3,592,792; July 13, 1971; assigned to Eastman Kodak Co.* produces poly-α-olefin films for use as agricultural ground cover which can be made heat absorbent to assist in plant development while preventing weed growth, and thereafter heat reflective to give increased crop production, and thereafter friable, by incorporating into the film composition certain fugitive heat absorptive colorants, light reflective opaqueing pigments, and prooxidants.

A process developed by *F.W. Stager and R.N. Minor; U.S. Patent 3,647,111; March 7, 1972; assigned to Biocor Corporation* is one yielding a plastic container which deteriorates after a prolonged exposure to moisture and oxygen in an atmospheric environment with the aid of biological agents. The container material has organic components that are nutrients to fungi and various bacteria which bring about the decomposition. The material is composed of an organic filler material, such as peat, and a phenolic resin impregnated with a metallic salt of a fatty acid. The decomposition process transforms the container material into a rich organic substance which enhances soil characteristics.

A process developed by *E.L. Cole, H.V. Hess and R.C. Pomatti; U.S. Patent 3,707,056; December 26, 1972; assigned to Texaco Inc.* is one in which the decomposition rate of polyolefin film used as an agricultural mulch is controlled by the incorporation therein of 3 to 20% of a cation exchanged sulfonated petroleum coke or cation exchanged wet air oxidized petroleum coke. The film also contributes ion exchange capacity to the soil, furnishes nutrients and improves the tilth of the soil. Other additives such as nematocides or fungicides can be incorporated in the film mulch.

A process developed by *L.J. Taylor and J.W. Tobias; U.S. Patent 3,797,690; March 19, 1974; assigned to Owens-Illinois, Inc.* involves the preparation of a degradable polymeric material by applying a degradation-sensitizing reagent to the surface of a polymeric material and then diffusing the reagent into the surface.

The plastic material may be selected from polyethylene, polypropylene, poly(4-methyl-1-pentene), polystyrene, and polyvinylchloride. The degradation-sensitizing reagent may be selected from organic peroxides, organic hydroperoxides, organic azo compounds, metallo-organic compounds, organic materials which undergo facile autoxidation, organic photosensitizers, and organic dyes.

A process developed by *B.F. Anderson; U.S. Patent 3,828,471; August 13, 1974; assigned to E.I. du Pont de Nemours and Company* is one in which films prepared from ethylene polymer-hydrocarbon elastomer blends are coated with petroleum derived rubber processing and extender oils. These films degrade more rapidly and provide for easier plant penetration than untreated films.

Water-Soluble Plastic Products

The next phenomenon which may be employed in environmentally disposable plastic products is solubility. The solubility of plastic materials varies considerably with plastic formation, temperature, solvent concentration, and solvent. However, for environmentally degradable plastics the only solvent to be considered is water. Therefore by solubility we mean completely soluble in water forming nontoxic homogenous solutions.

There are several plastics that are completely water-soluble and thermoplastic (23). Examples are polyvinyl alcohol (PVA), hydroxypropyl cellulose (HPC), and polyethylene oxide (PEO). Normally small amounts of plasticizers, internal lubricants, and antioxidants are added to improve processing and utility. These plastics can be pigmented and coated. They have been extruded, injection molded, thermoformed and blow molded.

As to the solubility, when an item of hydroxypropyl cellulose, for example, is immersed in water, it quickly forms a slippery gel on the outer surface. This gel layer must dissolve and wash away before the water can penetrate progressively deeper to dissolve the complete item. Thus additives such as fillers, plasticizers and lubricants which tend to wick or absorb water through the outer gel layer will speed up the overall rate of solution. Increasing the molecular weight or thickness, raising the water temperature, or adding hydrophobic modifiers such as waxes and oils increases the dissolving times.

The solubility of water-soluble plastics varies with formulations, molecular weight and temperature. Hydroxypropyl cellulose is insoluble in water above 45°C; polyethylene oxide at 65°C. Reversing these trends achieves faster solution times of course, except when freezing temperatures are reached. However, heat promotes the solubility of polyvinyl alcohol. Its solubility depends on degree of alcoholization. For example, completely alcoholized grades are hot water-soluble and cold water-insoluble. Partially alcoholized types, about 87%, are soluble in both hot and cold water.

Since these materials are water-soluble, the effect of high (above 80%) relative humidity on stability is important. Figure 100 shows the percent water absorption of polyethylene oxide increases considerably at over 80% relative humidity. Equilibrium moisture is a key to their performance. They will not decompose with high humidity.

They are nontoxic, edible, noncaloric, nonnutritive and wash through plumbing without damage or clogging. They have low biological oxygen demand and do not support mold or bacterial growth. In an incinerator they readily burn and eventually decompose to carbon dioxide, water, and residual carbon. They are resistant to grease, oil, and petroleum hydrocarbons and therefore lend themselves to the packaging industry.

FIGURE 100: MOISTURE SORPTION OF POLYOX AT VARIOUS HUMIDITIES

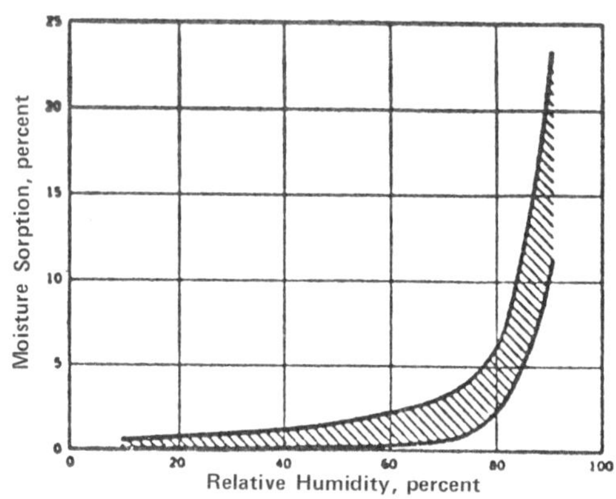

Source: J.B. Titus (23)

Photodegradable Plastic Products

The last and most promising technique for producing environmentally disposable plastic products is the utilization of photodegradation. Some plastic materials, depending on their molecular structure or additives are ultraviolet light sensitive. They absorb ultraviolet energy below 320 microns from the sun. When enough energy in the form of photons is absorbed, the bonds between carbon and hydrogen are broken and oxygen-reactive free radicals are formed. In the presence of oxygen these free radicals react with oxygen to produce peroxides and hydroperoxides which decompose further to produce carbonyl groups, hydroxyl groups, water and carbon dioxide. Figure 101 shows this reaction for a polystyrene (23).

These reactions cause a breakdown of a long chain backbone and reduce molecular weight. Once initiated, this reaction continues even in the dark. Thus the plastic loses its strength, becomes embrittled, and can be easily broken up by natural erosion forces such as rain and wind into small particles. These may become part of the soil or, if one of the materials previously discussed, can be attacked by microorganisms. As previously stated, low molecular weight alone will not constitute a biodegradable material. Photodegradation is time and temperature dependent and will vary with the polymer structure material thickness, and concentration and content of additives such as pigments, ultraviolet accelerators and promoters, ultraviolet absorbers and antioxidants. Such additives do not significantly affect the material's processing but would influence storage, coloration and physical properties.

The following materials are those which have been successfully degraded by this technique: polyethylene, polystyrene, polypropylene, polybutene, polybutadiene, acrylonitrile-butadiene-styrene, and polyvinyl chloride. The most successful materials are the linear, non-aromatic, molecular structured plastics. For example, unvulcanized syndiotactic polybutadiene is inherently degradable by ultraviolet light without sensitizers. The degree of crystallinity, normally between 15 and 30%, midway between an elastomer and a thermoplastic is the key to its degradability. Under direct sunlight it has disintegrated in periods ranging from one week to more than one year.

Several companies and universities are developing ultraviolet accelerators and promoters for initiating and controlling the photodegradation process. Since most of this work is proprietary, few of them have been chemically identified. However, some of these additives mentioned in the literature are: paraffin, oleic acid, iron compound, benzophenones, coal tar dyes, cobalt, zinc, copper and copper oxides. Most additives are unaffected by the usual fabricating methods and have little effect on plastic properties. They are inexpensive, usually nontoxic, increasing compound costs as little as ½ cent a pound.

FIGURE 101: PHOTODEGRADATION OF POLYSTYRENE

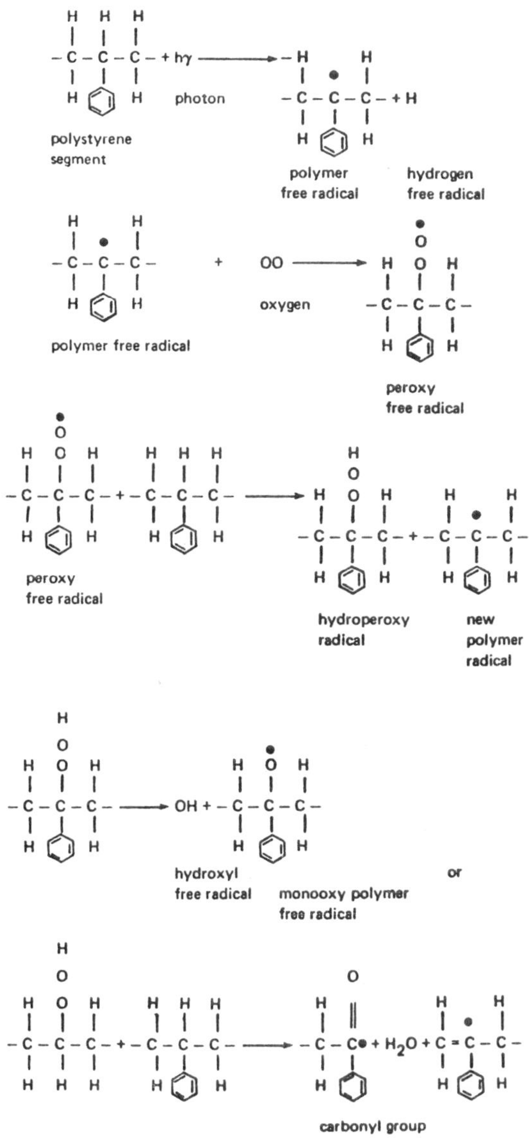

Source: J.B. Titus (23)

Photodegradable materials present the best candidates for all environments. These materials have the following advantages and disadvantages. Advantages: the best physical properties; the widest range of resin formulations; a variety of processing techniques; the best degradation rates; the ability to initiate degradation via artificial ultraviolet sources; and the ability to continually degrade without continuing ultraviolet exposure. The disadvantages are the rate of degradation is sensitive to pigmentation, coatings, ultraviolet intensity, formulation, and part thickness.

A process developed by *J.P. Pijst, U.S. Patent 3,590,527; July 6, 1971; assigned to N.V. Hollandsche Draad-en-Kabelfabriek, Netherlands* yields a sheet of foamed plastic material for protecting a vegetative plant, the material being air and water permeable and sensitive to light so as to be gradually disintegrated thereby.

In a typical example, a flexible polyfoam is prepared by uniformly and continuously mixing together 100 parts by weight of a triol of polypropylene oxide, 60 parts of about 80% 2,4-toluene diisocyanate and 20% 2,6-toluene diisocyanate, 4.8 parts of water, 0.15 parts of ethylene diamine, 0.50 parts of stannous octoate, 2.65 parts of silicon oil and 18 parts of monofluorotrichloromethane.

The mixture was formed into a continuous slab and the density of the material was 15 kg/m^3. The slab was sliced in sheets of a thickness of about 6 mm. In a field test, sheets of materials so produced were spread over vegetating bulbs. The effect of the sunlight on these samples and the atmospheric influences were evaluated by determining the loss of weight of the covering layer.

Bulbs during spring growth grew through a covering layer manufactured of the foam products, the material having completely disappeared at the time for lifting the bulbs. In selecting sheet material for use as a covering layer,.it is preferred that when the material is applied at the end of October or the beginning of November the sheet thickness willldecrease at the end of January to 60 to 80% of the initial value, and during the month of March it will decrease to 30 to 45% of the initial value.

Obviously, the rate at which the sheet material deteriorates will be selected on the amount of sunlight and temperature in the particular climate in which it is used, as well as on the particular species of vegetation to be protected. The range of density may vary between 10 and 30 kg/m^3 and the sheet thickness is preferred to be between 3 and 7 mm. Generally, for lower temperatures and increased amount of sun, sheet material having lower density and greater thickness will be employed.

A process developed by *T.H. Shepherd; U.S. Patent 3,590,528; July 6, 1971; assigned to Princeton Chemical Research, Inc.* yields a polybutene-1 film for use as an agricultural mulch. This film degrades and substantially disintegrates of its own volition within a growing season.

A process developed by *J.W. Henry; U.S. Patent 3,676,401; July 11, 1972; assigned to Eastman Kodak Co.* yields compositions comprising a copolymer of ethylene and 0.1 to 18% by weight of carbon monoxide, the percentage range being the proportion of carbon monoxide in the copolymer, and an ultraviolet light degradable amount of an organometallic compound or salt of a common metal, e.g., ferricine, cobalt acetyl acetonate, etc., whereby film, articles of manufacture and especially packaging means, upon exposure to ultraviolet rays emitted by the sun are capable of accelerated deterioration thereby causing such materials including packaging means to decompose forming dust or flakes which may become embedded in the soil or carried away by rain water under those circumstances where they are thrown away out of doors so as to be exposed to natural environmental conditions.

A process developed by *M.C. McGaugh; U.S. Patent 3,825,626; July 23, 1974; assigned to The Dow Chemical Company* is one in which the photodegradability of ethylene polymer containing a photosensitizing material is enhanced by incorporating therein a minor amount of polypropylene. The resultant photodegradable blends are useful as molding or extrusion compositions for the production of disposable plastic articles, e.g., wrappings, containers, etc., which degrade rapidly when exposed to sunlight and other ultraviolet light sources.

A process developed by *M.C. McGaugh; U.S. Patent 3,825,627; July 23, 1974; assigned to The Dow Chemical Company* is one in which the photodegradability of ethylene polymer containing a photosensitizing material such as tris(N,N-di-n-butyldithiocarbamato)iron (III) is enhanced by incorporating therein a minor amount of polypropylene. The resultant photodegradable blends are useful as molding or extrusion compositions for the production of disposable plastic articles, e.g., wrappings, containers, etc., which degrade rapidly when exposed to sunlight and other ultraviolet light sources.

A process developed by *D.E. Hudgin and T. Zawadski; U.S. Patent 3,830,764; August 20, 1974; assigned to Princeton Polymer Laboratories, Inc.* is one whereby a degradable composition is made from a polymer of a mono olefin having 2 to 3 carbon atoms or styrene and an additive comprising a derivative of an organic compound of a metal which has at least two valence states and a benzoyl derivative of an organic compound or a triazole.

A process developed by *A.R. Ingram; U.S. Patent 3,833,401; September 3, 1974; assigned to Arco Polymers, Inc.* is one whereby polyethylene film is caused to degrade in a predetermined pattern by the printing of the pattern on the surface of the film using printing ink containing a degradation promoter such as ultraviolet sensitizers, oxidation promoters, or embrittling agents.

A process developed by *R.A. White and S.A. Margosiak; U.S. Patent 3,847,852; November 12, 1974; assigned to De Bell & Richardson, Inc.* is one whereby thin hydrocarbon plastic articles are made more easily and readily degradable by incorporating within the hydrocarbon resin, prior to formation of the article, 0.1 to 15 parts by weight of a very finely divided additive which promotes ultraviolet degradation of the resin, per 100 parts by weight of the resin. The additive is preferably silica or a fatty acid metal salt.

A process developed by *J.E. Guillet and H.G. Troth; U.S. Patent 3,860,538; January 14, 1975* yields a polymeric composition which is photodegradable on exposure to ultraviolet radiation. This composition is made up of a major proportion of an addition polymer or copolymer of one or more vinylidene monomers represented by the general formula

$$CH_2{=}C\begin{matrix} \diagup X \\ \diagdown Y \end{matrix}$$

where X is selected from the groups hydrogen, lower alkyl and halogen, and Y is selected from the groups hydrogen, lower alkyl, aryl, substituted aryl, carboxylic acid, carboxylic acid lower alkyl ester, hydroxyl, lower acyloxy, halogen, cyano, alicyclic of not more than 8 carbon atoms, lower alkoxy, amido and lower alkenyl; and a minor proportion of a ketone copolymer of at least one unsaturated ketone of the general formula

$$R'\!-\!\overset{\overset{\displaystyle CH_2}{\|}}{C}\!-\!\overset{\overset{\displaystyle O}{\|}}{C}\!-\!R$$

where R' represents hydrogen or a lower alkyl group, and R represents an alkyl, alkaryl, alkenyl or aryl group, having from 1 to 9 carbon atoms, and at least one vinylidene monomer of the general formula

$$CH_2{=}C\begin{matrix} \diagup R'' \\ \diagdown Z \end{matrix}$$

where R" represents hydrogen, halogen or an alkyl group, and Z is a functional group selected from aryl, substituted aryl, carboxylic acid, lower alkyl carboxylate, lower alkyl, lower alkenyl, hydrogen, halogen, nitrile, lower acyloxy, lower alkoxy and amido; the amount of keto carbonyl groups in the composition being from about 0.01 weight percent to about 5 weight percent, based upon the total weight of addition polymer and ketone copolymer in the composition.

INCINERATION

One of the basic methods for destroying wastes is by burning or incineration. Volume reduction for most urban wastes after incineration ranges from 70 to 80%, and weight reduction is between 60 to 80% (16).

Incineration is basically a high temperature oxidation process. Thus, plastics, which contain primarily carbon and hydrogen, are oxidized to carbon dioxide and water vapor provided the conditions of time, temperature, and turbulence are right. When thermoplastics are heated, they become progressively softer and gradually melt; when the temperature is increased further, the viscosity of the melt decreases until it becomes a fluid. This generally occurs between 150° and 250°C.

In some cases, as with polystyrene, the polymer depolymerizes and often degrades or decomposes at or near its melting point. On the other hand, when thermosetting resins are heated, they generally do not deform but simply degrade. Because thermal decomposition starts from the outside, all materials with high ratios of surface to volume melt and decompose fairly rapidly. Conversely, the larger the article the more slowly it burns in the incineration process.

Most plastics have a high calorific value, often equal to that of high grade fuel oil. For example, polyethylene has a fuel value of 18,000 Btu per pound; in contrast, PVC, which has chlorine in addition to carbon and hydrogen, has a value of only 9,500 Btu per pound.

The polyolefins burn relatively easily and give off no noxious gases; however, they tend to drip, and the drops often continue to burn. When burned in the open, polystyrene, which has a self-ignition temperature of 900°F, burns with a yellow flame throughout and develops dense smoke. PVC, on the other hand, burns with difficulty; burning has to be sustained by an outside heat source. PVC softens and chars, and if no plasticizer is present, it does not drip. On burning, PVC generates 58 pounds of hydrogen chloride per hundred pounds of polymer.

Incineration is most commonly used in metropolitan areas where high population density often prevents the use of landfill, and where high total population results in the generation of large quantities of refuse. At present, about 20 to 25% of the commercial and domestic waste collected in the urban centers is treated by incineration; this amounts to about 10% of the total waste collected in the U.S. (urban and rural areas included). By the year 2000 it is anticipated the fraction of refuse incinerated in urban areas will increase to almost 30%.

Within the operating lifetime of many active municipal incinerator plants, municipal incineration technology has advanced from the early handstoked hearth units to the contemporary mechanically stoked water-wall boilers. The pace of design innovation has been greatly accelerated in recent years by federal assistance through demonstration grants. Nevertheless, looking ahead to the next few years, it is believed most incineration will still be carried out in units now in existence and in units greatly resembling those now in use. According to a recent survey by the Environmental Protection Agency, there are only about 3,000 municipal-type incinerators (capacities in the hundreds of tons per day) and of these only about 25% are rated as operable within current air and water pollution restrictions.

In the incinerator system, the handling of refuse begins with the delivery of materials to the site. In some cases compaction has occurred in the packer vehicle; such refuse has a density of 15 to 30 pounds per cu ft. The refuse is then stored temporarily in a pit where further densification can occur. Again, plastics are troublesome because they have low bulk density. However, because the pit and the associated transfer crane that is used to transfer the refuse to the incinerator constitute only 5 to 10% of the total incinerator plant costs, plastics only have a small effect on this aspect of plant economics.

From the pit the refuse is usually fed into the incinerator system and, in all but a very few old furnaces, the refuse then sits on a metallic grate where it is burned. These grates are fabricated from materials such as cast iron and various high-alloy stainless steels. In some portion of the burning process active burning begins at grate level, and the refuse is rapidly elevated to high temperatures. Under such conditions plastics (thermoplastics) have been known to fuse and close off the openings of the grate. If the plastic melts, it also can flow through the grates and cause undergrate fires with consequent grate damage.

In most plants, the furnace enclosure is constructed of refractory brick, but in the more modern plants in Europe and in a few places in the U.S. continuous water walls are used. Water-wall construction consists of steel tubes laid side by side and welded together to form panels which are in turn formed into walls of the furnace. In recent years, there has been considerable concern with the corrosive effects of the hydrogen chloride from PVC on the refractory wall materials but, more importantly, on the metal water walls. The alumina in the refractory wall furnaces can react with the hydrogen chloride to give aluminum chloride, which can volatilize under the temperature conditions of the furnace.

Following the initial combustion processes, which occur in the primary chamber, there is usually a secondary chamber to provide for the more complete burnout of combustible materials. Once again, the older plants use refractory construction in the secondary chamber, but some of the plants use water-filled metal tubes. In some European plants, where metal tubes are used, corrosion has been noted, and hydrogen chloride is suspect.

The next major unit of the incinerator is a gas-cooling device such as a water-spray evaporative cooling system. Again, aqueous hydrogen chloride can cause a corrosion problem. This problem is compounded if air pollution devices are used, because these often can include substantial areas of unprotected metal. Many problems associated with the scrubber water arise from its acidity, which is as low as pH 2 after a single pass. This acidity can rapidly degrade unprotected metallic surfaces, pipes, etc., and has caused catastrophic failure of scrubbers and fans. Furthermore, the treatment of wastewater from such scrubbers represents a substantial cost, and it can be a potential source for water pollution. Only in the past few years have true wet scrubbing systems, as differentiated from crude spray chambers or wet baffles, been applied to municipal incinerators; fewer than 10% of the municipal incinerators have them today.

At present, old incinerating systems that were not designed with the recognition of these corrosion problems suffer significant down-time maintenance charges. Unfortunately, however, such costs are difficult to document and are not only associated with the hydrogen chloride content of the flue gases but also reflect the action of sulfur and organic acids as well as improper operation.

Not withstanding what has been said above, there are large modern incineration plants in operation today which, after several years, have shown insignificant or no signs of corrosion. Often the absence of corrosion has been attributed to the partial removal of HCl from the combustion gases by its reaction with alkaline materials such as carbonates and silicates in the refuse.

Results of tests in which the normal refuse was supplemented with 3.25% PVC have been reported. It was noted in this case that only a small portion of the theoretical amount of HCl expected could be detected in the gases. It was also observed that if the incinerator operated at sufficiently high temperatures (above dew-point conditions), PVC presented virtually no risk of corrosion. On the other hand, the results of another study indicated that when industrial PVC waste was added to the urban refuse in an incinerator, inordinately high amounts of HCl were present in the flue gases (2,000 ppm). Also, severe corrosion occurred in the high temperature region of the incinerator.

A clear statement concerning the corrosive effect of HCl from PVC is difficult to make, because corrosion can be caused by poor operating conditions, uneven addition of plastics to the burning refuse, other acid producing wastes, and other artifacts. Recently Prof. Elmer R. Kaiser, in a study sponsored by the Society of the Plastics Industry, added double and triple amounts of plastics to the normal solid waste loads and burned them in a relatively recently constructed municipal incinerator. He added PVC, polystyrene, polyethylene, and polyurethane, and found that large amounts of plastics (up to 6%) in the solid waste mix did not have a corroding effect on the incinerator during the operating period. The short operating period of these experiments makes these findings questionable.

The incinerator was one that employed a conventional continuous feed system with oscillating grates and processed an average of 200 tons of mixed refuse a day. Prof. Kaiser also reported that the incineration of plastics in these large amounts caused no significant increase in smoke levels and did not increase pollution in the atmosphere. During the tests, the plastics did not drip through the incinerator grates. In fact, the higher burning temperatures of the plastics improved the incinerator performance, particularly when the normal refuse was wet.

As the discussion above indicates, PVC is the most troublesome plastic in an incinerator. Polystyrene also can be difficult, although it does not produce corrosive or noxious gases. Polystyrene burns easily in an incinerator if the rate of heat release, which is approximately 18,000 Btu per pound as opposed to 4,000 Btu per pound for normal refuse, is controlled. The very high rate of heat release must be retarded to insure complete combustion. Common methods to control this heat release are to cut the air supply and/or maintain a small ratio of polystyrene to normal refuse. The thick black smoke which can be generated in burning polystyrene is eliminated by using an efficient after burner.

As a treatment process, incineration is relatively expensive; incineration costs range from as low as two dollars per ton to as high as eight dollars per ton. The average is approximately $5.30 per ton. Incinerators require a large capital investment; construction costs vary from $1,500 to more than $6,000 a ton at a rated 24-hour burning capacity. Upkeep costs are also relatively high, and more skilled employees are required. Labor costs are a significant part of the operating costs of an incineration plant (Table 74). Plastic wastes can affect these economics by requiring more resistant materials of construction or heightened maintenance due to the more rapid corrosion of refractory and metal parts.

Increasing costs of collection and disposal compounded by increasing rates of refuse generation have led many industries, shopping centers, hotels, etc., to dispose of their own solid wastes by incineration. In many cases, the savings from reduced requirements for refuse storage and less frequent refuse collections together with the elimination of the aesthetic and unhealthful accumulation of waste are sufficient to justify the installation and operation of a private incinerator. While there are several types of incinerators that are effective for industrial solid waste, those which do not have grates that plug easily are more widely used where plastic wastes are involved.

In this case, rotary kiln types have been used successfully both in the U.S. and in Europe. They can incinerate solid waste containing relatively high proportions of plastics (up to 30% of the charged weight). The rotary kiln types are preferred because they afford excellent blending and mixing of the refuse during the combustion process. Any pool of molten plastics that forms under the refuse bed is continuously exposed to the combustion air by the rotation of the drum.

As might be expected, the incineration of industrial waste is more likely to produce pollutants than the incineration of domestic refuse; therefore, safeguards are required. Specially designed incinerators are available for burning scrap PVC, and in some instances, these allow the recovery of HCl for reuse.

Residues may be found in four locations in incinerator systems: on the grate, beneath the grate, in the secondary combustion chambers, and in the air pollution control device. Grate residue consists primarily of noncombustibles such as metal cans, wire, foils, etc., and slow burning refuse such as books, catalogs, and high moisture materials. Plastic solidified from the melt sometimes adheres on or adjacent to the grates; this must be broken with a hammer to be removed. Residue falling through the grates consists primarily of fine ash, aluminum, broken glass, etc., but it also may contain chunks of plastics that have solidified. Residues from the incinerator usually go directly to a final disposal site such as a landfill. In some instances, as mentioned above, metal cans are salvaged from these residues.

TABLE 74: ESTIMATED COST OF INCINERATION USING BATCH INCINERATION SYSTEMS (1969 dollars)

Reference location:	Boston, Massachusetts, mid-1969
Plant capacity:	400 TPD (two, 200-TPD furnaces)
Load factor:	250 days per year; 3 shifts/day
Annual consumption:	100,000 tons refuse/year
Capital investment:	$2,300,000

	Units	Dollars/Unit	Units/Ton Refuse Consumed	Dollars/Ton Refuse Consumed
Variable Costs				
Electricity	kwh	$0.015	22.3	0.33
Water	M gal	0.30	0.90	0.27
				$0.60
Semi-Variable Costs				
Labor				
1 crane operator per shift @ $5.00/hour				0.31
2 furnace operators (stoking and dumping) per shift @ $3.00/hour				0.38
1 charging floor man per shift @ $3.00/hour				0.19
2 residue removal men per shift @ $3.00/hour				0.38
1 supervisor @ $14,000/year				0.14
1 weigh scale operator @ $5,500/year				0.06
				$1.46
Overhead @ 40% of labor and supervision				0.58
Maintenance and repair @ 2.5% of capital investment annually				0.58
				$2.62
Fixed Costs				
Amortization @ 4.7% of capital investment annually				1.08
Interest on loan @ 3.0% of capital investment annually				0.69
				$1.77
Total Costs				$4.98

Source: Arthur D. Little, Inc. (16).

A device developed by *J.A. Boyd and D.E. Boyd; U.S. Patent 3,490,395; January 20, 1970; assigned to Washington Incinerator Sales & Service, Inc.* provides an improved incineration system for thermoplastic materials wherein a combustion chamber is provided with an inlet or charging port. A moving stream of air is generated outside the incineration combustion chamber and is directed toward and through the charging port into the interior of the combustion chamber. The thermoplastic materials to be incinerated are introduced into the air stream at a location remote from the charging port and are conveyed along and supported by the air stream and are thus introduced through the charging port into the combustion chamber where they are burned. By using the air stream for supporting and conveying the thermoplastic materials, such materials are maintained out of contact with the elevated temperature portions of the incinerator itself and hence will not melt and stick to surfaces of the incinerator.

Instead, the thermoplastic materials will be carried into the central portion of the combustion chamber where they are to be burned, and the air stream which carries the thermoplastic materials additionally creates turbulent flow conditions within the combustion chamber to promote and facilitate burning of the materials.

Another device developed by *J.A. Boyd and D.E. Boyd; U.S. Patent 3,495,555; Feb. 17, 1970; assigned to Washington Incinerator Sales & Service, Inc.* is an incinerator specially adapted for burning plastic materials in such a manner that the materials are substantially completely burned with a minimum of smoke and odors. Such plastic materials are introduced into a primary combustion chamber where they are subjected to initial burning, and thereafter, the products of combustion from the primary combustion chamber are directed to at least one secondary combustion chamber which includes a diffusion chamber means therein having at least one refractory wall with spaced openings therein. A burner in the secondary combustion chamber means heats the refractory wall to an incandescent condition and preheated air from the diffusion chamber means is mixed with the products of combustion to promote additional burning thereof. Turbulent heated air is introduced into the primary combustion chamber to promote combustion therein.

A device developed by *W.R. Fuller, B.H. Bieler and D.C. Morgan; U.S. Patent 3,556,024; January 19, 1971; assigned to The Dow Chemical Company* permits reducing the amount of halogen emitted during the incineration of halogen-containing plastics. The method involves applying an alkali to the plastic before it is burned. The method can reduce the emission of halogen by greater than 75% when properly employed.

A device developed by *R.F. Stockman; U.S. Patent 3,572,265; March 23, 1971; assigned to The Air Preheater Company, Inc.* provides for incinerating a meltable plastic material by means of first reducing the plastic to a liquid, heating the liquid to vaporization, and then burning the fumes that are given off from the surface of the liquid before they may be exhausted to the surrounding atmosphere.

Figure 102 shows details of the construction and operation of such a device. **10** refers to a housing or a container for the incineration of plastic material or the like. The housing has substantially vertical side walls **12** and an apertured top **14** with an inlet duct **16** which extends therethrough and terminates above the floor of housing **10** for adding solid plastic material into the container while an outlet port **17** at the top of the housing exhausts the fumes given off therefrom.

The port **17** is provided with an outlet duct **18** of cylindrical design which terminates a short distance above the housing **14** and is centrally positioned within the housing in a manner that permits a reasonable freedom from the congestion of fumes given off from the molten plastic.

The stack **22** exhausting from the container **10** may be of any approved construction, however one of modified venturi configuration performs best in mixing the fumes with air for combustion and then subjecting the mixture to a secondary process of combustion.

FIGURE 102: INCINERATOR FOR WASTE PLASTICS FEATURING MELTING AND COMBUSTION FUME INCINERATION

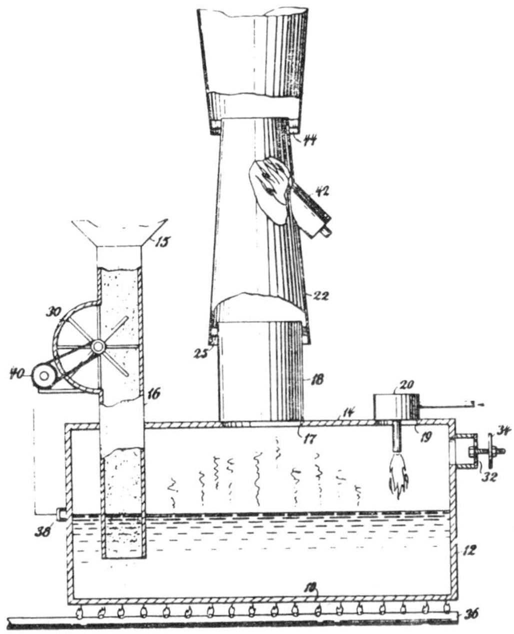

Source: R.R. Stockman; U.S. Patent 3,572,265; March 23, 1971

At the lower end of the stack air supply ports **25** are provided whereby the upward flow of fumes from the container **10** will induce the flow of air therethrough. As the air mixes with the fumes it is compressed slightly by passing through the converging section of the venturi shaped stack. Adjacent its point of greatest restriction a burner **42** including an independent source of gas and air is injected into the stack and ignited whereby the fumes from the plastic together with the air supplied through ports **24** are then subjected to combustion. As the gases ignite in the stack they expand and draw in still more air through the ports **44** whereby combustion is nearly complete in the remaining section of the stack before the spent gases are exhausted to the atmosphere.

The top **14** of the container includes a third opening **19** for a start up burner **20** which is inserted therethrough to initiate combustion of the plastic fumes. An air supply port **32** in the upper portion of the housing **12** includes an adjustable valve **34** therefor that may be opened a predetermined degree to partially burn the fumes being given off from the surface of the liquified plastic and produce sufficient heat in the combustion process to insure that the mass of plastic material in the housing is maintained in a liquid state.

In operation a mass of plastic material is loaded into the hopper **15** leading to container **10** through the inlet duct **16**. The size of the individual pieces of plastic will determine the size of duct **16**, thus for a mass of particle sized plastic an inlet duct **16** of smaller diameter is adequate while for larger chunk sized pieces of solid plastic an inlet of larger diameter is required.

A feeding device such as a star wheel **30** may be required to feed the plastic evenly and to preclude the back flow of excess quantities of vapor or fumes given off from the molten plastic which lies within the inlet tube **16**. A temporary source of heat such as a burner **36** is applied to the container **10** adequate to heat the contents thereof and thus reduce the solid plastic to a liquid state. As the plastic is heated it liquifies and its rate of vaporization increases so that it increasingly gives off fumes and vapors which are in turn ignited by the burner or other ignition means in the opening **20**. Fumes given off from the incomplete combustion of fumes in the housing **10** are exhausted through the oulet **18** to the stack **22**.

The fumes from the incompletely burned plastic are then mixed with air entering through ports **25**, mixed in the throat of the venturi and subjected to the incinerating action of the secondary burner **42**. Such incineration causes the inspiration of additional air through ports **44** so that combustion is complete before the gases are exhausted to the atmosphere. The step of injecting air and combustible fuel can be located at as many points along the stack as may be required to provide for the complete elimination of combustible products in the exhaust gas.

Heat from the plastic burning in the container **10** is sufficient to melt additional plastic material being added thereto so that auxiliary or temporary source of heat such as the burner **36** is not required for continuous operation. A liquid level sensor **38** on the container **10** may be made to control operation of the motor **40** and star wheel feeder **30** so that the level of plastic supplied through tube **16** to container **10** is maintained above the lower end opening to the inlet tube **16** so that little vapor from the molten plastic is allowed to escape up the tube **16** to the atmosphere. The star wheel **30** provides an additional barrier in tube **16** preventing still further the escape of vapor from the container.

The amount of combustion air permitted to enter the container **10** through supply valve **34** is controlled carefully to provide heat for the combustion of plastic within the container **10** adequate only for the continuous reduction to a molten state of the solid plastic being added through inlet **16**, it being understood that level of the plastic within the container is continuously controlled by the liquid level controller **20** so that it remains substantially constant at a point lying between the open bottom end of tube **16** and the top of the container.

Therefore the surface area of the body of molten plastic remains substantially constant, its temperature too is almost constant, and the air for its combustion is allowed to vary little so that an almost constant rate of combustion is permitted. Inasmuch as wide fluctuations in the amount of fuel available for burning are substantially eliminated, combustion is nearly complete and little smoke or obnoxious exhaust gas escapes to the atmosphere.

A device developed by *B. Faurholdt; U.S. Patent 3,670,667; June 20, 1972* is an incinerator for the combustion of waste products and particularly plastic materials comprising two combustion chambers, the primary combustion chamber comprising refractory ceramic elements located adjacent to one another so as to form inclined continuous guide surfaces and means for supplying primary combustion air to the primary combustion chamber opening into the chamber at different levels above the bottom thereof.

In the prior art incinerators the destruction of plastic materials presents serious problems because these plastic materials burn in a manner which is very different from that of household refuse.

The problems which are encountered during the incineration of plastic materials in prior art incinerators comprising a grate supporting the plastic materials are due to the fact that the grate quickly stops functioning in the normal manner because the holes in the grate through which the air is supplied become clogged. This clogging has the effect of initiating a destruction of the grate if it is made from commonly used materials, such as cast steel alloys.

Other problems are due to the high soot content in the flue gases leaving incinerators in which plastic materials are incinerated. This soot content which is due to an incomplete combustion strongly contaminates the atmosphere and is a nuisance in the area surrounding such incinerators.

The reason for these difficulties is that plastic materials possess the property of becoming soft and melting before they are ignitied and burn. If refuse containing plastic materials is introduced into and ignited in an incinerator in which the grate consists of perforated grate elements of cast iron or grate lamellae separated by passages for supplying air, the openings or passages in the grate become clogged whether it is constructed as a step grate, a plane grate, or as an inclined grate because of the melting of the plastic materials.

During the combustion which takes place on the basis of the combustion air which is supplied to the refuse without passing through the grate, very high temperatures are developed on the upper surface of the grate. Because of the clogging of the passages in the grate the latter is not subjected to the cooling which normally takes place during the passage of the cool combustion air through the grate passages. Thus, the grate, e.g., made from cast iron, is heated to very high temperatures and the heating causes an irreversible increase of volume.

Furthermore, the plastic materials possess the property that even when heated to a moderate temperature large amounts of gases are developed and the amount of the gases increases exponentially with increasing heating temperatures. Thus, if refuse containing plastic materials is introduced into and ignited in the above mentioned incinerators, large amounts of gases containing carbon particles are quickly formed in the zone above the refuse layer and the carbon particles escape from the combustion chamber together with the flue gases. If it is attempted to avoid the formation of these carbon particles by introducing more combustion air in the combustion chamber, the increased amount of heat resulting from the secondary combustion causes a generation of still greater amounts of gases and sooty carbon particles.

Thus, the problems are progressively increased until the gases and carbon particles have been completely burned away. At this time the supply of a large amount of secondary air effects a strong cooling of the combustion chamber, thus decreasing the combustion temperature of the material which has not been completely burned out. This temperature decrease also leads to the formation of soot which leaves the incinerator together with the flue gases.

As mentioned above the rate at which the generation of volatile gases and soot particles takes place within the primary combustion chamber depends on the generation of heat and consequently the temperature within the primary combustion chamber. However, by effecting the combustion in two thermally separated chambers it can be prevented that the generation of heat and the increase of temperature resulting from the secondary combustion influence the combustion process in the primary combustion chamber and the generation of gases and sooty particles within the chamber.

According to this development there is provided an incinerator having a primary combustion chamber which comprises inclined continuous guide surfaces consisting of guide elements made from a refractory ceramic material and in which the means for supplying primary combustion air to the primary combustion chamber open into the chamber at different levels above the bottom thereof.

When incinerating plastic materials in the incinerator according to the process, the plastic material while melting flows down along the inclined guide surfaces and during the movement the temperature is increased to a value such that a gasification takes place. Due to the fact that the primary combustion air is supplied at different levels above the bottom of the chamber, there is effected an intimate mixing of the primary combustion air and the compounds formed during the gasification which compounds subsequently burn. Figure 103 shows the essentials of this incineration design.

In the figure **1** is a primary combustion chamber containing inclined guide elements **2** of a refractory ceramic material. These guide elements **2** are supported by hollow metal pipes **3** provided with holes and connected to means for supplying primary combustion air to the combustion chamber. The primary combustion chamber is connected to a secondary combustion chamber **6** through a duct **4** and a flow-turning chamber **5**. In an opening in the wall of the secondary combustion chamber **6** there is provided an oil burner **7** and a large number of air supply nozzles **8**.

FIGURE 103: TWO-CHAMBER INCINERATOR WITH MULTIPLE INCLINED REFRACTORY GRATES

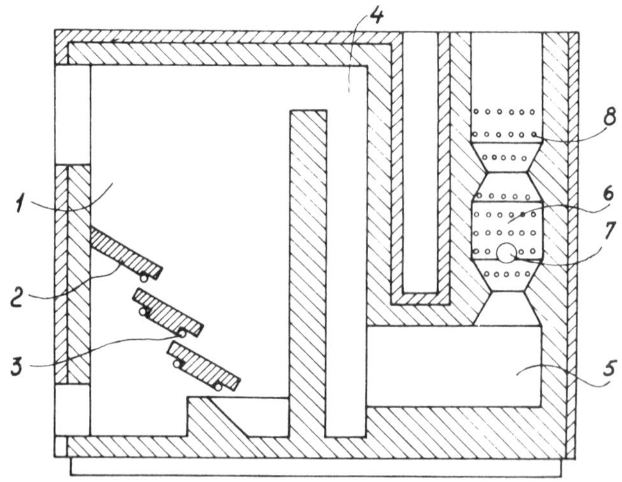

Source: B. Faurholdt; U.S. Patent 3,670,667; June 20, 1972

A device developed by *R.C. Pryor; U.S. Patent 3,680,500; August 1, 1972; assigned to Phillips Petroleum Company* is one in which effective incineration of waste products including plastic products of various types is obtained by means of a two-stage grateless incinerator including a first stage vaporization zone with burner and forced air supply, and a second stage combustion zone with burner and forced air supply. The two-stage vaporization-combustion incinerator provides effective oxidative incineration, with gaseous products being smokeless, and any nonvaporized noncombusted solid matter is essentially retained in the first stage vaporization chamber.

A device developed by *J.G. Roy; U.S. Patent 3,710,739; January 16, 1973; assigned to Union Carbide Corporation* is a means for the self-supportive incineration of thermoplastic materials wherein the thermoplastic material serves as the fuel and wherein the heat from the incineration is utilized to melt the thermoplastic material to permit rapid gravity flow onto an essentially vertical heat conductive surface maintained at the vaporization temperature of the thermoplastic material. This method and apparatus are claimed to provide a solution to the ecological problem of bulk waste disposal of thermoplastic materials.

A process developed by *M. Shigaki, S. Kido and Y. Chiba; U.S. Patent 3,716,339; Feb. 13, 1973; assigned to Takuma Kikan Mfg., Co.* is a device for the disposal of scrap vinyl polymers containing chlorine and hydrogen. The apparatus includes a rotary kiln that is adapted to receive the scrap plastic and heat the same to a temperature at which the plastic partially decomposes, and the major portion of the chlorine escaping therefrom as hydrogen chloride gas, which gas together with any chlorine which results from the decomposition is recovered in a confined space for subsequent usage. The partially decomposed scrap plastic is sequentially removed from the rotary kiln in a dried state and crushed. The crushed plastic is discharged by a blower onto a grate in a furnace where the scrap plastic is burned. The hot gases of combustion from the burning together with fines and dust arising from the burning operation are conducted through a passage to heat air that is subsequently used in the heating of the rotary kiln. The hot gases of combustion, together with entrained solid materials, are discharged into a dust collector to remove the fines and dust therefrom. The gases after being freed of solid materials are subsequently discharged through a chimney to the ambient atmosphere. The gases discharged to the ambient atmosphere are substantially free of chlorine and hydrogen chloride gases.

A process developed by *W.D. Jones; U.S. Patent 3,719,471; March 6, 1973; assigned to American Metal Climax, Inc.* is a system for recovering copper from insulated copper scrap containing electrical insulation of the type comprising plastic, rubber and the like which, upon being thermally decomposed and partially combusted, forms a waste gas which may pollute the atmosphere. The scrap is thermally treated in a cupola to form blister copper and the thermally decomposed insulation is separated as a waste gas and passed through an after burner where it is substantially completely combusted and the effluent gas therefrom then scrubbed and passed through an electrostatic precipitator following which it is vented to the atmosphere as relatively clean gas.

A process developed by *J.E. Perry; U.S. Patent 3,744,779; July 10, 1973; assigned to Horizons Research Incorporated* relates to the recovery of metal values in scrap materials in which the metal values are associated with nonmetallic materials, such as insulation on scrap wire.

Among the methods which have been suggested for recovering the metal values separately from the nonmetal associated therewith are incineration or burning of combustible material, such as plastics, rubber, resin or paper, carried on the metal, which it is desired to reclaim. Such methods require an oxidizing atmosphere and usually are carried out at relatively high temperatures. As a consequence, substantial amounts of the metal are oxidized with consequent loss of metal product salvaged as metal.

Another method which has been suggested involves heating metal scrap and the nonmetallic materials found with it in a retort or closed chamber. Such a procedure requires the treatment of small batches of product and is inherently more expensive than a process in which the material is processed continuously in an orderly manner through a suitable apparatus.

These and other suggested procedures in which the materials are exposed to relatively high temperatures are objectionable for at least two reasons. First, because the metal is contaminated by oxidation and second because the physical properties of the metal which is later recovered are usually adversely affected by the overheating, and are, therefore, not as good as those possessed by the metal before it was processed for removal of the undesired organic or inorganic materials which may be considered as contaminating the metal to be recovered.

One object of the process is to provide a process in which waste materials associated with metal scrap are continuously removed from the metal by a simplified, continuous process which avoids the smoke generated and overheating accompanying combustion processes, and in which the metal is recovered in a relatively uncontaminated state while it retains its physical properties and in which the major portion of the contaminating nonmetallic material is recovered as a volatile or condensible gaseous product which is useful as a fuel or heating means.

Briefly, this method consists of heating the contaminated scrap, in the absence of oxygen, to a temperature at which all of the volatile components are distilled off, using a continuous feed and removal of residue. At the temperatures involved, some volatile components are formed by destructive distillation. The volatile components, liquid and gas fractions, may be collected for reuse or they may be ignited and burned with a smokeless, odorless flame.

A process developed by *E.O. Ohsol and A. Perlmutter; U.S. Patent 3,750,600; August 7, 1973; assigned to American Cyanamid Company* provides for the disposal of thermoplastic materials utilizing a fuel oil as a solvent for the thermoplastic, by dissolving from about 1 to 20%, by weight, of a thermoplastic material or a mixture of the materials in from 99 to 80%, by weight, of a fuel oil, in the presence or absence of a cracking catalyst, usually from 1 to 10%, by weight, at elevated temperatures under atmospheric or superatmospheric pressure and, thereafter, burning the fuel oil containing thermoplastic material with resultant minimal residues and other solid, liquid, or gaseous pollutants.

A process developed by *E.B. Johnson; U.S. Patent 3,754,743; August 28, 1973* is a system for the rendering of silver from photographic waste in which the film or other waste is ignited and brought up to a satisfactory burning temperature in a primary chamber and maintained at this temperature until the film or waste is burned by controlling the amount of oxygen supplied to the primary chamber. During the ignition and burning of the photographic waste the smoke and particulate matter is directed to a secondary chamber where it is subjected to burning at a temperature that is higher than that in the primary chamber. After the photographic waste is burned, the silver bearing ash is cooled and removed from the primary chamber. The ash is then processed by conventional techniques to refine the silver. This apparatus permits the economical rendering of silver from photographic waste by burning with the elimination or reduction of smoke and other air pollutants.

A process developed by *F. Ensslin, G. Frenzel and E. Löbermann; U.S. Patent 3,776,147; December 4, 1973; assigned to Preussag AG-Metall, Germany* is a means for eliminating plastic waste and recovering the metals and metal compounds contained in the waste. The plastics waste, whether in a solid, pasty or liquid form, is converted into a combustible fluid-like form and is directed through a burner into a combustion chamber. The combustion of the waste results in combustion gas and fly ash which is cooled and directed through a separator to obtain the metals and metal compounds from the fly ash in the form of oxides.

A process developed by *Z.J. Przewalski; U.S. Patent 3,777,679; December 11, 1973* utilizes an incinerating system for recovering copper wire following the burning of the insulation therefor as combustible waste while simultaneously insuring against wire overheating. The process includes the steps of: first burning the insulated wire in an ignition chamber, second preheating the secondary chamber, and third exploiting the fuel generated in the ignition chamber for maintaining the temperature of the secondary chamber while passing the smoke generated in the ignition chamber through the secondary chamber preliminary to passage to the atmosphere.

A process developed by *E.H. Swartz; U.S. Patent 3,821,026; June 28, 1974* is one in which plastic-coated scrap metal is heated within a vented combustion chamber to cause decomposition of the plastic without adversely affecting the metal core with respect to its subsequent separation from the decomposed plastic and recovery as a substantially pure metal product. A fuel gas mixture is fed at a controlled rate to the combustion chamber to initially elevate the temperature of the chamber to an optimum value that is then maintained constant. The products of combustion form a nonoxidizing atmosphere in the chamber within which the desired decomposition of the plastic occurs.

A process developed by *R.F. Atkin; U.S. Patent 3,828,701; August 13, 1974; assigned to Pyrocom, Inc.* utilizes an incinerator which is capable of disposing of combustible waste containing a substantial percentage of plastic materials without creating attendant atmospheric pollutants. This is accomplished by provision of a controlled, stabilized process of combustion within the incinerator, wherein the solid wastes are volatilized at a controlled rate in the presence of a flame, the resultant combustion gases are combined with excess oxygen and are burned in an oxidizing flame; the remaining combustion products are then thoroughly mixed with the remaining excess oxygen at an elevated temperature and then accelerated into another oxidizing flame where the residue of the plastic materials is completely consumed.

A process developed by *H.L. Wentworth and K.H. Seelandt; U.S. Patent 3,841,240; October 15, 1974; assigned to Sola Basic Industries, Inc.* provides a continuous fluid bed furnace system which can be used to remove insulation from wire or other material.

COMPOSTING

Another treatment that also is a way of utilizing solid waste in composting. Composting is essentially the biodegradation of refuse or a portion of it to produce a humus-like material. Although plastics are not compostible, because they are not susceptible to rapid biodegradation, they have not been reported to cause significant problems if the refuse is shredded. Fine plastic particles merely behave as any other inert material like grit, small stones, or slate; therefore, they would not be expected to have any detrimental effects on the root development and growth processes of plants.

Composting has not been an effective way of disposing of municipal solid waste in this country. Despite its very early use as a soil conditioner and fertilizer in Europe, composting has made little progress in the U.S. This lack of progress has been due to the low cost and wide use of chemical fertilizers and the high operating and distribution cost of compost. Thus, the number of successful composting operations in the United States has been limited and municipalities have been reluctant to enter into this type of commercial operation.

LANDFILLS

Plastics make a suitable component of sanitary landfill. Since they do not decompose readily, they release no odors, gases or liquids that pollute surrounding land, air or water. The Los Angeles County Sanitation District Office has given the following statement in a letter to the Society of the Plastics Industry dated December 7, 1970.

> "The District staff does not subscribe to the opinion or suggestion that sanitary landfill is not a proper disposal method for plastic material because they are nonbiodegradeable. If we consider for discussion purposes a sanitary landfill in which only inert materials have been placed, the settlement anticipated would occur only through mechanical means. Nonbiodegradeable plastic wastes disposed of in such a landfill would seem to be as suitable as dirt, broken concrete, bricks and such like materials which also do not undergo decomposition. Since plastic materials do not degrade, they can be considered the equivalent of any other inert material in this respect. The economics of plastics products versus decomposable materials in a sanitary landfill are insignificant."

Compacted plastic wastes or residues from incineration are normally delivered to a landfill or, in a few instances, transferred to barges for subsequent ocean dumping. More than 90% of the household, commercial, and industrial solid wastes are handled by some type of landfill. Costs include capital investment of about $2,000 per ton and one to two dollars per ton as operating costs.

The practice of land fill can range from an open burning dump to a truly sanitary landfill operation. This latter operation is basically one of spreading refuse on the ground; in some instances, the refuse is densified by burning or by mechanical means using a compaction vehicle, and afterward covering the refuse with a layer of clean earth. In the case of the open burning dump, many plastics are undoubtedly subjected to some measure of volume and mass degradation with an accompanying emission of air pollutants. Where compaction vehicles are used to densify the refuse, many plastic items are ruptured and reduced in volume.

Plastic materials generally do not decay or rot even after prolonged soil burial and, therefore, they represent some of the most persistent components in the fill. For example, plastic bottles are not easily squashed, and only those in the lower strata of the refuse become distorted, flattened, or compressed by the overlying weight. Those with less weight or overburden are likely to remain intact and can, therefore, contribute to the springiness of the top layer. Accordingly, whatever process is used to compress refuse on-site or off-site, the results will be a reduction, if not a complete elimination, of the deleterious effects that plastics may have in landfill. If plastic bottles are properly squashed, they lose much of their springiness, although some shapes tend to unsquash more than others.

The public has often attacked plastics because they are not biodegradable. However, as recently pointed out by some officials of the county sanitation district of Los Angeles. there are no meaningful differences between plastics and decomposible materials when used in a sanitary landfill. In fact, they state that because plastics can be considered the equivalent of an inert material such as broken concrete, plastics are better suited than decomposible materials for landfill disposal, because they provide immediate stability. Here again, plastics are most suitable if they are in a pulverized form.

Biodegradation is dependent on time and the geometry of the article in the disposal area. In a study of plastics that were buried in a dump in Sweden for five years, it was found that while heavy plastic parts were found, thin plastic films had already disappeared. It was also reported that after five years the molecular weight of the polyethylene in refuse was one-quarter of the original, whereas the molecular weight of PVC in refuse was two-thirds of the original.

DISPOSAL OF RUBBER PRODUCTS

Tires are one of the most difficult consumer wastes to dispose of properly. In sanitary landfills, whole tires cannot be effectively compacted and tend to work their way up to the landfill surface. They also consume more landfill space per unit weight than other items and are not biodegradable (19). Tires on the surface of landfills or dumps or littered in urban or rural areas provide nesting places for rodents, flies, and mosquitoes.

Shredding, splitting, or otherwise reducing the size of tires overcomes the problem of landfilling to a large degree. Shredding is the most feasible method of physically altering the tires but is presently employed only in a few locations. Cost is the primary deterrent, and neither public nor private entities are likely to pay the price of processing in the absence of disposal regulations.

A small number of tires can be combined with large quantities of refuse in incineration as long as they do not comprise more than about 5% of the charge. Large percentages of tires result in damage to furnace walls and also require flue gas control. Tires represent only about 1.5% of the municipal waste stream but are often delivered to a disposal site in large batches rather than in an even day-to-day flow. Shredding of the tires would allow the percent of tires in the incinerator charge to be increased above 5% if the shredded rubber is thoroughly mixed with other waste materials.

INCINERATION

Incineration of tires for steam generation in special furnaces has been examined by some of the major tire manufacturers and appears to be a promising disposal and recovery method. One such plant has been built. When in full operation, the plant will handle about 1 million tires per year. The incinerator is reported to be completely odorless and pollution free. The major drawback appears to be the high capital cost of the equipment (19).

A device developed by *M.E.P. Hill; U.S. Patent 3,565,021; February 23, 1971* is a combustion apparatus particularly aimed to dispose of materials which are normally difficult to completely burn, such as used engine oil and used rubber tires. The apparatus has a refractory structure forming the combustion chamber into which the material to be burnt is fed. The material is subject to a current of forced air. Such a device may take the form shown in Figure 104. The combustion chamber indicated at **40** is of a generally cylindrical shape and is formed of firebrick **41**, bearing externally a layer of insulation **42** consisting of vermiculite and cement fondu, and is housed in a metal casing **43**. The chamber is provided with an inlet **44** in the general form of a slot in an end wall **45** and with a flue exit **46** in the opposite end wall **47**. The floor of the chamber departs from the cylindrical form of the rest of the chamber and descends by steps **48** into a wall **49** constituting an ash pit, to which access is provided through a door **50** in the metal casing. Forced air is provided by a compressor **51** through three inlet passages (shown in outline) in the combustion chamber, one passage **52** providing forced air in the ash pit, one passage **53** at the right-hand side of the chamber as viewed from the inlet side of the installation, in the floor of the chamber directing forced air segmentally upwards therein, and one passage **54** in the left-hand side of the chamber as viewed from the inlet side of the installation directing forced air horizontally across the chamber inlet so that immediately the leading part of a tire is received through the inlet slot **44** it is subjected to the current of air from passage **54**.

FIGURE 104: INCINERATOR DESIGN SUITABLE FOR DISPOSAL OF AUTOMOBILE TIRES WITH THE PRODUCTION OF USEFUL HEAT

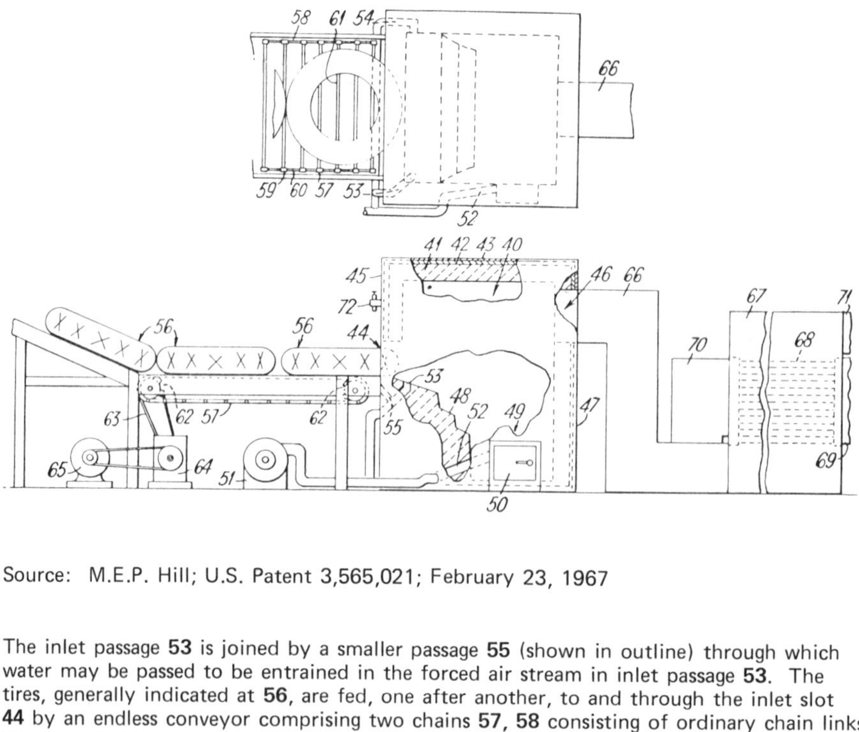

Source: M.E.P. Hill; U.S. Patent 3,565,021; February 23, 1967

The inlet passage **53** is joined by a smaller passage **55** (shown in outline) through which water may be passed to be entrained in the forced air stream in inlet passage **53**. The tires, generally indicated at **56**, are fed, one after another, to and through the inlet slot **44** by an endless conveyor comprising two chains **57, 58** consisting of ordinary chain links **59** connected by metal straps **60** to which are secured bars or slats **61** to extend between the two chains; the chains being carried round sprockets **62** one of which is driven by a belt **63** from a reduction-gear box **64** in turn driven by a belt drive from an electric motor **65**, to give a forward speed of the conveyor such that the tires are introduced at a rate equal to the rate at which they are burned.

The flue exit of the combustion chamber leads into a pipe **66** to conduct the effluent gases to a tubular head exchanger **67** which receives material to be heated, by passage through the tubes **68** indicated in outline, through an inlet pipe **69**. The material passing through the tubes **68** is discharged through pipe **70** and the cooled effluent gases are discharged through pipe **71**. Water may be fed onto the tires before entry to the combustion chamber from the mains supply through a tap **72** although if water is to be introduced with the tires any other means may be used for depositing water thereon. In the operation of the above described tire-burning installation, combustion may readily be started by dousing the leading part of the first tire with an easily ignited fuel, e.g., petrol and igniting the fuel. The forced air supply to the combustion chamber not only provides a current of forced air for the tire as it enters the chamber but also gives rise to a turbulent or vortex movement of the gas stream in the combustion chamber thereby increasing the residence time of the gaseous materials and particles of the tire entrained therein.

As a result combustion is substantially complete by the time the combustion products are discharged from the chamber into the flue pipe **66**. In one experiment a combustion chamber was employed of dimensions of some 5 ft in axial length from the inner surface of one end wall to the like surface of the other end wall and some 1½ ft in internal radius giving a height of some 5 ft from the floor of the ash pit to the interior surface of the roof of the combustion chamber. Water was supplied at a rate of 12 gal/hr mostly by deposition on and within the tires and the compressor supplied air at approximately 3,200 ft^3/min under a pressure of some 2¾ psi gauge pressure to provide approximately one third this amount in each inlet passage. Tires were burned at about 720 lb/hr to give an effluent temperature of from 800° to 850°C of the combustion gases effluent from the combustion chamber and about 100°C of the combustion gases effluent from the heat exchanger.

A device developed by *J. Brion; U.S. Patent 3,779,183; December 18, 1973; assigned to Commissariat a l'Energie Atomique, France* is an incinerator designed to burn whole tires representing 100% of the charge. A hearth which is smaller in width than the diameter of the tires to be burned has longitudinal vertical walls provided with lateral orifices for the admission of combustion air. A charging lock-chamber having internal dimensions slightly larger than the dimensions of a vertical tire has a sloping floor along which the tire rolls into the hearth and remains vertical. The combustion chamber of the incinerator is fitted with nozzles for the injection of air in order to induce high turbulence.

A device developed by *K.W. Stookey; U.S. Patent 3,785,304; January 15, 1974;* is a device for thermally reducing rubber or plastic materials in which the rubber or plastic material is comminuted to a predetermined size and it is then subjected to primary combustion in the lower portion of a first chamber with the gases and distilled combustible vapors in the upper portion of the first chamber being subjected to a secondary combustion and with the still only partially oxidized gases then being conducted to a second chamber and admixed with further air for tertiary combustion and the combustion thereof completed and with these last hot gases being used as the heat input for a boiler, such as a waste heat boiler. Automatic controls are provided to control the temperature in the upper portion of the first mentioned chamber to within predetermined limits and a further control is provided to insure substantially complete combustion in the second mentioned chamber, and with a bare minimum of excess air.

CHEMICAL RECOVERY BY PYROLYSIS

Destructive distillation, carbonization, and hydrogenization are thermal conversion processes that may be used to recover the chemical constituents of tires. The first two processes are both forms of pyrolysis, i.e., thermal decomposition in a low oxygen atmosphere; hydrogenization is a process of chemical synthesis involving addition of hydrogen. The primary product of carbonization (high-temperature destructive distillation) is carbon black, a major tire raw material. Lower temperature destructive distillation yields a mixture of oils, gases, and carbon residue with significant fuel value. Hydrogenization yields hydrocarbon products that can be used in tire manufacturing.

Pilot-plant-scale carbon black processes utilizing scrap rubber have been tested by some of the major rubber companies. It was reported that the processes are feasible but econimically unattractive at present, with costs of about three to four times that of carbon black production by conventional means from petroleum (19). Destructive distillation of tires has been examined by the Bureau of Mines in conjunction with a major rubber company. The products consist primarily of heavy oils, light oils, gases, and char. It has been reported that the value of these products from 100 pounds of tires is approximately $1.50 (i.e., $30 per ton of tires processed). It has been further suggested that even at this value the process is uneconomical if shipping, handling, storage, and preparation are included (19). Hydrogenation of the products of destructive distillation has been examined by Hydrocarbor Research Inc. at the pilot plant scale, and preliminary projections were made that such a plant could operate at a profit if tires could be obtained for $5/ton (19).

A process developed by *C.E. Scott; U.S. Patent 3,700,615; October 24, 1972; assigned to Cities Service Company* is one in which waste vulcanized rubber is depolymerized in a liquid hydrocarbon medium by means of agitation, heat, free-radical initiators, and molecular oxygen. The resulting solution of rubber-modified hydrocarbon can then be utilized in the manufacture of end-use products, thereby disposing of the waste rubber without pollution of the ecological environment. The rubber-modified hydrocarbon can, e.g., be vulcanized to produce a moisture barrier or insulation coating. It can also be mixed with asphalt to provide rubberized asphaltic compositions. In still another application, the rubber-modified hydrocarbon can be thermally decomposed to produce carbon black.

A process developed by *S.B. Alpert; U.S. Patent 3,704,108; November 28, 1972; assigned to Hydrocarbon Research, Inc.* is one in which waste natural and synthetic rubbers are converted to useful products in the presence of a catalyst and in a hydrogen environment by passing a particulate rubber slurry and hydrogen upwardly through a reactor maintained under temperatures in the 700° to 850°F range and pressures in the 500 to 3,000 psig range. The particulate rubber is converted in the reactor to hydrocarbon liquids while unconvertible solids originally compounded with the rubber leave unchanged in the product. The hydrocarbon stream can have a wide range of properties depending on the objectives of the user of the process. Figure 105 shows a suitable form of the apparatus.

As shown in Figure 105, waste rubber is continuously fed in line **1** to shredder **5** and then passes in line **9** to grinding zone **13**. The rubber is ground to a size preferably in the range of about ⅛ inch to 100 mesh USS. The particulate rubber then passes in line **17** to metals separation zone **21** and the metals leave in line **25**. This zone **21** can be a magnetic separation device, a shaking table and the like. The essentially metals-free particulate rubber passes in line **29** to feed preparation zone **33**.

In this zone **33** a caustic solution in **27** can be used to remove any fibrous materials with the spent caustic leaving in line **35** or the fibers may be left in the particulate rubber which leaves zone **33** in line **37** for slurrying in zone **45** with a suitable hydrocarbon oil such as aromatic decant oils, catalytic heavy gas oil, virgin heavy gas oil fed to **45** in line **41**. The rubber slurry then passes in **49** to enter, by upflow, ebullient reaction zone **57**. The rubber slurry, in passing through **49**, is heated to 500° to 800°F in heater **50**. A hydrogen-containing gas is fed to reaction zone **57** in line **53**. The gas contains preferably greater than 60% hydrogen.

The slurry enters the reaction zone at a space velocity of between about 10 and about 400 $lb/hr/ft^3$ reactor and at a temperature between about 700° and about 850°F. The hydrogen enters the reaction zone at a temperature between about 700° and about 1000°F at a rate of about 10,000 to 50,000 standard cubic feet (scf) of hydrogen per ton of waste rubber.

The reaction zone **57** contains a catalyst suitable for desulfurization and hydroconversion such as nickel molybdate on alumina, cobalt molybdate or silica-alumina or cobalt molybdate on alumina. The reaction zone **57** is operated at a pressure between 300 and 3,000 psig. It is to be understood that reaction zone **57**, although described as a single stage, can be a series of two or more stages. A recycle stream in line **105** can be fed to the reaction zone to provide additional velocity for the reactants in the reactor.

The total space velocity of the solid waste rubber entering the reaction zone is between 10 and 400 $lb/hr/ft^3$ of reactor space. The reactor velocity is sufficient to maintain the catalyst and particulate rubber in random motion in the liquid with no carryover of catalyst but complete carryover of fine inorganic compounds. The velocity is sufficient to carry compounds out of the reaction zone **57** in the liquid effluent in line **61** because they are small in size relative to the size of the catalyst particles. The catalyst will have a size range between 100 USS mesh (149 microns) and 1/16 inch whereas the inert compounds will be smaller than 50 microns in diameter.

FIGURE 105: PROCESS FOR HYDROCONVERSION OF WASTE NATURAL AND SYNTHETIC RUBBERS TO USEFUL PRODUCTS

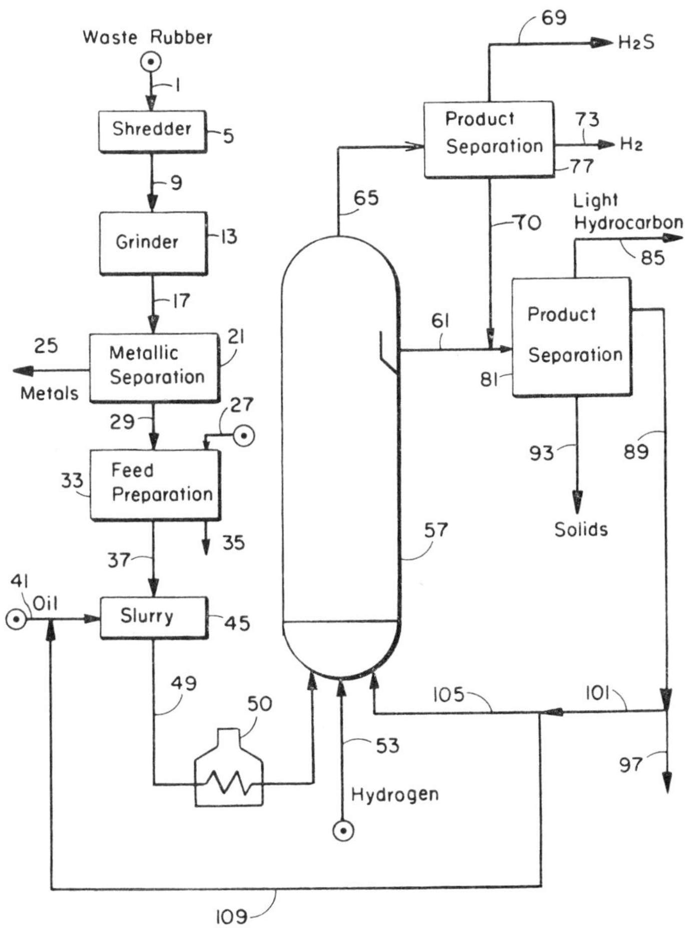

Source: S.B. Alpert; U.S. Patent 3,704,108; November 28, 1972

The gaseous effluent leaves reaction zone **57** in line **65**. This effluent enters separator **77** and the H_2S is removed in line **69** while the hydrogen is removed in line **73**. The hydrogen in **73** can be recycled to supplement the hydrogen feed in line **53** or the entire stream in **65** can be sent to a refinery for recovery. Light hydrocarbons in **70** can be combined with **61** or can pass out of the system in combination with **85**. The liquid effluent leaves reaction zone **57** in line **61** carrying with it any fine small size compounds. Line **61** enters separation zone **81** wherein a light hydrocarbon stream is removed in line **85** and a heavier hydrocarbon stream is removed in line **89**. Solids leave **81** in line **93** and are suitable for recovery and reuse for compounding rubber, water treatment, pigments, filler and insulation or for burning as a solid carbonaceous material.

A liquid hydrocarbon product is removed from line **89** in line **97**. This product is suitable for use as a low sulfur fuel oil with a sulfur content of less than 0.1% or it may be further upgraded in a subsequent separate catalytic operation. A recycle of liquid hydrocarbons is taken from line **89** in line **101** and passes as direct recycle to reaction zone **57** in line **105** or part can be passed in line **109** to serve as slurry oil in zone **45**. When the reaction zone **57** is operated to produce low sulfur liquid and solids the preferred conditions are at a pressure of about 1,000 psig, a space velocity of 30 to 200 pounds of rubber per hour per cubic foot of reactor, and a temperature in the range of 750° to 850°F. Hydrogen circulation rates of 20,000 to 40,000 cubic feet per ton are preferred.

This waste rubber process may also be integrated with an existing operating refinery. The process takes in a rubber suitably reduced in size by shredding and grinding operations to permit slurrying and pumping the waste rubber up to the pressure of the system. Extraneous foreign metals such as nails, and bits of glass are removed by conventional techniques familiar to rubber reclaiming operators.

The ground and prepared waste rubber is then slurried with a solids-free oil. The oil stream is either generated from the hydroconversion of the waste rubber or is supplied by the refinery. The slurry of waste rubber and oil is then mixed with hydrogen. The hydrogen gas is obtained from a hydrogen generation source inside the refinery. The source may be any hydrogen production technique such as natural gas reforming, hydrogen from catalytic reforming, or partial oxidation hydrogen generating processes conventionally used in hydrogen manufacture.

If desired a gas oil, preferably boiling above 600°F, is made available as a slurrying oil from the refinery to carry the waste rubber into the reaction system. The ground rubber is reacted, and the products sent to the refinery for recovery. In the refinery, the liquid and gas products are recovered, and the inorganic constituents such as carbon black, zinc oxide, pigments, etc, are separated. These latter compounds are reusable.

A process developed by *F. Schaub and S. Speth; U.S. Patent 3,787,166; January 22, 1974; assigned to Ruhrchemie AG, Germany* is one in which wastes of the group comprising synthetic plastics and rubber are thermally disposed of by cracking at temperatures between 250° and 450°C. The cracking may be carried out in the presence of an auxiliary medium which is in the liquid state at the reaction temperature and capable of dissolving the waste and/or the cracking products at least partially. The thermal decomposition may be accelerated by the addition of catalysts.

OTHER WASTE RUBBER UTILIZATION TECHINIQUES

Road Building

Road building and repair offers a large potential outlet for old shredded tires. Rubber that has been ground into small particles can be added to asphalt as a binder. The rubber is reported to enhance the resilience of the road in cool weather and reduce the flowing characteristics in hot weather. Rubber can reduce the tendency of asphalt to bleed to the road surface where it presents a skidding hazard, and the rubber in the mix allows more asphalt to be added for better aging and reduced raveling tendencies.

However, estimated costs of adding rubber to asphalt mix (usually in concentrations of 3 to 8%) range from $1.50 per ton to $2.50 per ton of mix. The justification for this cost will depend on whether the cost can be offset by lower maintenance costs or longer road life. Tests of rubberized roads to quantify benefits are presently underway in several locations (19). The State of New York is already using reclaimed rubber with hot asphalt to seal cracks and joints, and Arizona has laid rubberized asphalt on streets and airport aprons.

Reef Building

Tires have been used to construct artificial reefs for fish spawning on the East and Gulf coasts in response to increasing interest in sportsfishing. It is reported that water action on the West coasts is too strong for construction of artificial reefs. EPA does not presently have extensive data on the economics and long-range potential of this method of tire reuse.

FUTURE TRENDS

Air pollution problems of the plastics industry are assuming greater importance as more becomes known about the toxicity of monomers such as vinyl chloride, vinylidene chloride and styrene. Stricter standards are being set by OSHA design and control techniques are being improved to permit conformance to these standards.

Water pollution problems of the plastic industry are being given increased attention by EPA. Wastes are being more strictly characterized and treatment methods adapted to meet zero discharge standards. Solid waste problems of the plastics industry are problems both of reuse and disposal. As summarized by DeBell and Richardson in a report for the Manufacturing Chemists Association (24), some recommendations for the solid waste management of plastics are as follows.

[1] For various reasons, incinerators are still being designed and installed in United States communities that fail to incorporate all of the known modern incinerator technology and may therefore be expected to give occasional difficulties in operation, especially with intermittent, relatively high concentrations of plastic wastes in the material to be incinerated. Unskilled labor is a contributing factor in incinerator operating inefficiency. The basic considerations for proper incineration design and operation appear to be:

(a) The feed material should be as uniform as possible in composition and physical form.

(b) Combustion should be performed in such a way that the evolved gases are always in an oxidizing atmosphere.

(c) Temperatures in the furnace proper should be above the dew point and below 1000°C (1832°F).

(d) Retention time must be sufficient to ensure complete combustion of organic material in the flue gases.

(e) The gases should be cooled to approximately 250°C before cleaning to remove particulate matter.

(f) Where heat exchangers are used to recover the fuel value, corrosion can be avoided by limiting superheater temperatures to a maximum of 350°C, and preferably 250°C.

(g) Materials of construction and fabrication methods must be properly chosen to withstand corrosive atmospheres containing sulfur and chlorine compounds.

[2] Reuse of plastics waste is as yet of little commercial value but deserves continuing research and economic evaluation for possible development in the future.

[3] Pyrolysis, theoretically attractive and under intensive development in some quarters, has yet to become established as a practical alternate to proper incineration or sanitary landfill.

[4] Research thus far has not produced plastic packaging materials which are truly biodegradable, but continuing and expanded research effort is desirable.

[5] The operation of high-pressure compaction and baling of solid wastes (especially after presorting and comminution) appears to be a long-term attractive method to alleviate the growing shortage of available sanitary landfill areas.

[6] The plastics industry has a continuing obligation to present the facts concerning plastics in solid waste to all parties, i.e., the Federal and Municipal governments, the industry members themselves, the legislators and municipal authorities, and the public in general.

[7] The plastics industry should continue to support meaningful research and development work, in collaboration with equipment designers and manufacturers, both in the United States and overseas, to complete our knowledge in this area, and refine our approaches for certain remaining areas of concern.

Solid waste from the rubber industry is a massive problem and tires constitute a substantial part of that problem. Motor vehicle tires are relatively difficult to dispose of by landfill and incineration, and many tires are disposed of inadequately, left as litter, or piled on open ground. The existing markets for recycling and reuse of old tires are the retreading industry, the rubber reclaimers, and the tire splitters. The latter two markets are very small relative to the quantity of tires discarded, and the retreading market has been declining in recent years.

Some research and development of new options for tire recycling has been carried out by both private industry and the Federal Government. Most of these techniques are more costly than disposal (which is often inadequate from an environmental or aesthetic point of view). Research and development, demonstrations, technical assistance, regulations, and fiscal subsidies are all possible mechanisms for instituting tire processing or recovery. These mechanisms need to be evaluated and the technical and economic feasibility of tire recycling needs to be further analyzed before recommendations can be made.

LITERATURE REFERENCES

(1) Ethyl Corporation, *Plastics: Present and Future–An Environmental View,* Baton Rouge, Louisiana (1974).

(2) National Industrial Pollution Control Council, *Plastics in Solid Waste,* Washington, D.C. (March 1971).

(3) Darnay, A. and Franklin, W.E., Midwest Research Institute under contract to U.S. Environmental Protection Agency, *Salvage Markets for Materials in Solid Wastes,* Publication 5W-29c, Washington, D.C. (1972).

(4) Cross, J.A., Welch, R.O., Hunt, R.G. and Park, W.R., Midwest Research Institute for Manufacturing Chemists Association, *Plastics: Resource and Environmental Profile Analysis,* Washington, D.C. (June 28, 1974).

(5) Danielson, J.A., Ed., U.S. Environmental Protection Agency, *Air Pollution Engineering Manual, 2nd Edition,* Publication Number AP-40, Research Triangle Park, N.C., Office of Air and Water Programs (May 1973).

(6) Carpenter, B.H., *Vinyl Chloride–An Assessment of Emissions Control Techniques and Costs,* Report Number EPA-650/2-74-097, Washington, D.C., U.S. Environmental Protection Agency (September 1974).

(7) U.S. Environmental Protection Agency, *Development Document for Proposed Effluent Limitations Guidelines and New Source Performance Standards for the Synthetic Polymers Segment of the Plastics and Synthetic Materials Manufacturing Point Source Category,* Report EPA-440/1-74/036, Washington, D.C. (Sept. 1974).

(8) U.S. Environmental Protection Agency, *Addendum to Development Document for Proposed Effluent Limitations Guideline and New Source Performance Standards for the Synthetic Resins Segment of the Plastics and Synthetic Materials Manufacturing Point Source Category,* Report EPA 440/1-74/036-2, Washington, D.C. (September 1974).

(9) U.S. Environmental Protection Agency, *Development Document for Proposed Effluent Limitations Guidelines and New Source Performance Standards for the Synthetic Resins Segment of the Plastics and Synthetic Materials Manufacturing Point Source Category,* Report EPA 440/1-73/010, Washington, D.C. (Aug. 1973).

(10) B.F. Goodrich Chemical Co., *Wastewater Treatment Facilities for a Polyvinyl Chloride Production Plant,* Report 12020 DJI 06/71, Washington, D.C., U.S. Environmental Protection Agency (June 1971).

(11) U.S. Environmental Protection Agency, *Economics of Proposed Effluent Guidelines: The Plastics and Synthetic Rubber Industry (Viscose Rayon, Cellophane, Cellulose Acetate and Triacetate Fibers),* Report EPA-230/1-73/022, Washington, D.C. (September 1973).

(12) U.S. Environmental Protection Agency, *Economic Analysis of Proposed Effluent Guidelines: The Plastics and Synthetics Industry, Phase II,* Report EPA 230/1-74/044, Washington, D.C. (September 1974).

(13) U.S. Environmental Protection Agency, *Development Document for Proposed Effluent Limitations Guidelines and New Source Performance Standards for the Tire and Synthetic Segment of the Rubber Processing Point Source Category,* Report EPA 440/1-73/013, Washington, D.C. (September 1973).

(14) Howarth, J.T., Carter, J.A. and Sidman, K.R., *Economic Analysis of Effluent Guidelines: Rubber Processing Industry,* Report PB 235,691, Springfield, Virginia, National Technical Information Service (August 1974).

(15) U.S. Environmental Protection Agency, *Economic Analysis of Proposed Effluent Guidelines: The Rubber Processing Industry,* Report EPA 230/1-73-024, Washington, D.C. (September 1973).

(16) Milgrom, J., *Incentives for Recycling and Reuse of Plastics,* Report PB-214,045, Springfield, Virginia, National Technical Information Service (1972).

(17) U.S. Environmental Protection Agency, *Development Document for Proposed Effluent Limitations Guidelines and New Source Performance Standards for the Fabricated and Reclaimed Rubber Segment of the Rubber Processing Point Source Category,* Report EPA 440/1-74/030, Washington, D.C. (August 1974).

(18) U.S. Environmental Protection Agency, *Economic Analysis of Proposed Effluent Guidelines: The Rubber Processing Industry (Phase II),* Report EPA-230/1-74-046, Washington, D.C. (August 1974).

(19) U.S. Environmental Protection Agency, *Second Report to the Congress—Resource Recovery and Source Reduction,* Publication Number SW-22, Washington, D.C. (1974).

(20) Holman, J.L., Stephenson, J.B. and Adams, M.J., *Recycling of Plastics from Urban and Industrial Refuse,* Report R.I. 7955, Rolla, Missouri, U.S. Bureau of Mines (1974).

(21) Pettigrew, R.J. and Roninger, F.H. of Uniroyal, Inc. under contract to U.S. Environmental Protection Agency, *Rubber Reuse and Solid Waste Management,* Publication SW-22c, Washington, D.C. (1971).

(22) Humpstone, C.C., Ayres, E., Keaney, S.G. and Schell, T., International Research and Technology, Corp., *Tire Recycling and Reuse Incentives,* Report PB 234,602, Springfield, Virginia, National Technical Information Service (1974).

(23) Titus, J.B., *Environmentally Degradable Plastics: A Review,* Report AD-760,718, Springfield, Virginia, National Technical Information Service (February 1973).

(24) Warner, A.J., Parker, C.H. and Baum, B., *Solid Waste Management of Plastics,* Washington, D.C., Manufacturing Chemists Association (December 1970).

(25) Smith, W.E. and La Shelle, J.R., *Characterization of Atmospheric Emissions from Polyurethane Resin Manufacture,* Report PB-237,420, Springfield, Virginia, National Technical Information Service (October 1974).

PATENTEE INDEX

NOTICE

RECLAIMING RUBBER AND OTHER POLYMERS 1973

by J. A. Szilard

Chemical Technology Review No. 5

As the title indicates, two basically different types of materials are considered. In the case of rubber, both natural and synthetic, reclaiming consists of devulcanizing and depolymerizing processes. This requires the use of heat, the addition of acids or alkalis, and the use of softeners and plasticizers. The resulting reclaimed rubber is a new raw material, different from the starting material.

In contrast, the reclaiming processes for plastics usually seek to recover the polymer, or a prepolymer, or at least the monomers from which the polymer was formed originally.

Some of the newer reclaiming processes are continuous and designed to reduce air or water pollution, and to render the solid or liquid waste materials capable of being assimilated by conventional sewage disposal systems.

127 patent-based processes. A partial and condensed table of contents follows. Numbers in ( ) indicate the number of patents per topic. Chapter headings are given, followed by examples of important subtitles.

ISBN 0-8155-0476-4

310 pages

RECYCLING AND RECLAIMING OF MUNICIPAL SOLID WASTES 1975

by F. R. Jackson

Pollution Technology Review No. 17

More and more cities, large and small, are exploring new technological systems for recycling solid municipal wastes.

Naturally the greatest interest centers upon those recycling systems that permit profitable reclamation of saleable items and materials from the town's refuse.

Articles made of metal can be cleaned, washed, compressed, and sold to the scrap metals industry. Glass can be comminuted and even separated as to color.

Paper and cardboard is a promising target for recycling, because paper fiber is the largest single component of municipal wastes and potentially very valuable.

Water-insoluble organic matter, such as plastics and wood, can be reclaimed and processed for conversion into energy in a variety of ways.

As these new technologies are developed, they will add to the alternatives available to a community, when it decides to recycle some of its wastes as part of its total solid waste management program. Numerous alternate ways and means are needed, because local conditions vary widely, and no single line of approach is capable of meeting every community's needs.

This Pollution Technology Review surveys information available up to the middle of 1974, and is based primarily upon studies conducted by industrial and engineering firms under the auspices of the EPA and other government agencies. Forward-thinking municipal administrators will find this book most useful, because it collates comprehensively the results of governmental and industrial research.

A partial and condensed table of contents follows here.

ISBN 0-8155-0560-4

342 pages